Wildlife and Wind Farms, Conflicts and Solutions

Dedication

This work is dedicated to my family: my wife Eleanor, with whom I share a vision of a better future for our planet; my children Merlin & Phoenix who are still young enough to wonder, and Morgan & Rowan, who took their place in society as women somewhere along the way; and my Mum and Dad. My Mum was taken from us before these books were completed and is acutely missed.

Wildlife and Wind Farms, Conflicts and Solutions

Volume 3
Offshore: Potential Effects

Edited by
Martin R. Perrow

Pelagic Publishing | www.pelagicpublishing.com

Published by Pelagic Publishing
www.pelagicpublishing.com
PO Box 874, Exeter, EX1 9YH, UK

Wildlife and Wind Farms, Conflicts and Solutions
Volume 3 Offshore: Potential Effects

ISBN 978-1-78427-127-5   (Pbk)
ISBN 978-1-78427-128-2   (ePub)
ISBN 978-1-78427-130-5   (PDF)

Colour reproduction of this book was made possible thanks to sponsorship by Vattenfall Wind Power Limited™. For more information visit https://corporate.vattenfall.com/

Cover images:

Top: Sandwich Terns *Thalasseus sandvicensis* (and a single Lesser Black-backed Gull *Larus fuscus*) resting on Scroby Sands, a dynamic sandbar that increased in height to once again become emergent at all states of tide soon after the construction of the Scroby Sands wind farm (UK). (Martin Perrow)

Left: Bottlenose Dolphins *Tursiops truncatus*, a species of cetacean predicted to be of increasing importance in impact studies as wind farms expand around the world. (Martin Perrow)

Middle: A jack-up vessel preparing to place T-pieces over the monopiles previously installed at a wind farm in the shallow waters in the Greater Wash, UK. (Martin Perrow)

Right: The hard substrates provided by turbine bases and scour protection provide new habitat for colonising hard substratum fauna in what is often a mainly sedentary benthic environment in the North Sea. (WG Ecosystem Functions, Alfred Wegener Institute, Helmholtz Centre for Polar and Marine Research)

Contents

Contributors

Lothar Bach Büro Bach Freilandforschung, Hamfhofsweg 125b, D-28357 Bremen, Germany

Richard J. Berridge ECON Ecological Consultancy Ltd, Unit 7 Octagon Business Park, Little Plumstead, Norwich NR13 5FH, UK

Göran Broström Department of Marine Sciences, University of Gothenburg, Box 461, 40530 Göteborg, Sweden

Jennifer Dannheim Alfred Wegener Institute, Am Handelshafen 12, 27570 Bremerhaven, Germany

Steven Degraer Royal Belgian Institute of Natural Sciences, Operational Directorate Natural Environment, Aquatic and Terrestrial Ecology, Marine Ecology and Management, Gulledelle 100, 1200 Brussels, Belgium

Michael Elliott Institute of Estuarine and Coastal Studies, University of Hull, Hull HU6 7RX, UK

Joris Everaert Research Institute for Nature & Forest/Instituut Natuur- en Bosonderzoek (INBO), Havenlaan 88, Box 73, 1000 Brussels, Belgium

Andrew B. Gill PANGALIA Environmental, UK and Centre for Offshore Renewable Energy and Engineering, School of Water, Energy, and Environment, Cranfield University, Cranfield MK43 0AL, UK

Andrew J. P. Harwood ECON Ecological Consultancy Ltd, Unit 7 Octagon Business Park, Little Plumstead, Norwich NR13 5FH, UK

Reinhold Hill Avitec Research, Sachsenring 11, D-27711 Osterholz-Scharmbeck, Germany

Ommo Hüppop Institut für Vogelforschung "Vogelwarte Helgoland", An der Vogelwarte 21, D-26386 Wilhelmshaven, Germany

Helen Jameson Vattenfall Wind Power Ltd, 1 Tudor Street, London EC4Y 0AH, UK

Adrian D. Judd Centre for Environment, Fisheries and Aquaculture Science (Cefas), Pakefield Road, Lowestoft NR33 0HT, UK

Sue King	Sue King Consulting Ltd, The Coach House, Hampsfell Road, Grange-over-Sands LA11 6BG, UK
Bjarke Laubek	Vattenfall Wind Power Ltd, 1 Tudor Street, London EC4Y 0AH, UK
Elke Ludewig	Institut für Meereskunde, Universität Hamburg, Bundesstr. 53, 20146 Hamburg, Germany
Bianca Michalik	Institut für Vogelforschung "Vogelwarte Helgoland", An der Vogelwarte 21, D-26386 Wilhelmshaven, Germany
Georg Nehls	BioConsult SH GmbH & Co. KG, Schobueller Str. 36, D-25813 Husum, Germany
Steven K. Pelletier	Stantec, 30 Park Drive, Topsham, ME 04086, USA
Martin R. Perrow	ECON Ecological Consultancy Ltd, Unit 7 Octagon Business Park, Little Plumstead, Norwich NR13 5FH, UK
Thomas Pohlmann	Institut für Meereskunde, Universität Hamburg, Bundesstr. 53, 20146 Hamburg, Germany
Jon M. Rees	Centre for Environment, Fisheries and Aquaculture Science (Cefas), Pakefield Road, Lowestoft NR33 0HT, UK
Emilie Reeve	The Renewables Consulting Group Ltd, Gilmoora House, 57–61 Mortimer Street, London W1W 8HS, UK
Jan T. Reubens	Flanders Marine Institute/Vlaams Instituut voor de Zee (VLIZ), InnovOcean site, Wandelaarkaai 7, 8400 Ostend, Belgium
Anja Schneehorst	Bundesamt für Seeschiffahrt und Hydrographie (BSH), Bernhard-Nocht-Straße 78, 20359 Hamburg, Germany
Heike Sittel	Vattenfall Wind Power Ltd, 1 Tudor Street, London EC4Y 0AH, UK
K. Smyth	Institute of Estuarine and Coastal Studies, University of Hull, Hull HU6 7RX, UK
Eric W.M. Stienen	Research Institute for Nature & Forest/Instituut Natuur- en Bosonderzoek (INBO), Havenlaan, 88, Box 73, 1000 Brussels, Belgium
Nicolas Vanermen	Research Institute for Nature & Forest/Instituut Natuur- en Bosonderzoek (INBO), Havenlaan, 88, Box 73, 1000 Brussels, Belgium
Magda Vincx	Marine Biology Research Group, Ghent University, Campus Sterre S8, Krijgslaan 281, B-9000 Gent, Belgium
Dan Wilhelmsson	Wilmaco, Skördevägen 23, 135 43 Tyresö, Sweden
Jennifer C. Wilson	Wood plc, Partnership House, Regent Farm Road, Gosforth, Newcastle-upon-Tyne NE3 3AF, UK

ECON
Ecological Consultancy Ltd

Preface

Wind farms are seen as an essential component of global renewable energy policy and the action to limit the effects of climate change. There is, however, considerable concern over the effects of wind farms on wildlife, especially on birds and bats onshore, and seabirds and marine mammals offshore. On a positive note, there is increasing optimism that by operating as reefs and by limiting commercial fishing activity, offshore wind farms may become valuable in conservation terms, perhaps even as marine protected areas.

With respect to any negative effects, Environmental Impact Assessment (EIA) adopted in many countries should, in theory, reduce any impacts to an acceptable level. Although a wide range of monitoring and research studies have been undertaken, only a small body of that work appears to make it to the peer-reviewed literature. The latter is burgeoning, however, concomitant with the interest in the interactions between wind energy and wildlife as expressed by the continuing Conference on Wind Energy and Wildlife Impacts (CWW) series of international conferences on the topic. In 2017, 342 participants from 29 countries attended CWW 2017 in Estoril; similar to the numbers of both at Berlin CWW in 2015. There is also evidence of increasing interest within individual countries. For example, the Wind Energy and Wildlife seminar (Eolien et biodiversité) in Artigues-près-Bordeaux, France on 21–22 November 2017, relating to both onshore and offshore wind, attracted over 400 participants.

Even with specific knowledge of the literature and participation in meetings, I reached the conclusion some years ago (which is still maintained today) that it was difficult for an interested party to judge possible effects on wildlife, and especially the prospects of ecosystem effects generated by ecological interactions between affected habitats and their dependent species, or between species one or more of which could be affected by wind farms. In other words, there was a clear need for a coherent, overarching review of potential and actual effects of wind farms, and perhaps even more importantly, how impacts could be successfully avoided or mitigated. Understanding the tools available to conduct meaningful research is also clearly fundamental.

A meeting with Nigel Massen of Pelagic Publishing in late 2012 at the Chartered Institute of Ecology & Environmental Management Renewable Energy and Biodiversity Impacts conference in Cardiff, UK crystallised the notion of a current treatise and the opportunity to bring it to reality. Even then, the project could not have been undertaken without the significant financial support of ECON Ecological Consultancy Ltd. expressed as my time.

At the outset of the project I did not imagine the original concept (one volume for each of the onshore and offshore disciplines) would morph into a four-volume series, with onshore and offshore each having a volume dedicated to (i) documenting current knowledge of the effects – the conflicts with wildlife; and (ii) providing a state-of-the-science guide to the available tools for monitoring and assessment, and the means of avoiding, minimising and mitigating potential impacts – the solutions. I also did not imagine that the gestation time to produce the volumes would be six years or so, or that the offshore volumes would run nearly two years behind the onshore volumes. The offshore industry has developed rapidly in the last few years, and this has meant many potential authors becoming swamped by their workloads within various roles within the industry. Perhaps inevitably, several authors fell by the wayside, and although replaced by others, this caused delays and some stop and start in the process. However, I believe this has meant that the books have benefited by being able to document the rapid progress in the last few years and now having a particularly active team of authors at the top of their game.

In this Volume 3, the concept was to cover as wide a taxonomic spread as possible, starting with *Seabed communities* of mainly invertebrates, and then covering *Fish, Marine mammals, Migratory non-seabirds and bats*, before dealing with *Seabirds* in two separate chapters on displacement and collision. All chapters were to outline potential as well as realised effects and possible impacts. To accompany the chapters on the taxonomic groups, the scene was set by an introductory chapter on the *Nature of wind farms*, with a further chapter on *Physical effects* (and to a lesser extent *Chemical effects*) of wind farms and *Atmosphere and ocean dynamics*, outlining what may prove to be a critically important field in terms of its ecological consequences. A *Synthesis of effects and impacts* upon all the taxonomic groups and encompassing the physical effects is then provided in a stand-alone chapter at the end of the volume. As it was completed after all other chapters, this summary chapter had the benefit of incorporating more recent information published in the intervening period.

To promote coherence within and across volumes, a consistent style was adopted for all chapters, with seven sub-headings: Summary, Introduction, Scope, Themes, Concluding remarks, Acknowledgements and References. For ease of reference, the latter are reproduced after each chapter. The carefully selected sub-headings break from standard academic structure (i.e. some derivative of Abstract, Introduction, Methods, Results and Conclusions) in order to provide flexibility for the range of chapters over the two volumes, many of which are reviews of information, whereas others provide more prescriptive recommendations or even original research. Some sub-headings require a little explanation. For example, the Summary provides a ~300-word overview of the entire chapter, whilst the Concluding remarks provide both conclusions and any recommendations in a section of generally ~500 words. The Scope sets the objectives of the chapter, and for the benefit of the reader describes what is, and what is not, included. Any methods are also incorporated therein. The Themes provide the main body of the text, generally divided into as few sub-heading levels as possible. Division between effects during construction and operation was generally avoided as this increased the number of sub-headings and led to an unwieldy structure. Any clear differences in effects between different stages of windfarm construction and operation are incorporated into specific sub-headings.

As well as being liberally decorated with tables, figures and especially photographs, which are reproduced in colour courtesy of sponsorship by Vattenfall, most chapters also contain Boxes. These were designed to provide particularly important examples of a particular point or case, or to suffice as an all-round exemplar and 'stand-alone' from the text. In a few cases, these have been written by an invited author(s), on the principle that it is better to see the words from the hands of those involved rather than paraphrase published studies. My sincere thanks go to all 27 chapter authors and 3 box authors for their contributions. I take any deficiencies in the scope and content in this and its sister volume to be my responsibility, particularly as both closely align to my original vision, and many authors have patiently tolerated and incorporated my sometimes extensive editorial changes to initial outlines and draft manuscripts.

Finally, it needs to be stated that with the current epicentre in northwestern Europe in the North and Baltic Seas, the coverage of this book could not be global. However, as the offshore wind industry develops at almost breakneck speed in a great range of countries, I hope the information and experiences gleaned from the pages of this book can be applied in a global context; with the proviso that applying any lessons learned to marine systems elsewhere on the planet would need to be accompanied by specific research to account for any inevitable differences in the ecological structure and functioning of those systems. Hopefully, this book is a further step towards the sustainable development of wind farms and the ultimate goal of win–win[1] scenario for renewable energy and wildlife.

Martin R. Perrow
ECON Ecological Consultancy Ltd
24 August 2018

---

1 Kiesecker, J.M., Evans, J.S., Fargione, J., Doherty, K., Foresman, K.R., Kunz, T.H., Naugle, D., Nibbelink, N.P. & Niemuth, N.D. (2011) Win–win for wind and wildlife: a vision to facilitate sustainable development. *PLoS ONE* 6: e17566.

CHAPTER 1

The nature of offshore wind farms

HELEN JAMESON, EMILIE REEVE, BJARKE LAUBEK and HEIKE SITTEL

Summary

Offshore wind has come of age since the turn of the millennium and is now a mainstream source of low-carbon electricity, at least in Europe. Decoupling from subsidy dependence remains key, but rapid technological development and exploitation of ever more favourable generating conditions have vastly improved competitiveness in a very short period. This chapter draws on the personal experience of the authors to address some basic questions: What makes an offshore wind farm (OWF)? What are the key components and activities involved? And how is it legislated and regulated? This chapter documents the development and consenting process and outline issues for consenting in a range of different countries in Europe and in key developing markets elsewhere in the world. By necessity, a high-level approach is adopted as the reality is an assortment of site-specific and conflicting complexities. The majority of offshore wind developments are sited in European waters, with more than 3,500 turbines concentrated in the North and Baltic Seas contributing close to 14 GW of the 16 GW global total. While, at the time of writing in 2017, the UK is the global leader, Denmark and Germany follow closely behind. Deployments in China and more recently Japan, Taiwan, the Republic of Korea and the USA are increasing. Industry knowledge and experience in project delivery and regulation have had to improve rapidly to facilitate the expansion of offshore wind. Selecting an appropriate site to deploy an OWF is a critical step as it can potentially avoid or reduce potential impacts on sensitive species, particularly seabirds and marine mammals, as well as other marine users. The monitoring of wildlife at operational wind farms has helped to reduce some of the uncertainties regarding environmental interactions. The lessons learned must be shared as new markets evolve to avoid curtailment of offshore wind's contribution to decarbonisation.

Introduction

The movement away from reliance on fossil fuels and the global challenge to decarbonise electricity generation continue to gain momentum. The expansion of offshore wind in the twenty-first century has achieved unprecedented levels of competitiveness, making it a mainstream supplier of low-carbon electricity (Hundleby *et al.* 2017). Historically, wind farms began to move offshore in the early 1990s with a number of limited-capacity, nearshore projects such as Vindeby (4 MW, 1991) and Tunø Knob (5 MW, 1991) in Denmark; Irene Vorrink (16.8 MW, 1996) in the Netherlands; and Blyth (4 MW, 2000) in the UK. These projects were early demonstrators and deployment did not truly take off until the turn of the millennium. Yet, at that time, only a handful of European countries incentivised offshore wind and so opened the door to the next generation of large-scale projects, such as Utgrunden (2000, 10 MW) and Middelgrunden (40 MW, 2000) in Swedish and Danish waters, respectively; North Hoyle (60 MW, 2003), Scroby Sands (60 MW, 2004) and Kentish Flats (90 MW, 2005) in the UK; and Egmond aan Zee (108 MW, 2006) in the Netherlands. This move was principally driven by technological developments allowing the exploitation of more favourable offshore wind resources, and an increasingly favourable political climate surrounding offshore wind.

The majority of offshore wind developments are sited in European coastal waters, with the North and Baltic Seas contributing close to 14 GW of a total of nearly 16 GW worldwide at the time of writing in 2017 (Figure 1.1). In fact, the most recent statistics show Europe's cumulative offshore wind capacity actually reached 15.78 GW from 4,149 grid-connected turbines in 92 wind farms across 11 countries (Wind Europe 2018). The UK is the global leader in offshore wind, followed by Germany and then Denmark, the Netherlands and Belgium. Outside European offshore wind markets, significant deployments in China and more recently Japan, Taiwan, the Republic of Korea and the USA are gathering pace (see *Consenting process and issues below*) and taking an increasing share (Figure 1.1). Advances in the commercialisation of floating offshore wind are enabling countries with deep water, such as Japan and the USA (on the west coast and California in particular) to consider a new frontier for deployment.

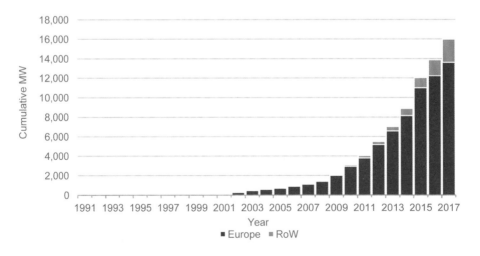

Figure 1.1 Cumulative European Union (EU) and rest of the World (RoW) or non-EU installed offshore wind capacity in megawatts from 1991 to 2017. (Renewables Consulting Group 2017)

The main benefit of offshore wind energy is the ability to deploy larger turbines in greater numbers, where wind resource is more favourable than onshore. There are, however, drawbacks when it comes to some installation and operational activities that have required adaptation owing to the harsher conditions experienced offshore. Moreover, as projects become located farther offshore, the cost escalates as larger foundations are required for deeper water, transit times to and from the wind farm increase, the weather windows within which installation vessels can operate become smaller, and there is reduced availability of suitable installation vessels and longer export cables. Technologies deployed must be able to withstand the often extreme conditions of the marine environment, and remain reliable in terms of their electrical output throughout their operational lifetime.

Offshore wind deployment still remains largely dependent on government subsidies, although significant industry progress has been made on cost reduction and technological advances, which is yielding positive results (Offshore Wind Programme Board 2016). In 2017, the German offshore wind market granted power purchase agreements to four projects for a total of 1,490 MW at a record-low weighted average strike price of €4.40 per MWh, with at least one project coming in at €0.00 per MWh, that is, at a zero subsidy. Such a ground-breaking strike price was enabled by a change in government policy in Germany to move towards a competitive auction process challenging developers to seek the most cost-effective options. Zero-subsidy projects have since been seen in the Netherlands. Whether similar strike prices can be achieved elsewhere remains to be seen, as there are varying factors that contribute to the proposed strike price. For example, in the UK, developers to date have covered not only the cost of development and delivery of the wind farm, but also transmission assets such as offshore platforms, export cables and onshore electrical infrastructure, and costs associated with the grid connection itself. Despite this, in 2017 the UK also experienced lower than expected strike prices for delivery year 2022/23, as part of the Contracts for Difference (CfD) Allocation Round 2, equating to a price drop of around 50% since February 2015 (see *Consenting process and issues* below).

Innovation in offshore wind technology is critical to the effectiveness of cost reduction measures and improved reliability during operation. However, before a project is built, a relatively long development and consent process is required to ensure that all potential impacts pass rigorous environmental and planning procedures, governed by legislation and in accordance with best practice. This process can take up to a decade before consent is awarded (Freeman 2014). From a developer's perspective, this represents significant risk and uncertainty as future innovation in technology, which could significantly reduce costs and environmental impacts, could become available by the time a project receives consent. Consequently, a project envelope (or Rochdale envelope) approach has been successfully adopted into UK, Dutch and Danish offshore wind development programmes to provide the necessary flexibility to future-proof or hedge against state-of-the-art components and systems such as larger and more efficient wind turbines, by the time major contracts are placed. The project envelope is highly iterative and allows developers to work towards finding an optimal design that is both technically feasible and commercially viable at the time services are procured from the supply chain. During the impact assessment stage, developers are required to define a set of project design parameters based on expected technical options that allow the realistic worst case to be assessed. These are mainly for the 'big-ticket' items such as wind turbines, installation vessels, cables and foundations, installation techniques and the like, but can also include bespoke solutions for project-specific environmental or technical issues. Without this flexibility built into planning, the alternative could result in a suboptimal site design and higher cost per megawatt and, ultimately, prevent the project from delivering the low-cost electricity generation necessary to compete in a subsidy auction.

Fortunately, project envelope designs have so far proven to be sufficiently wide enough to ensure that projects are future-proofed. However, a potential shortcoming to the envelope approach is the need for developers to seek the widest possible envelope design to insulate the project from any future uncertainty. Given that developers rarely build out their maximum project envelope design, the significance of cumulative impacts predicted during the planning rarely reflects the reality once the wind farm is built. This can negatively impact the ability of a neighbouring wind farm developer to achieve consent for their future development. In 2012, the UK Government refused permission to consent to Docking Shoal offshore wind farm (OWF) (540 MW) based on predicted cumulative impacts on Sandwich Tern *Thalasseus sandvicensis*. At the time, two other nearby OWFs, Race Bank and Dudgeon, were also seeking consent to build. The then Department for Energy & Climate Change (DECC) determined that if all three wind farms were built, there would be unacceptable and significant cumulative impacts on the local, internationally important Sandwich Tern population (DECC 2012a). Ultimately, the regulator refused Docking Shoal consent. Yet, once the other projects had eventually been constructed, one built out at 402 MW rather than the 560 MW that had been consented to, arguably providing capacity or 'headroom' for at least a part of Docking Shoal to have been constructed (Freeman & Hawkins 2013). More importantly, monitoring of another site nearby, Sheringham Shoal, has shown that Sandwich Terns avoided built turbines even before they became operational, to a greater extent than had been predicted (Harwood *et al.* 2017). The availability of rigorous monitoring data is of clear benefit in reducing environmental uncertainty and consenting risk.

The acquisition of monitoring and environmental baseline data to inform impact assessments is more challenging offshore than onshore, in part because of the need for specialist equipment and vessels to access remote and hostile areas of the sea. Reduced visibility and challenging weather conditions also make detecting and quantifying marine life difficult. Moreover, there are relatively few OWFs compared to their onshore counterparts, and so a smaller body of data exists.

Methodologies employed for monitoring may vary, making extrapolation and applicability to future projects uncertain. As a result, assessments of impact significance and the parameters on which they are based may be highly precautionary. A good example of this is the accepted methodology for assessment of collision risk for birds. The estimated number of collisions that may occur at a wind farm is calculated using a collision risk model. The model takes into account a number of assumptions associated with species' flying behaviour, turbine size, number of turbines, and various other factors such as swept area and rotor speed. These models are based on theoretical assumptions, as monitoring actual seabird collision events is extremely difficult. Direct observation requires the use of specialist equipment that can not only detect a collision, but also provide sufficient information to identify the species and be able to withstand exposure to the offshore marine environment. Although gathering empirical evidence on bird collisions is complicated, research is being undertaken to understand the true collision risk and so better inform collision risk models. Until sufficient evidence is available, however, the precautionary approach is adopted when considering collision risk to seabirds.

Significant progress is being made in understanding the environmental impacts of OWFs (e.g. Huddlestone 2010; Skeate *et al.* 2012; BSH & BMU 2014; Shields & Payne 2014; Köppel 2017). Researchers have targeted a number of key themes, including the establishment of baseline conditions, behavioural responses of species to the installation and operation of OWFs, and the development of methods to model and measure impacts in a more meaningful way. In addition, several collaborative research programmes have been set up to advance and improve the understanding of environmental impacts of offshore wind, including the registered charity Collaborative Offshore Wind Research Into the Environment (COWRIE) and the Offshore Renewables Joint Industry Programme (ORJIP), which involves several governmental organisations, offshore wind developers, statutory nature conservation bodies, academics and leading experts. Research outputs from both programmes are part of a growing body of evidence that, collectively, will better inform consenting decisions and improve the scientific knowledge base.

The technical chapters in Volume 4 of this series provide further detail on the understanding regarding the monitoring and mitigation of offshore wind environmental impacts. The more typical aspects that require assessment during the various stages of construction, operation and decommissioning are outlined in Table 1.1. The rest of the chapters in this volume provide further detail on the potential effects arising from offshore wind deployment upon coastal processes (Chapter 2), ocean dynamics (Chapter 3), and a range of marine life including benthos or seabed communities (Chapter 4), fish (Chapter 5), marine mammals (Chapter 6), migratory birds and bats (Chapter 7) and seabirds (Chapters 8 and 9).

Scope

The aim of this chapter is to provide an introduction to offshore wind as the basis for subsequent technical chapters. Specifically, this chapter will aim to provide a description of the physical components of an OWF, the installation and operational requirements, the key legislative tools employed in the consent process and the basics of Environmental Impact Assessments (EIAs). The breadth of engineering required to build a wind farm is sufficient for a weighty tome of its own. Therefore, detail in this chapter is limited to the basics required to understand what exactly developers are seeking permission for and where opportunities for interaction with environmental receptors may arise. As legislation can vary between countries, this chapter will only provide examples from a range of countries active in offshore wind, to enable the reader to understand the variances between countries. Specific examples will be provided for the UK, Germany, the Netherlands, China, Japan, Taiwan and the USA.

It should be noted that this chapter does not provide a systematic review of all available evidence. It is based mainly on the working experiences of the authors, drawn upon from European and Scandinavian waters, where offshore wind has the longest track record and experience from pre-, during and post-construction studies.

Themes

What makes an offshore wind farm?

An OWF involves a number of uniquely designed components working together to transform wind energy into electricity that is then transported to shore via export cables, and connected to the electricity grid via an onshore connection point. As demonstrated in Figure 1.2, a wind farm is made up of multiple wind turbines, supported by foundations and connected to other turbines and any offshore platforms via inter-array cables. If required, one or more offshore substations will step up the voltage of the power generated offshore before electricity is transmitted via one or more export cables to the onshore substation. Transmission may be alternating current (AC) or direct current (DC); in the latter case, offshore converter stations will convert AC electricity to DC before export. High-voltage

Table 1.1 Overview of the environmental effects of the varying activities involved in construction, operation and decommissioning of an offshore wind farm that typically require assessment.[a]

	Seismic survey	Wind turbines	Foundations	Cables	Vessels/helicopters	Aviation/navigational lighting	Onshore infrastructure[b]
Birds	–	Collision, migratory displacement, habitat loss	Attraction of resting birds	–	Disturbance, displacement	Attraction, collision, disturbance	Habitat loss, disturbance
Marine mammals	Noise	Operational noise/vibration	Piling noise/vibration	–	Disturbance, displacement	–	–
Bats	–	Collision, migratory displacement	–	–	–	Attraction, collision	Habitat loss, disturbance
Fish	Noise	Potentially reduced fishing within array	Piling noise/vibration	Electromagnetic fields	–	–	–
Benthos	–	–	Habitat loss	Habitat loss	–	–	–
Invasive species	–	–	Colonisation of structures	–	–	–	Spread during construction
Chemical/physical	–	Oil/solvent use	Hydrology/coastal process change, oil/solvent use	–	Oil/solvent use	–	Oil/solvent use, land take

a This list is not exhaustive and does not attempt to quantify the scale of effects and possible impacts.
b Onshore constructions typically cover, for example, cables and substation.

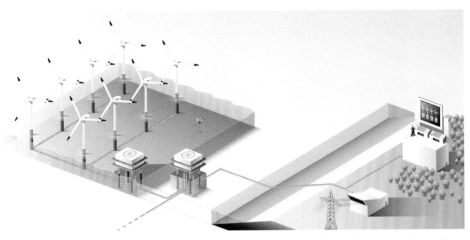

Figure 1.2 Diagrammatic illustration of the interacting components of an offshore wind farm. (Jorg Block 2011)

direct current (HVDC) transmission may be preferable when an OWF is sited far from shore, as electrical losses over distance are reduced. The onshore substation will further transform the electricity generated offshore to a voltage and form suitable for entry into the national transmission or distribution network. Depending on the location and purpose of each wind-farm component, most will have some potential for interaction with environmental receptors.

Wind turbine generator

The turbines are the most visible and instantly recognisable element of any wind farm. The most commercially established technology is the three-blade, horizontal-axis structure (Figure 1.3), comprising:

- **Rotor:** a hub and three blades.
- **Nacelle:** housing the generator and various components required to convert mechanical energy from the rotor into electrical energy.
- **Tower:** a cylindrical steel tube supporting the structure.

Aside from these primary components, turbines contain drives controlling blade pitch and yaw (positioning relative to wind direction), cooling and control systems, and anemometry.

Since the early days of offshore wind, when 0.45 MW turbines were installed at Vindeby OWF, offshore turbines have become much larger than their onshore counterparts. Two factors have driven the increase in size; the first being the availability of space offshore and the reduced need to consider other structures that make planning challenging, and the second being a significantly more favourable wind resource. However, the offshore environment brings additional complexities for wind turbines, including the need to manage saltwater corrosion and access difficulties associated with operations and maintenance (O&M) activities when exposed to the harsh conditions offshore. Many newly consented projects are proposing turbines of 8 MW-plus with blades measuring over 80 m in length, rotor diameters measuring up to 180 m and tip heights rapidly approaching 200 m (Figure 1.4). However, even as the first 8 MW turbines were being installed (for example at Burbo Bank Extension, near Liverpool, UK, in 2016), turbine designers were already planning machines capable of generating 10–15 MW and above.

Figure 1.3 A 3.6 MW wind turbine at DanTysk offshore wind farm in Germany, showing the rotor, nacelle and tower. (Paul Langrock)

Although the three-bladed horizontal-axis structure is the most commonly used technology to capture wind energy, there are alternatives. The most significant, but still much less common, is the vertical-axis wind turbine, with blades that spin around a central tower. As more innovative approaches to offshore wind generation come to the fore, this introduces additional uncertainty for both developers and those assessing applications for consent, as these technologies lack the body of empirical data needed to understand relative environmental impacts and support the permitting process.

Foundations

The foundation, or support structure, secures the turbine to the seabed. Foundations must be strong enough and deep enough to stabilise the entire construction under the

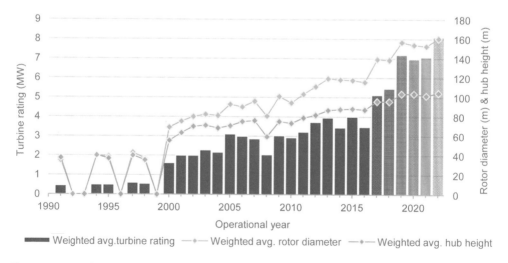

Figure 1.4 Weighted average turbine rating, weighted average hub height and weighted average rotor diameter from 1991 to 2017 and predicted in the future. (Renewables Consulting Group 2017)

constant battering of wind, waves, tide and current. Offshore foundation design is highly specialised, owing to marked differences in water depth, ground and metocean conditions between sites. These foundations are typically also much larger than their onshore equivalents, with standard monopiles ranging up to 6 m in diameter and extra-large monopiles exceeding 7 m in diameter (Figure 1.5). Offshore foundations must account for increased mechanical stresses and greater distance between the base of the tower and the seabed than their onshore counterparts. In the case of the monopile foundation, the height from the tower base to the full depth below the seabed can easily be as great as that from tower base to blade tip. Considering that the largest deployed turbines now approach a blade tip height of 200 m, this is a significant feat of engineering.

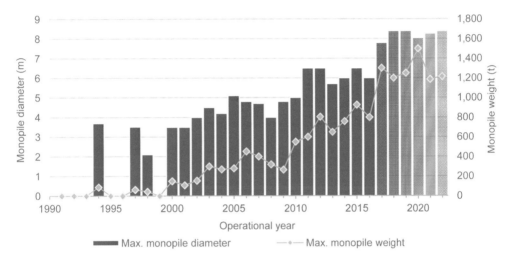

Figure 1.5 Maximum monopile diameter and maximum monopile weight from 1994 to 2017 and predicted in the future. (Renewables Consulting Group 2017).

Although numerous new designs are moving towards commercial viability, including anchoring for floating turbines at deeper water sites, the most common foundation types remain as follows:

- **Monopile.** These relatively simple structures comprise cylindrical steel tubes (Figure 1.6) driven into the seabed to a depth governed by site conditions. They are an economical choice in shallower waters but less suitable in deeper waters or for larger turbine models. The installation process for monopiles usually requires the use of a jack-up vessel, which is raised on four steel legs to ensure stability and the precise positioning of the vessel. The jack-up vessel will have a heavy-lift crane and offshore pile-driving hammer on board. The heavy-lift crane is used to lift the monopile into place with the driving hammer positioned above the pile, then a hydraulic system releases a heavy weight down on to the pile, driving it into the seabed. The forces required to drive the pile vary depending on the size of hammer required and will emit varying levels of peak noise when hammering. This noise is of particular concern as it can be harmful to marine mammals, damaging their hearing, either temporarily or permanently, depending on their distance from the source (Table 1.1).

- **Jacket.** Jackets are lattice-like structures of tubular steel, typically on four legs, each of which is fixed to the seabed using a pin-pile (Figure 1.7). They are highly resistant to force and therefore well suited to deeper water and strong tides or currents. Jackets

Figure 1.6 Offshore wind substructure designs according to water depth. From left (shallow) to right (deep): monopile, four-leg jacket, three-leg 'twisted' jacket/tripod, semi-submersible platform floating, tension leg floating and spar buoy floating. (From NREL 2015 – courtesy Josh Bauer, National Renewable Energy Laboratory)

are generally a more expensive alternative to monopiles, although innovations in design using less steel, such as the twisted jacket, are making them more economical. The initial installation process for a jacket is similar to that of the monopile, using a jack-up vessel and pile-driving hammer; however, the pin piles being used to pin the jacket foundation to the seabed are much smaller (around 1.2–2.6 m) than for a monopile and therefore require a smaller hammer size and reduced hammer energy. Once the pin piles have been secured into the seabed the jacket is lowered and secured on to the piles.

- **Tripod.** As the name suggests, tripods comprise three legs extending from a central column and fixed in place using pin-piles. These extremely stable structures are well suited to deep waters and require minimum seabed preparation. Again, they tend to be significantly more expensive than monopiles. The installation process for tripod foundations is very similar to that undertaken for the jacket foundation; however, pin piles in this instance tend to be even smaller and may therefore require a different hammer size.

- **Gravity base.** Unlike other designs, the gravity base rests directly on the seabed, relying on concrete and ballast weight to hold it in place. Significant seabed preparation is required for installation and handling can become difficult farther offshore, with deeper water requiring a larger and heavier structure. Depending on the design of the structure, the gravity-base foundation either can be floated out to position to be placed on to the seabed, or requires the use of a heavy

Figure 1.7 Jacket foundations ready for installation at Ormonde Offshore Wind Farm on the English west coast. (Ben Barden)

lift crane to position it in place. Installation of the gravity-base foundation does not require any piling and therefore will not emit noise from piling activities.

- **Suction bucket/caisson.** This cylindrical steel bucket-shaped structure is upturned so the open end is facing down. The suction bucket is only suitable for use on soft sediments and is a relatively new foundation type for offshore wind, but has been used extensively in the oil and gas industry. The installation process involves the steel bucket being placed on the seabed with the use of pumps, to pump the water out of the bucket, lowering the pressure inside the bucket causing the foundation to sink into the seafloor. This process does not require the use of pile-driving hammers and therefore will not emit noise from piling activities. Vattenfall's European Offshore Wind Deployment Centre, located in Aberdeen Bay and under construction at the time of writing, is deploying an innovative new foundation concept involving a jacket structure secured by suction caisson anchoring, negating the need for any piling operations at all during installation.

- **Floating.** Floating offshore wind technology is rapidly gaining commercial viability and is expected to become more common within the next few decades as interest increases in developing offshore wind projects in deeper waters. Although there are over 30 floating wind concepts under development at the time of writing, each with its own strengths and weaknesses depending on the country and conditions within which they are being deployed, it is expected that those concepts reaching full commercialisation will be relatively few. Most floating wind foundation designs fall into the following categories: semi-submersibles, spar-buoys, tension-leg platforms, multiturbine platforms and hybrid wind–wave devices.

The typical sequence of installation will see the foundation installed first, followed by the transition piece, then the tower and finally the turbine hub comprising nacelle and blades. The transition piece is a cylindrical steel tube placed over the top of a foundation and sealed into place to provide a connection between the foundation and the turbine base. Installation vessels will have to be capable of not only transporting the weight of constituent parts to site before installation, but also lifting and positioning the various components into place with precision (Figure 1.8). In some locations, turbulence caused by placing a solid structure into the seabed can result in erosion of sediments at the interface between the foundation and seabed surface, known as scour (Rees & Judd, Chapter 2). To protect the structure from becoming unstable, scour protection, typically in the form of rock or concrete mattresses, may be introduced.

Figure 1.8 A rotor being loaded on to a specialist vessel in port. (Ben Barden)

Cabling

Various cable types are required to transmit the electricity generated offshore to the onshore substation, including:

- **Inter-array cables** between individual wind turbines and between turbines and offshore platforms use AC of medium voltage, typically 33 kV, although higher voltage designs such as 66 kV are beginning to be used more often. These are advantageous in terms of reducing electrical losses within the array and reducing cable lengths required within the wind farm, as a larger number of turbines can be accommodated on a single circuit. The use of higher voltage inter-array cables can, in some cases, remove the need for offshore transformation, and therefore the need for offshore substation platforms, as the inter-array cables can essentially transport electricity from the wind-farm array directly to shore.
- **Connecting cables** between offshore platforms.
- **Subsea export cables** between offshore platforms and landfall may be high-voltage alternating current (HVAC) or HVDC. HVAC remains most common but as wind farms increase in size and move farther offshore, HVDC technology becomes increasingly cost effective.
- **Onshore export cables** are jointed directly to offshore export cables at landfall and then conduct electricity to the onshore substation.

Offshore cables may be laid directly on the seabed or buried using jetting, ploughing or trenching methods. Maintaining burial may be difficult because of the dynamic nature of the seabed. Therefore, additional measures may be used such as rock placement, concrete mattresses or ducting along sections with a higher risk of exposure. At landfall, subsea export cables are brought ashore for connection to onshore cables. This may involve trenching within the intertidal area or horizontal directional drilling underneath it (Figure 1.9). Subsea and onshore cables are then jointed together within an excavated bay typically set back from the landfall. Subsea cables are very large and expensive, and are designed specifically for subsea conditions. Therefore, it is advantageous to limit the total length required by jointing to onshore cables as close to landfall as possible.

Figure 1.9 Intertidal cable installation at Ormonde on the English west coast. (Tony West)

Offshore platforms

Voltage must be stepped up between the generating station and grid connection point, as onshore transmission and distribution networks typically operate at voltages upwards of 132 kV. Most OWFs will have one or more offshore substation platforms housing transformers and other electrical equipment. The number of platforms is often dependent on the size of the wind farm and the distance to shore. In cases where a wind farm is located close to shore (meaning several kilometres as opposed to tens of kilometres) and electrical losses over distance are minimised, offshore transformation may not be required at all and electricity can be directly transmitted to shore at the inter-array cable voltage. For example, Vesterhav Nord and Syd in Denmark and the Aberdeen Offshore Wind Farm in the UK are taking this approach. Sites utilising HVDC transmission have generally had more platforms, as they require converter platforms offshore as well as collector platforms. Platforms may also house accommodation for personnel working offshore and include the means of transfer by helicopter (Figure 1.10) and/or transfer vessels. Again, this tends to be for sites farther offshore. Each site has unique requirements and therefore the number and configuration of platforms will be site specific.

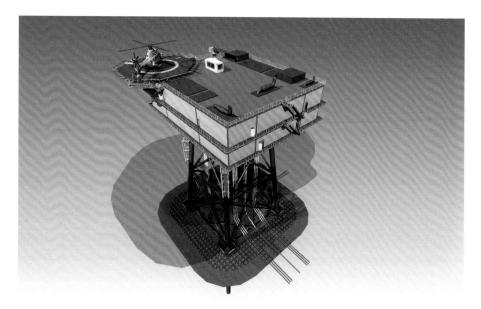

Figure 1.10 Diagrammatic illustration of the offshore substation platform at DanTysk, a 288 MW site in 31 m deep waters near Sylt Island in Denmark. (Strukton)

Maritime operations and logistics

It is important to note that in addition to the main structural components, the logistical requirements of an OWF during installation and operation are extremely complex, involving different vessel and equipment types, as well as the skilled technicians required to operate them. During the installation phase a variety of vessels will be working at the site to install wind-farm components, including:

- **Foundation installation vessels** are typically large jack-up vessels of 100–150 m in length, with four to six movable legs that are hydraulically winched or jacked down into the seabed, thus lifting the hull of the vessel clear of the water to create a stable platform from which to lift components.

- **Cable installation vessels** of varying types may carry out any pre-lay works such as jetting or grapnel run, as well as the installation of the inter-array and export cables. The installation operation often requires the laying of several sections of cable that are then jointed together. Pre-lay and installation operations may be carried out by the same vessel or by several vessels with different capabilities.

- **Offshore platform installation vessels** responsible for transporting and lifting the offshore substation platform, transformers and associated infrastructure into position. Often also jack-up vessels, they must be capable of transporting and lifting extremely heavy loads with precision.

- **Turbine installation vessels** are required to lift the tower and hub into place once the supporting foundation is in place and may be the same vessel as that used to install the foundations.

Once the wind farm has been installed and fully commissioned, the activity on site changes from construction to O&M. This usually requires far smaller vessels than construction, unless large-scale works such as major component exchanges are required, although this is relatively infrequent. Crew transfer vessels (CTVs) will usually transfer up to 12 technicians from the port to the OWF to undertake general maintenance of the structures. Depending on the size of the site, the wind farm may require several CTVs and occasionally other vessels for routine operations.

Legislation and regulation

Advances in the understanding of environmental issues associated with offshore wind began to be translated into a growing body of environmental legislation in the 1960s and 1970s. Since then, this body has become vast and a comprehensive guide to all relevant global environmental protection legislation with relevance to offshore wind is not possible in this chapter. Rather, the focus here is on key global agreements and examples of how nature conservation and environmental assessment obligations are delivered on a regional or national scale.

Global conventions

At its highest level, ratification of international conventions on the subject of nature and biodiversity has shaped regional and national legal frameworks and the regulation of developments with the potential to affect the natural environment, including offshore wind. The following are key examples:

- **Convention on Wetlands of International Importance especially as Waterfowl Habitat**, otherwise known as the Ramsar Convention. This convention came into force in 1975 and introduces specific protection for important wetland habitats and their natural and economic benefits through the establishment of a global network of 'Ramsar sites'.

- **Convention on the Conservation of European Wildlife and Natural Habitats**, otherwise known as the Bern Convention. This convention, which came into force in 1982, requires signatories to meet specific obligations to conserve and protect designated species and habitats.

- **Convention on the Conservation of Migratory Species of Wild Animals**, otherwise known as the Bonn Convention. This convention, which came into force in 1985, introduced requirements for the conservation of migratory species and their habitats via a number of legally binding agreements such as the African–Eurasian Migratory Waterbird Agreement (AEWA) and Agreement on the Conservation of Small Cetaceans in the Baltic, North East Atlantic, Irish and North Seas (ASCOBANS).

- **Convention on Biological Diversity (CBD)**. This convention came into force in 1993, and acknowledges the complexity of biological life and natural processes sustaining it, while emphasising the need for sustainable exploitation of natural resources.

- **United Nations Convention of the Law of the Sea (UNCLOS)**. This convention came into force in 1982 and defines the rights and responsibilities of nations with respect to their use of the world's oceans, establishing guidelines for businesses, the environment and the management of marine natural resources.

- **Convention for the Protection of the Marine Environment of the North-East Atlantic (OSPAR)**. This convention came into force in 1992 and is the current legislation regulating international cooperation on environmental protection in the

North-East Atlantic. It combines the 1972 Oslo Convention on dumping waste at sea and the 1974 Paris Convention on land-based sources of marine pollution.

- **Convention on the Prevention of Marine Pollution by Dumping of Wastes and Other Matter**, otherwise known as the London Convention. This convention came into force in 1972 and defines an agreement to control pollution of the sea by dumping, and to encourage regional agreements supplementary to the convention, such as the Oslo Convention.

European Union legislation

The European Union (EU) delivers the requirements of international conventions through a series of European Directives, including several that are key to the development of offshore wind, particularly in countries such as the UK, Germany and the Netherlands. As the majority of global offshore wind capacity has been installed within EU waters, it is the EU legislation that has framed the context within which offshore wind has been developed. For example:

- In the UK, Strategic Environmental Assessment (SEA) 2001/42/EC and Environmental Impact Assessment (EIA) 2014/52/EU Directives (as amended). SEA is the assessment of environmental consequences of public plans or programmes, whereas EIA is the assessment of public and private projects considered likely to have significant environmental impacts. Each EU member state has a different approach to the EIA screening process; some will consider EIA necessary owing to the number of proposed turbines, while others will base the decision on the energy generation potential of the project.

- Council Directive 2009/147/EC on the conservation of wild birds ('Birds Directive') and Council Directive 92/43/EEC on the conservation of natural habitats and of wild fauna and flora ('Habitats Directive'). These are key pieces of legislation in the regulation of offshore wind in order to meet the obligations of the Bonn and Bern Conventions. The implementation of these directives led to the establishment of the Natura 2000 network of sites, comprising Special Protection Areas (SPAs) for birds and Special Areas of Conservation (SACs) for habitats (EC 2010). EIA of an offshore wind project must detail any likely significant effect on the integrity of a designated site.

- In the Netherlands, the Nature Conservation Act states that an appropriate assessment has to be conducted for developments such as an OWF located within or deemed likely to have significant effect on the Natura 2000 area in order to evaluate the likely effect of the development on the conservation objectives of Natura 2000 areas. If the negative effects are considered to be 'not significant' a permit can be granted. If considered significant despite any proposed avoidance or mitigation measures, then according to Article 6(4) of the EU Habitats Directive, consent can only be granted if there is no acceptable alternative solution and there are imperative reasons of overriding public interest and compensatory measures have been secured. If protected species present are judged unlikely to be harmed significantly, an exemption can be provided and development may be consented.

- Water Framework Directive (WFD) 2000/60/EC and Marine Strategy Framework Directive (MSFD) 2008/56/EC introduced targets for the improvement and sustainable exploitation of aquatic ecosystems and wetlands, and marine waters, respectively.

Responsibility for transposing the requirements of each piece of EU legislation into national law lies with the individual member state. This can lead to conflicting interpretations between countries. Within the southern North Sea, for example, a significant region for offshore wind, there are multiple international marine boundaries within a relatively small area, and the nations within these boundaries each have their own strategy for offshore wind development and interpretation of the EU Directives. As projects grow larger and move closer to these boundaries, transboundary issues may become more significant, with potential consequences for neighbouring nations.

Non-European Union legislation

As the development of offshore wind increases globally, legislation in non-European countries is also starting to frame the context of offshore wind deployment. However, as OWFs are relatively new to these countries, some adaptation is required to ensure that the legislation can account for the development of wind energy within the marine environment. Depending on the country, the legislation may differ and can be illustrated by the differences in legislative provision among a number of countries such as Japan, China, Taiwan and the USA. These nations are becoming more prominent in the offshore wind market and will be discussed in greater detail within the *Consenting process and issues* theme, later in this chapter.

It should be noted that all projects requiring loans from international banks also have to conform to World Bank Equator principles and International Finance Corporation performance standards. This is not formally a legislative requirement but does in many countries, especially developing nations, set a rather high standard for environmental studies and assessments compared to Europe and North America for example, and may be more stringent than local legislation.

A final note on non-EU countries is reserved for the, soon to be ex-EU, UK. At this time, an EU Withdrawal Bill is being debated by the UK Government and it is not yet clear how exactly the UK will extricate itself from the jurisdiction of European law. It is likely that a large body of existing legislation will be transferred directly to UK domestic law, as this is most straightforward approach, and would result in little discernible change to current consenting regimes, at least in the short term. However, from the formal 'Brexit' date onwards, the UK will no longer be obliged to comply with EU requirements. Within the arena of environmental regulation, there has been concern that independent administration of neighbouring countries can have a detrimental effect on the assessment and management of environmental impacts, especially given that environmental stressors such as noise and emissions to air and water, as well as the receptors they affect, such as water masses, birds and marine mammals, do not respect national boundaries.

Development and consenting process

Planning approaches to consenting to offshore wind projects and the regulatory agencies involved differ between nations. In European countries where offshore deployment has matured, the regulatory systems of governance share common ground in adopting European legislation for the protection of wildlife and climate change abatement through binding renewable energy targets.

Often, the first stage in developing a national strategy for offshore wind deployment is a strategic-level assessment or master plan to evaluate the potential wind resource against possible physical, biological and socio-economic constraints over a large geographic area.

The task of defining and assessing an individual project may lie with a governmental body or fall to the developer, depending on the consenting framework in place in that particular country. Under a public tender process, the government is responsible for delivering an EIA and obtaining consents. Once this is complete, developers are invited to tender for the site and attempt to define an optimal project design, or design options, while selling themselves as the best party to deliver it. Denmark is an example of such a system, as is the Netherlands.

The public tender process is advantageous for the developer as it carries less risk in that the onus is on the government to deliver early development and obtain consent, and the cost of the EIA will only be payable if the bid is successful. In developer-focused systems such as that in the UK, the risk and pre-consent cost are shifted on to the developer, who must conduct all environmental assessments and early engineering work when there is no guarantee of a consented project or CfD subsidy award at the end of it (see the next theme, *Consenting process and issues*).

Once a project has been defined, within the EU at least and regardless of who is responsible for delivery, adoption of the EIA Directive and other associated legislative tools has resulted in the development of a broadly consistent approach to EIA for large-scale developments. The steps in this procedure are as follows:

- **Site selection and feasibility study.** The earliest stages of a project will involve both environmental characterisation to highlight issues likely be significant during consenting, and acquisition of reliable data on geotechnical and metocean parameters, which will govern the feasibility of installation and operation in that particular area. Traditionally, these early stages involve little engineering input, although there are significant advantages to a higher level of input in reducing development risk. In simple terms, the higher the level of certainty of physical and environmental site conditions, the greater the confidence that the project has been designed appropriately according to those conditions and is ultimately buildable. Water depth, seabed conditions, marine and coastal processes, wind resource, distance from shore, and suitable cable landfall sites and port facilities are all important factors to consider, both in terms of financial feasibility and in helping to predict the likely level of impact that the prospective development may have on environmental receptors.

- **Screening and scoping.** These terms are defined within the EIA Directive and associated national regulations. Screening is the process whereby the outline detail of the project in question is presented to the appropriate regulatory body, which then advises on whether the project requires an EIA. Screening is not mandatory under the EIA Directive, but is useful to ensure that the developer's understanding is in line with that of the authorities. The next stage of the process, scoping, is mandatory if a development has been 'screened' as requiring EIA. Scoping determines the coverage and content of assessment, setting out the evidence that the competent authority and its advisors wish to see in support of an application for consent.

- **EIA.** Impact assessment begins with definition of the project parameters for which a consent is sought, establishment of baseline conditions and characterisation of environmental receptors with the potential to be affected by the development. The impact assessment proper uses standard methods to estimate the significance of impacts on each environmental receptor based on what is known about that particular receptor's sensitivity, as well as the likely magnitude and duration of impacts. The results of the EIA are presented in the form of an Environmental Statement (ES). Within the ES, reasonable alternatives must be considered, and all conceivable direct and indirect effects and impacts considered, across all topic areas defined during scoping and during both construction and operation of the development. Some EIAs will also include consideration of decommissioning; however, the timescales involved often necessitate a further assessment of impacts associated with project termination to be undertaken closer to the time of decommissioning to account for developments in technology and best practice. Where impacts are considered unacceptable, appropriate mitigation measures must be proposed. All technical data must be presented in the form of appendices. An ES for an OWF is invariably an extremely large document or series of documents (which for some projects have reached over 12,500 pages) as a result of the breadth and depth of the information provided on a wide range of physical, chemical and biological receptors (Table 1.1).

- **Consultation.** Regulatory agencies and various advisory bodies within a specific country, and neighbouring countries if applicable, will be provided with the opportunity to analyse the data presented and provide a formal response, which must be considered by the developer during the consenting process. At the same time, the ES must be made publicly available so that anyone with an interest in the project can access the information and provide comment if they so wish. The number and length of responses from all stakeholders can be significant, and all must be considered.

- **Determination.** After consultation, the regulator will consider all evidence and comments from stakeholders and will work with senior officials to determine whether the project proposed can be granted a licence. Once a decision has been made by the regulator on whether to grant a consent or consents to develop a particular site, this decision and the justification for it must be made public. There will be a certain length of time, sometimes undefined, during which a consent decision is subject to legal challenge. Although a legal challenge is rare, it is not unheard of and can take some time to resolve, as demonstrated by the Royal Society for the Protection of Birds (RSPB) mounting a legal challenge against the consent to four Scottish offshore wind sites, which took just under 3 years to resolve (https://www.telegraph.co.uk/business/2017/11/07/three-year-legal-war-local-birds-won-2bn-offshore-wind-farm/).

- **Consent compliance.** Even under the most favourable regime, the process from early stage feasibility to irrevocable consent can take many years. However, any granting of consent will be accompanied by a number, perhaps even a very large number, of conditions requiring discharge before construction can begin. These will include the various environmental monitoring and mitigation measures thought to be appropriate according to the EIA and subsequent discussions, as well as a suite of management plans to control various aspects of construction and operational activities. Failure to meet any of these many requirements would result in a breach of consent and potentially the revocation of one or more permits.

Consenting process and issues

Consenting processes and the issues encountered vary between countries. To illustrate, this section provides a brief description of key legislation, identifies the principal authorities responsible and their roles, and outlines specific requirements in several countries in north-western Europe with an established OWF industry, as well as case studies from emerging markets elsewhere in the world.

The UK

Offshore wind in the UK has undergone six leasing rounds since 2000 and, by design, has played a significant role in the rapid expansion of the emerging offshore market by incentivising a competitive leasing process for large-scale commercial projects. This step change in offshore wind demonstrated the benefits of a strategic approach to deployment by providing developers with the means to gain technological, economic and environmental expertise at a scale not previously witnessed. Round 3 aimed to deliver up to 25 GW by 2020 and, against all expectations, 96 UK and international companies and consortia registered for a piece of seabed real estate termed 'Round 3 Zones'. Unlike previous leasing rounds, The Crown Estate (TCE) invested £100 million in the development stages and worked closely with developers towards solutions to address consenting challenges.

In 2013, London Array Offshore Wind Farm (630 MW) came on line as the largest OWF in the world, although this title was stolen by the 659 MW Walney Extension in 2018 for a brief time until the 1.2 GW Hornsea Project One, planned to become operational in 2020. The latter project won its subsidy via the UK CfD regime, which has seen a spectacular reduction in the levelised cost of energy (LCoE) of offshore wind in less than 3 years. The UK CfD regime is a government-supported mechanism whereby the generator of low-carbon energy is paid the difference between the contractual strike price and the wholesale market price for electricity by the Low Carbon Contracts Company, a government-owned company. In the event that wholesale electricity prices exceed the agreed strike price, the generator is obliged to pay the difference back. The system allows the generator to have more certainty on their investment return during the 15-year CfD lifetime, while keeping the cost to consumers down.

The 2017 CfD Allocation Round 2 saw strike prices for offshore wind plummet to £57.50, an almost 50% reduction from the lowest bid price in the 2015 Allocation Round 1 of £114.39. The challenge put forward to the industry by the DECC to reach an LCoE of £100 per MWh by 2020 (DECC 2011) has been delivered and exceeded well ahead of schedule. One of the biggest enablers for this UK record low strike price has been development of larger turbines that are soon expected to exceed 10 MW, as they are more efficient in converting kinetic energy from wind into electricity and can utilise a greater range of wind speeds to generate electricity.

The UK has a relatively mature consenting regime, with offshore wind growth in excess of 1 GW each year. In England and Wales, the key legislation governing offshore wind projects above 50 MW is the Planning Act 2008 (as amended by the Localism Act 2011). Developers are required to obtain a Development Consent Order from the relevant Secretary of State, and in order to do so must fulfil many pre-application requirements, including detailed EIA and consultation, before undergoing a strictly timetabled examination process via the 'PINS' process, overseen by the Planning Inspectorate (PINS). Although this system is heavily front loaded and resource intensive, it has advantages in terms of planning as, once the application has been accepted, there is a definitive date for determination.

The Scottish system is somewhat different, as there is no set timetable for determination. Consent must be obtained under Section 36 of the Electricity Act 1989 and a Marine Licence awarded under the Marine (Scotland) Act 2010 before a project can be constructed. Developers must also submit an ES in support of their application to meet the requirements of the EIA regulations.

The main consenting issue for many sites within UK waters remains ornithology, and with the number of sites increasing, cumulative ornithological impact is a growing concern. As a result, much attention has been focused on ornithological issues through a variety of research initiatives to improve our understanding of environmental effects. The forerunner of these was the COWRIE initiative. The number of projects and topic areas supported was very wide, including historic environment, coastal processes, fish and seabed communities, and the potential impact of electromagnetic fields (Huddleston 2010). Many projects focused on birds, including the development of standardised survey techniques (Camphuysen *et al.* 2004) and developing guidance on cumulative impact assessment (King *et al.* 2009), as well as research on individual species or groups, including assessing the displacement of Common Scoter *Melanitta nigra* (Kaiser *et al.* 2002), tracking swans on migration routes (Griffin *et al.* 2010) and quantifying the use of coastal waters occupied by wind farms by terns (Perrow *et al.* 2010).

After the closure of COWRIE, the Strategic Ornithological Support Services (SOSS) projects were established, aimed at providing ornithological advice to inform decision making for offshore wind. Among other projects, SOSS sought to understand how different species are displaced following offshore wind construction in an area, and provided guidance on methods to assess bird collision risk and monitor collision events. In 2012, industry and government stakeholders created the ORJIP, which jointly funded research into the impacts of offshore wind on the marine environment. The three most notable projects undertaken through ORJIP are:

- **The Bird Collision Avoidance study** monitored how birds behaved within and around the operational Thanet Offshore Wind Farm, with the aim of better understanding their avoidance behaviour and risk of collision (Skov *et al.* 2018).
- **The Acoustic Deterrent Devices study** aimed to determine the efficacy of acoustic deterrent devices in deterring seals, Harbour Porpoise *Phocoena phocoena* and Common Minke Whale *Balaenoptera acutorostrata* from offshore wind construction zones (Sparling *et al.* 2015; McGarry *et al.* 2017).
- **The Piling and Installation Restriction study** aimed to understand the impacts of piling during the Atlantic Herring *Clupea harengus* spawning season (Boyle & New 2018).

These significant research projects have considerably improved our understanding of the impacts of offshore wind on the marine environment. However, further research is still required, particularly in relation to cumulative impacts, which could eventually become a barrier to further development in UK waters. But, as it stands, only two sites have been rejected as a result of an unacceptable predicted impacts upon particular bird species: Shell Flats in Liverpool Bay and Docking Shoal in the Greater Wash (DECC 2012b). In Scotland, the initially successful judicial review brought by the RSPB against four large-scale offshore wind projects in Scottish waters was overturned by the Inner House of the Court of Session following a lengthy legal battle, and the previously quashed consents reinstated. A further appeal by RSPB to the UK Supreme Court was rejected in November 2017 (https://www.supremecourt.uk/news/permission-to-appeal-decisions-07-november-2017.html).

Germany

The key pieces of legislation covering the protection of nature and biodiversity are the Federal Nature Conservation Act (BNatSchG) and Offshore Installations Ordinance (SeeAnIV). The BNatSchG sets guidelines regarding protected areas, including Natura 2000 sites, biotopes and species. SeeAnIV regulates procedure throughout the lifetime of an OWF, from authorisation to construction and decommissioning. The competent authority is the Federal Maritime and Hydrographic Society of Germany (BSH), and the German Standard for Environmental Impact Assessments (StUK), published by the BSH, sets out mandatory requirements for EIA and monitoring of offshore wind installations. The highest priority for the German regulatory system is to minimise the impact on the natural environment (BSH & BMU 2014), which includes prescriptive requirements for protection of certain species that must be adhered to by law. For example, while the consenting regime in most countries requires piling noise during foundation installation to be minimised as far as possible to protect marine mammals, in Germany there is a specific requirement to stay below 160 dB sound exposure level at 750 m from the source. To achieve this, a number of experimental noise-mitigation techniques have been tested and continue to be tested to develop a solution that can achieve these specific requirements, such as the AdBm noise-abatement system, which was tested at the Butendiek wind farm in 2014.

A further example of the commitment to monitoring is provided by the Alpha Ventus project, originally named Borkum West OWF, the pioneering project in German waters that was approved in 2001. A grant of €5 million from the German Federal Ministry for the Environment, Nature Conservation and Nuclear Safety (BMU) enabled the German Offshore Wind Energy Foundation to buy the rights to the wind farm in 2005. The site was subsequently leased to Deutsche Offshore-Testfeld und Infrastruktur, a consortium of the energy utilities EWE, Vattenfall and E.ON. A five-year research project accompanied the wind farm's construction and operation to evaluate the BSH StUKplus. The results of this research on oceanographic processes, benthic communities, fish, seabirds, migratory birds and marine mammals are presented in a book (BSH & BMU 2014) and reflected elsewhere in this volume (see Broström *et al.*, Chapter 3; Dannheim *et al.*, Chapter 4; Hüppop *et al.*, Chapter 7; King, Chapter 9; Perrow, Chapter 10). The Alpha Ventus research programme provides something of a model of assessment for developing markets.

The Netherlands

The authority responsible for the seabed in the Netherlands is the Rijkswaterstaat, also known as the Directorate-General for Public Works and Water Management. The Dutch approach to gaining consent is similar to that in Denmark, where the government takes on the cost and risk of site development including consenting. In July 2015, the Offshore Wind Energy Act came into force whereby the Dutch Ministries of Economic Affairs and of Infrastructure and the Environment were made responsible for the execution of the law and took over responsibility from offshore wind developers regarding spatial planning arrangements and environmental assessment of proposed projects. In addition, the government took responsibility for the offshore grid connection, with developers taking responsibility to ensure that all necessary permits are obtained for the realisation of the project. As part of their responsibilities, the Dutch Government will decide which sites can be developed and will undertake all relevant environmental data collection. This information is released publicly and made available to developers interested in bidding for the site. Therefore, in contrast to the UK approach, developers bidding for a site in the Netherlands are only required to consider, and demonstrate, the financial and technical feasibility of the project.

The European requirements of the Habitats and Birds Directives were transposed into Dutch law via the Natuurbeschermingswet 1998 (Nature Conservancy Act 1998). In January 2017, the Nature Conservation Act replaced the Flora and Fauna Act, 1998

Nature Conservancy Act and the Forestry Act, combining legislation from all three for application beyond 12 nautical miles offshore and covering the entire Dutch continental shelf.

Significant research has been undertaken at Dutch OWFs to understand the impacts of offshore wind on the marine environment. In a similar vein to Alpha Ventus in German waters, research at the OWF Egmond aan Zee, built by Noordzeewind (a joint venture of Nuon Duurzame Energie and Shell Wind Energy), aimed to evaluate the economic, technical, ecological and social effects of OWFs through the development of a site monitoring and evaluation programme. The site was constructed in the summer of 2006 and has been in operation since January 2007. Monitoring was again undertaken across a range of trophic levels from bivalves (Bergman 2010), to both demersal (Lambers & Hofstede 2009) and pelagic (Ybema *et al.* 2009) fish, with further special studies on Atlantic Cod *Gadus morhua* and Sole *Solea vulgaris* (Winter *et al.* 2010), as well as birds (Krijgsveld *et al.* 2011; Poot *et al.* 2011) and marine mammals (Brasseur *et al.* 2012; Scheidat *et al.* 2012). The general effects for all groups over the short term, with a focus on the operational phase, are summarised by Lindeboom *et al.* (2011).

China

China is rapidly becoming a major player in the offshore wind industry, having exceeded the installed capacity of a number of European countries in recent years (Figure 1.1). At the time of writing in 2017, China has the greatest offshore wind installed capacity of all non-European countries (Global Wind Energy Council 2015) following the setting of ambitious targets by the government of 5 GW installed by 2015 and 30 GW by 2020 (Carbon Trust 2014a). However, achieving these targets has proved more difficult than anticipated, with only 1.6 GW installed by 2016. The main factors driving the slower pace of development were the lack of a complete local supply chain and the considerable investment and risk associated with developing offshore wind in a new market, which discouraged investors.

The consenting process is much faster in China than in many other nations, taking around 2 years to obtain the various permits required from government agencies. Under the Chinese EIA Law (1999), developers must conduct an EIA for construction projects that have a potential of having a significant environmental impact and submit an Environmental Impact Report to the State Environmental Protection Administration. In 2013, the National Energy Administration (NEA) devolved powers to provincial governments to try to accelerate consenting, although this had limited effect in the absence of clear implementation guidance.

Historically, without clearly defined roles within government agencies, conflicts between the NEA and other government agencies have arisen. In an attempt to address these conflicts, the NEA and State Oceanic Administration released a set of regulations and frameworks for offshore wind that allocate responsibilities between agencies as well as introducing rules, such as a minimum distance of 10 km from shore for all new projects, to avoid conflict with other marine users and thereby facilitate the consenting process.

Japan

In Japan, national policy has shifted in favour of renewable energy following the Fukushima nuclear incident (International Atomic Energy Agency 2015) and in response to public opposition to onshore wind on the basis of visual and noise disturbance (Carbon

Trust 2014b). Japan has significant scope for offshore wind development, with the Japanese Wind Power Association ambitiously proposing that a possible 37 GW of capacity could be installed by 2050, although installed capacity stood below 60 MW in February 2017.

Japan also has one of the most favourable subsidy schemes of any country, although its grid infrastructure requires significant upgrade, and this will temper progress for some years to come. Japan faces particular technical and environmental challenges, being located in an earthquake zone and experiencing regular tropical storms. Coastal waters are deep even close to shore, and as a result floating turbine projects feature heavily on the future agenda (GWEC 2014). The environmental impacts of floating turbines are being investigated at the Kabashima floating turbine demonstration project (Goto Floating Offshore Wind Turbine 2012), and it is hoped this will provide a model for future EIAs. Additional research projects at the Choshi and Kitakyushu offshore demonstration projects are looking at the impacts of fixed turbines, with a similar view to streamlining the consenting process. The Ministry of Land, Infrastructure and Transport is responsible for permitting and the Ministry of Environment for regulating EIA in offshore wind installations and elsewhere.

Taiwan

Taiwan has abundant wind resources along its west coast, with the installed capacity of 8 MW in 2017 coming from demonstration sites (see Hu 2013). To meet the requirements of Taiwan's nuclear phase-out policy framework, the government has set a development target of at least 3.5 GW available for feed-in tariff and a maximum 6.5 GW of offshore wind power by 2025. In the short term, by 2020, the aim is to see 520 MW built out in Taiwanese waters. A high feed-in tariff price draws the attention of heavyweight offshore wind developers; however, those policy targets also come with multiple challenges that increase project uncertainty. Once the permitted capacity is over the 3.5 GW available for the feed-in tariff, developers will be invited to participate in a bidding competition.

According to the rules announced by the Bureau of Energy, all potential developers must go through an EIA process to secure the rights to develop a specific wind farm. The main environmental receptor of note in Taiwan is the Chinese White Dolphin *Sousa chinensis*, and the need to limit the impacts on this species has led to the downscaling of most offshore wind projects. Developers also need to go through a selection process with a certain percentage of local content as a requirement. The Taiwanese government has set a high local industry development strategy of 71%, but the details are unclear. In addition, a lack of health, safety and environmental standards, working vessels, substation sites and grid connections may increase the burden of offshore wind development in Taiwan.

Taiwan has the same environmental threats as Japan, namely earthquakes and tropical cyclones that frequently visit the island. The highest standards of design, manufacturing and installation are necessary to ensure that both the developers' assets and the country's electricity supplies are protected under such conditions. Grid connection is another challenge, as generation tends to be focused in particular geographic areas, leading to network constraints. Most projects are crowded along the coastal areas of Chang Hua County, while most of 20 GW photovoltaics projects are located around Yun Lin County, Chang Hua County and Chia Yi County nearby. The main grid operator, Taipower, is exploring several solutions to ease the load tension between both wind and solar development in central Taiwan and ensure that energy is distributed to high-demand areas.

The USA

The first offshore wind energy development project in the USA was Cape Wind. In 2001, the developer, Cape Wind Associates, submitted an application to the Army Corps for a project in Nantucket Sound, Massachusetts, on the Outer Continental Shelf (OCS), and in 2002 was granted a permit for a meteorological tower, which was installed in the same year. The developer then applied for a commercial lease from the Department of the Interior's Minerals Management Service in 2005.

The OCS is regulated by the federal government and begins, for most states, 3 nautical miles offshore and extends for at least 200 nautical miles to the edge of the Exclusive Economic Zone. Under the Energy Policy Act of 2005, offshore wind leasing on the OCS became the responsibility of the Bureau of Ocean Energy Management (BOEM), part of the Department of the Interior, which is responsible for all leases, easements and rights of way for renewable energy development activity. The mission of BOEM's Office of Renewable Energy Programs is to regulate environmentally responsible offshore renewable energy development activities. BOEM carries out an EIS, required under the National Environmental Policy Act, as well as technical review of the Construction and Operation Plan (COP). Once the COP has been approved, the lessee has to submit a Facility Design Report and a Fabrication and Installation Report before construction of a project can commence.

For Cape Wind, the final EIS was published in 2009, followed by the Environmental Assessment and Finding of No New Significant Impact in April 2010. In October 2010 at the American Wind Energy Association conference in Atlantic City, the then Secretary of the Interior, Ken Salazar, signed the first commercial offshore wind lease in the USA for Cape Wind for a 33-year term comprised of a five-year assessment term and 28-year operations term. In 2015, the developer submitted a request for a two-year suspension of the operations term, which BOEM approved in July 2015. A further application for suspension was made in June 2017. Unfortunately, after this extremely lengthy process it was announced in December 2017 that the developer had decided to terminate the lease.

Despite this setback, the 'Smart from the Start' wind energy initiative for siting, leasing and construction of new projects in the Atlantic OCS launched in 2010 is starting to prove to be a successful process for US offshore wind. Smart from the Start is a coordinated approach to spatial planning and leasing, not too dissimilar to TCE's zonal development process for UK Round 3. Priority Wind Energy Areas (WEAs) are identified for potential development, and data are gathered by BOEM from key federal agencies, for example on environmental constraints, physical characteristics and other sea users, which are then made available to potential bidders for lease auctions. The key objective of this process is to accelerate development and permitting through better coordination and siting of offshore wind areas that reduce the risk of conflict with wildlife and other marine users. The Bureau of Land Management had previously used such a strategy to site solar projects in the western USA.

In 2012, BOEM reached agreement on a commercial lease, the first awarded under the Smart from the Start initiative, with prominent US developer Bluewater Wind for an area of the OCS off Delaware. Following a Determination of No Competitive Interest, a lease was executed with Bluewater Wind in November 2012. BOEM has so far identified WEAs off Massachusetts, Rhode Island/Massachusetts, New Jersey, Delaware, Maryland and Virginia. Competitive lease sales have already been concluded for two WEAs. On 31 July 2013, BOEM held the first ever competitive lease sale for renewable energy in federal waters. Deepwater Wind, backed by New York investment company the D.E. Shaw Group, won the first two leases (North and South) in the Massachusetts/Rhode Island WEA with a

$3.8 million bid. The auction reportedly lasted for 11 rounds. A commercial lease sale was held for the Virginia WEA in September 2013; Dominion Virginia Power was the winner with a $1.6 million bid in an auction lasting for six rounds.

Concluding remarks

The expansion of offshore renewable energy generation is a response to the world energy trilemma of carbon emission reduction, security of energy supply and reduction in energy costs (World Energy Council 2016).

Those delivering OWFs face unique technical and operational complexities due to the rapid pace of development and the challenging environments in which these structures are constructed. Since the first commercial offshore wind turbines were installed at Vindeby in 1991, the pace of technological development for all offshore wind components has continued to increase, with the development of more efficient, more robust and, in some cases, larger components, from wind turbines and foundations, inter-array and export cables, to platforms and construction equipment, to improve the productivity of wind farms. Moreover, innovators are developing technology to take the interactions with the marine environment into consideration and to address and reduce potential environmental impacts.

The interactions between offshore wind technology and environmental receptors are continuing to be better understood. Notably, the majority of OWFs installed within European waters to date have been of a similar nature. However, as offshore wind begins to expand into Asia and the USA, and beyond, there will be differences in both the ecosystems in which wind farms are installed and the species within them, with regard to their physiology, behaviour and sensitivity to disturbance.

The development of floating offshore wind structures will soon enable some countries to achieve 100% power from renewable energy and others to reduce significantly their dependence on fossil fuels. With the increased adoption of offshore wind and technological development allowing greater efficiency and dependency of renewable energy generation, the price of offshore wind is expected to reach parity with fossil fuels, as demonstrated by the recent round of UK CfD subsidy allocation, the results of which were announced in September 2017.

An international collaborative effort between academia, industry and regulation is required first to address the gaps in our baseline understanding and then to develop effective methods for quantifying the impacts of structures and operations on the marine environment. There are signs of progress in this area but further work, and the funding required to deliver it, is essential to understand the real impacts of offshore wind and replace reliance on models and precautionary estimates with real empirical data. If done effectively, this will achieve the twin aims of gaining a better understanding of the global marine environmental baseline and the subsequent development of suitably robust analytical techniques for assessment and monitoring environmental impacts. This will then facilitate standardisation of international approaches to assessment and regulation, thereby streamlining the consenting process and allowing the continued expansion of offshore wind without compromising our natural environment and resources.

Acknowledgements

The authors would like to thank the following friends and colleagues for contributing their thoughts and time to this chapter: Piers Guy, Andy Paine, Kathy Wood, Mary Thorogood, Eva Philipp, Pietr van Zijl, Judith Jehee, Nicole Kadagies, Kazuaki Hosokawa, Ole Bigum Nielsen, Willem van Dongen, Sam Park, Hannah Hendron and Michael Chang. Steve Freeman and Gareth Lewis of the Renewables Consulting Group Limited assisted in completing the editorial revision of an earlier draft of this chapter.

References

Bergman, M., Duineveld, G., Van't Hof, P. & Wielsma, E. (2010) Final Report. Impact of OWEZ Wind farm on bivalve recruitment. Retrieved 12 June 2018 from http://www.noordzeewind.nl/wp-content/uploads/2012/02/OWEZ_R_262_T1_20100910-Benthos-Recruitment-T1.pdf

Boyle, G. & New, P. (2018) ORJIP Impacts from Piling on Fish at Offshore Wind Sites: Collating Population Information, Gap Analysis and Appraisal of Mitigation Options. Final report – June 2018. The Carbon Trust. United Kingdom. Retrieved 19 September 2018 from https://www.carbontrust.com/media/676435/orjip-piling-study-final-report-aug-2018.pd

Brasseur, S., Aarts, A. *et al.* (2012) Habitat preferences of harbour seals in the Dutch coastal area: analysis and estimate of effects of offshore wind farms. Retrieved 12 June 2018 from http://www.noordzeewind.nl/wp-content/uploads/2012/04/OWEZ_R_252_T1_20120130_harbour_seals-2.pdf

BSH & BMU (2014) *Ecological Research at the Offshore Windfarm* Alpha Ventus*: Challenges, Results and Perspectives*. Federal Maritime and Hydrographic Society of Germany (BSH) and Federal Ministry for the Environment, Nature Conservation and Nuclear Safety (BMU). Weisbaden: Springer Spektrum.

Camphuysen, C.J., Fox, A.D., Leopold, M.F. & Peterson I.K. (2004) Towards standardised seabirds at sea census techniques in connection with environmental impact assessments for offshore wind farms in the UK. Retrieved 12 June 2018 from https://tethys.pnnl.gov/sites/default/files/publications/Camphuysen-et-al-2004-COWRIE.pdf

Carbon Trust (2014a) Detailed appraisal of the offshore wind industry in China. London: Carbon Trust. Retrieved 12 June 2018 from https://www.carbontrust.com/media/510530/detailed-appraisal-of-the-offshore-wind-industry-in-china.pdf

Carbon Trust (2014b) Appraisal of the offshore wind industry in Japan. London: Carbon Trust. Retrieved 12 June 2018 from https://www.carbontrust.com/media/566323/ctc834-detailed-appraisal-of-the-offshore-wind-industry-in-japan.pdf

Department of Energy & Climate Change (DECC) (2011) UK renewable energy roadmap. London: DECC. p. 42. Retrieved 13 November 2017 from https://www.gov.uk/government/uploads/system/uploads/attachment_data/file/48128/2167-uk-renewable-energy-roadmap.pdf

Department of Energy & Climate Change (DECC) (2012a) Record of the appropriate assessment undertaken for applications under Section 36 of the Electricity Act 1989. Projects: Docking Shoal Offshore Wind Farm (as amended), Race Bank Offshore Wind Farm (as amended), Dudgeon Offshore Wind Farm. London: DECC. Retrieved 11 December 2017 from https://itportal.beis.gov.uk/EIP/pages/projects/AAGreaterWash.pdf

Department of Energy & Climate Change (DECC) (2012b) Offshore Wind Cost Reduction Task Force Report 2012. London: DECC. Retrieved 12 June 2018 from https://www.gov.uk/government/groups/offshore-wind-cost-reduction-task-force

European Commission (EC) (2010) Wind energy developments and Natura 2000: EU guidance on wind energy development in accordance with EU nature legislation. Luxembourg: Publications Office of the European Union. Retrieved 12 June 2018 from http://ec.europa.eu/environment/nature/natura2000/management/docs/Wind_farms.pdf

Freeman, S. & Hawkins, K. (2013, summer) Pushing the 'envelope'. Special Report. *Offshore Wind Engineering* pp. 6–7.

Freeman, S.M. (2014) Consenting preparedness of offshore wind stakeholders: survey and recommendations. RenewableUK industry report.

Global Wind Energy Council (GWEC) (2012) Global wind report: annual market update 2012. Brussels: Global Wind Energy Council. Retrieved12 June 2018 from https://www.gwec.net/wp-content/uploads/2012/06/Annual_report_2012_LowRes.pdf

Global Wind Energy Council (GWEC) (2014) Global wind report: annual market update 2013. Brussels: Global Wind Energy Council. Retrieved12 June 2018 from http://www.gwec.net/wp-content/uploads/2014/04/GWEC-Global-Wind-Report_9-April-2014.pdf

Global Wind Energy Council (GWEC) (2015) Global wind statistics 2014. Brussels: Global Wind Energy Council. Retrieved 12 June 2018 from http://www.gwec.net/wp-content/uploads/2015/02/GWEC_GlobalWindStats2014_FINAL_10.2.2015.pdf

Goto Floating Offshore Wind Turbine (2012) Technology. Retrieved 12 June 2018 from http://goto-fowt.go.jp/english/about/tech/

Griffin, L., Rees, E. & Hughes, B. (2010) Whooper Swan migration in relation to offshore wind farms. *BOU Proceedings – Climate Change and Birds*. Retrieved 12 June 2018 from http://www.bou.org.uk/bouproc-net/ccb/griffin-etal.pdf

Harwood, A.J.P., Perrow, M.R., Berridge, R., Tomlinson, M.L. & Skeate, E.R. (2017) Unforeseen responses of a breeding seabird to the construction of an offshore wind farm. In Köppel, J. (ed.) *Wind Energy and Wildlife Interactions. Presentations from the CWW2015 Conference*. Cham: Springer International Publishing. pp. 19–42.

Hu, S.-Y. (2013) Policy and promotion of offshore wind power in Taiwan. Thousand Wind Turbines Project. Retrieved 12 November 2017 from https://www.mofa.gov.tw/Upload/RelFile/2508/111034/25bcd458-67d7-4ed4-994b-128a7ba49d17.pdf

Hundleby, G., Freeman, K. & Barlow, M. (2017) Unleashing Europe's offshore wind potential: a new resource assessment. WindEurope–BVG Associates. Retrieved 12 November 2017 from https://windeurope.org/wp-content/uploads/files/about-wind/reports/Unleashing-Europes-offshore-wind-potential.pdf

Huddlestone, J. (2010) *Understanding the Environmental Impacts of Offshore Windfarms*. COWRIE.

International Atomic Energy Agency (IAEA) (2015) The Fukushima Daiichi accident: report by the Director General. Vienna: International Atomic Energy Agency. Retrieved 12 June 2018 from https://www-pub.iaea.org/MTCD/Publications/PDF/Pub1710-ReportByTheDG-Web.pdf

Kaiser, M.J. *et al.* (2002) Predicting the displacement of Common Scoter *Melanitta nigra* from benthic feeding areas due to offshore windfarms. Retrieved 12 June 2018 from https://tethys.pnnl.gov/sites/default/files/publications/Kaiser%20et%20al.%202002.pdf

King, S., Maclean, I.M.D., Norman, T. & Prior, A. (2009) *Developing Guidance on Ornithological Cumulative Impact Assessment for Offshore Wind Farm Developers*. COWRIE. Retrieved 12 June 2018 from https://tethys.pnnl.gov/sites/default/files/publications/King-et-al-2009.pdf

Köppel, J. (ed.) (2017) *Wind Energy and Wildlife Interactions. Presentations from the CWW2015 Conference*. Cham: Springer International Publishing.

Krijgsveld, K.L., *et al.* (2011) Effect studies: Offshore Wind Farm Egmond aan Zee. Final report on fluxes, flight altitudes and behavior of flying birds. Retrieved 12 June 2018 from http://www.noordzeewind.nl/wp-content/uploads/2012/03/OWEZ_R_231_T1_20111114_2_fluxflight.pdf

Lambers, R.H.R. & Hofstede, R. (2009) Refugium effects of the MEO-NSW Windpark on fish: progress report 2007. Retrieved 12 June 2018 from http://www.noordzeewind.nl/wp-content/uploads/2012/02/OWEZ_R_264_T1_20091110_demersal_fish.pdf

Lindeboom, H.J., Kouwenhoven, H.J., Bergman, M.J.N., Bouma, S., Brasseur, S., Daan, R., Fijn, R.C., de Haan, D., Dirksen, S., van Hal, R., Hille Ris Lambers, R., ter Hofstede, R., Krijgsveld, K.L., Leopold, M. & Scheidat, M. (2011) Short-term ecological effects of an offshore wind farm in the Dutch coastal zone; a compilation. *Environmental Research Letters* 6(3).

McGarry, T., Boisseau, O., Stephenson, S., Compton, R. (2017) Understanding the Effectiveness of Acoustic Deterrent Devices (ADDs) on Minke Whale (*Balaenoptera acutorostrata*), a Low Frequency Cetacean. ORJIP Project 4, Phase 2. RPS Report EOR0692. The Carbon Trust, United Kingdom: 107pp. Retrieved 19 September 2018 from https://www.carbontrust.com/media/675268/offshore-renewables-joint-industry-programme.pdf

National Renewable Energy Laboratory (NREL) (2015) 2014–2015 Offshore wind technologies market report. Retrieved 12 June 2018 from https://www.nrel.gov/docs/fy15osti/64283.pdf

Offshore Wind Programme Board, Offshore Renewable Energy Catapult (2016). Cost reduction monitoring framework 2016. Retrieved 13 November 2017 from http://crmfreport.com/wp-content/uploads/2017/01/crmf-report-2016.pdf

Poot, M.J.M, *et al*. (2011) Effect studies Offshore Wind Egmond aan Zee: cumulative effects on seabirds. A modelling approach to estimate effects on population levels in seabirds. Retrieved 12 June 2018 from http://www.noordzeewind.nl/wp-content/uploads/2012/02/OWEZ_R_212_T1_20111118_Cumulative_Effects-20111123-reduced.pdf

Perrow, M.R., Gilroy, J.J., Skeate, E.R. & Mackenzie, A. (2010) Quantifying the relative use of coastal waters by breeding terns: towards effective tools for planning and assessing the ornithological impacts of offshore wind farms. ECON Ecological Consultancy Ltd, Report to COWRIE. ISBN: 978-0-9565843-3-5.

RenewableUK (2011) Consenting lessons learned. An offshore wind industry review of past concerns, lessons learned and future challenges. London: RenewableUK. Retrieved 12 June 2018 from http://www.renewableuk.com/en/publications/index.cfm/Offshore-Wind-Consenting-Lessons-Learned.

Scheidat, M., Aarts, G., Bakker, A., Brasseur, S., Carstensen, J., van Leeuwen, P. W., Leopold, M., van Polanen Petel, T., Reijnders, P., Teilmann, J., Tougaard, J. & Verdaat, H. (2012) Assessment of the effects of the Offshore Wind Farm Egmond aan Zee (OWEZ) for Harbour Porpoise (comparison T0 and T1). Retrieved 12 June 2018 from http://www.noordzeewind.nl/wp-content/uploads/2012/03/OWEZ_R_253_T1_20120202_harbour_porpoises.pdf

Shields, M. & Payne, A. (eds) (2014) *Marine Renewable Technology and Environmental Interactions*. Dordrecht: Springer Science Business Media.

Skeate, E.R., Perrow, M.R. & Gilroy, J.J. (2012) Likely effects of construction of Scroby Sands offshore wind farm on a mixed population of harbour *Phoca vitulina* and grey *Halichoerus grypus* seals. *Marine Pollution Bulletin* 64: 872–881.

Skov, H., Heinänen, S., Norman, T., Ward, R., Méndez-Roldán, S. & Ellis, I. (2018) *ORJIP Bird Collision and Avoidance Study*. Final Report – April 2018. The Carbon Trust. Retrieved 20 June 2018 from https://www.carbontrust.com/media/675793/orjip-bird-collision-avoidance-study_april-2018.pdf

Sparling, C., Sams, C., Stephenson, S., Joy, R., Wood, J., Gordon, J., Thompson, D., Plunkett, R., Miller, B. & Götz , T. (2015) *The Use of Acoustic Deterrents for the Mitigation of Injury to Marine Mammals during Pile Driving for Offshore Wind Farm Construction. ORJIP Project 4. Stage One of Phase Two*. The Carbon Trust. Retrieved 19 September 2018 from https://www.carbontrust.com/media/675291/orjip-add-study-final-report-stage-1-phase-2.pdf .

Wind Europe (2018) Offshore Wind in Europe: Key Trends and Statistics 2017. Retrieved 27 July 2018 from https://windeurope.org/wp-content/uploads/files/about-wind/statistics/Wind-Europe-Annual-Offshore-Statistics-2017.pdf

Winter, H.V., Aarts, G. & van Keeken, O.A. (2010) Residence time and behaviour of sole and cod in the offshore wind farm Egmond aan Zee (OWEZ). Retrieved 12 June 2018 from http://www.noordzeewind.nl/wp-content/uploads/2012/02/OWEZ_R_265_T1_20100916_Residence_time_cod_sole_OWEZ.pdf

World Energy Council (2016) World Energy Trilemma Index: Benchmarking the sustainability of national energy systems. London: World Energy Council. Retrieved 26 November 2017 from https://www.worldenergy.org/wp-content/uploads/2016/10/Full-report_Energy-Trilemma-Index-2016.pdf

Ybema, M.S, Gloe, D. & Lambers, R.H.R. (2009) OWEZ – pelagic fish, progress report and progression after T1. Retrieved 12 June 2018 from http://www.noordzeewind.nl/wp-content/uploads/2012/02/OWEZ_R_264_T1_20091110_pelagic_fish.pdf

CHAPTER 2

Physical and chemical effects

JON M. REES and ADRIAN D. JUDD

Summary

This chapter examines how offshore wind farms (OWFs) may cause changes to the prevailing hydrodynamic and coastal processes, including changes to seabed morphology and sediment transport and to water and sediment quality. Physical effects can be attributed to the physical presence of OWF infrastructure and the techniques used to install these structures. Chemical effects can be attributed to the substances such as biocides, corrosion inhibitors and cements used in construction and maintenance. Both physical and chemical effects have the potential to affect habitats and species. Here, evidence from both grey and peer-reviewed literature is used to describe the nature and scale of any effects, both individually and cumulatively. Different foundation designs using monopile and gravity-based structures provide two extreme scenarios. Despite a limited body of evidence, understanding of the physical and chemical impacts of monopile and gravity-based structures of individual OWFs is reasonably sound. Material is divided between themes exploring physical processes and local changes to seabed conditions, including scour and scour protection, cabling and cable protection and chemicals. Effects characterised from observations, models and measurements may be split into construction, operation and decommissioning phases. This evidence shows that physical effects are predominantly in the near field for tides, sediment transport and chemical factors; that is, within one spring tidal ellipse of the structure supporting the turbine. In contrast, wave modelling studies have shown that impacts can extend beyond the near field. Insufficient evidence is available to draw conclusions on the cumulative physical and chemical effects; that is, overlapping effects within one tidal ellipse or the aggregated effect from effects within multiple tidal ellipses. Observations and measurements have yet to identify any significant effects of chemicals from OWF construction or operation.

Introduction

Structures in the marine environment have the potential to change hydrodynamic and coastal processes (Carroll *et al.* 2010). Hydrodynamic and coastal processes are defined as physical processes on water and the substrate. These physical processes include:

- **Tidal elevations.** Periodic rise and fall of sea level caused by gravitational and centrifugal forces from the sun, moon and rotation of the earth.

- **Waves.** Horizontal movement of water generated by surface winds, also influenced by the fetch and depth of the water body.

- **Currents.** Horizontal movement of water influenced by the rise and fall of the tide, wind and temperature differences in surface waters.

The energy of tides, waves and currents can suspend seabed sediments into the water column and transport them as bedload or suspended load in the direction of flow. When the water flow increases, sediment particles can be lifted into suspension. Conversely, when water flow slows down, sediment particles can no longer be supported and are deposited on the seabed or coastlines. The hydrodynamic effects of artificial structures will impede or enhance water flows, in that increased flow speeds could lead to sediment erosion (scour) and decreased flow speeds could lead to sediment deposition. Waves and currents, and hence sediment transport induced by these forces, can vary by several orders of magnitude and are strongly influenced by prevailing physiographic conditions and consequently climate change, owing to the variability of tidal elevations, waves, winds and temperature.

The combination of variables influencing tides, waves and currents as hydrodynamic and coastal processes can impact on local sediment transport in the near field of the 'structure' concerned, whether this is a rock outcrop on the seabed or the foundations of a wind turbine or offshore substation or accommodation platform, or at sites farther afield along the coastline. In this context, near field is defined as within the array licensed boundary with a buffer of the local tidal ellipse centred on the turbine structures (spring tidal excursion), whereas far field is likely to be either at the outer limits of the tidal excursions within which the structure is located or, if there is an active connection of the processes, within adjacent overlapping tidal excursions, where there is the potential for effects to spread beyond a single spring tidal excursion. While the coastlines can naturally be highly dynamic, impacts from offshore structures under certain conditions have the potential either to reduce or to increase the wave climate inshore, thus modifying coastline erosion or deposition patterns. This also applies to habitats within a wind farm and along prevailing wave and tidal axes.

A typical offshore wind farm (OWF) consists of three main elements: the foundations and associated turbine, the electrical intra-array, and export cabling. Some developments will also have an offshore substation (or substations) with foundations in the seabed, especially if located some distance from the electrical grid. Different foundation options are available for OWFs, including monopiles, micropiles, gravity or caisson bases, jackets, suction buckets and floating structures (Box 2.1). The towers for offshore wind turbines vary in size depending largely upon the generating capacity of the turbine and local ground conditions. Monopile foundations typically increased from 4 m diameter in 2004, such as the Scroby Sands wind farm in the UK (Rees 2006), to 6.5 m diameter by 2015, such as the Riffgat wind farm in Germany (ASCOBANS 2014). Monopiles of 8.1 m diameter have been proposed in Environmental Statements, such as the Hornsea 1 OWF in the UK with 7 MW turbines, with future projections of 15 m diameter for planned 15 MW turbines. The foundations and towers ultimately support the nacelle and horizontally rotating blades, although vertical-axis turbines have been tested (Jameson *et al.*, Chapter 1).

Box 2.1   Foundation types for offshore wind farms

Different foundation options are available for turbines within offshore wind farms (OWFs). The most common foundation to date is the monopile, but other types include tripods and lattice structures, gravity bases/caissons and floating structures (see Figure 1.6 in Jameson *et al.*, Chapter 1).

In more detail, a monopile foundation consists of a large-diameter cylindrical steel tube (typical pile diameter 4–6.5 m) with a transition piece connecting the pile to the turbine tower (Figure 2.1a). Depending on the soil characteristics, monopiles are predominantly driven into the seabed and are suitable for shallow water up to 25–35 m deep relative to mean sea level (MSL). Some sites will also require drilling to enable the monopile to reach the design penetration depth (typically 20–30 m). They can be installed in deeper water, but that increases the cost of deployment. A 5 m diameter monopile without scour protection will have a 20 m² footprint on the seabed. An existing variant of monopile foundations for deep water is guyed monopile towers, allowing the monopile to be stabilised with tensioned guy wires (Malhotra 2011). Potential impacts include scour pits, scour wakes and sediment disturbance from constructional activity, including ploughing, jetting for cables and 'spudcan' seabed indentations due to jack-up vessels.

Tripod, jacket or lattice foundations sit on the seabed anchored by smaller 'pin' piles up to 2 m in diameter, with shallower sinking depths than monopiles (Figure 2.1b). Potential impacts are similar to monopiles but reduced, as the size of the individual members of the structure is correspondingly smaller.

(a)   (b)

Figure 2.1 (a) Monopile foundations are by far the most common foundation installed; (b) jacket lattice-type foundations are better suited to deeper waters with strong tides or currents. (Martin Perrow)

Gravity-base structures typically consist of a steel or concrete substructure which rests on the seabed. A wide variety of gravity-base and caisson-type foundations have been either designed or installed. The main characteristics include a wide, typically 50 m diameter base made from concrete, typically 5–10 m high, with a tower extending to the surface. The design depends on the application, hydrodynamic regime, water depth (normally shallow to deep water <70 m) and seabed characteristics. The installation of gravity-base structures may require the dredging of large volumes of material for seabed preparation to create a level and stable surface. Furthermore, the footprint of the gravity-base structure is large, which can cause a significant impact in terms of habitat loss. A 16 m diameter gravity-base foundation without scour protection will have a 200 m² footprint on the seabed. Peire *et al.* (2009) describe the ground preparation and installation methods for gravity-base foundations as deployed at the Thornton Bank OWF in Belgium and as shown in Figure 2.2, as follows:

- Foundation pits are dredged to create a level surface on the seabed (each pit measures 50 m wide by 80 m long).
- A two-layer gravel bed is installed on which the gravity base is placed.
- A heavy-lift vessel manoeuvres the gravity base into position.
- The foundation pits are backfilled.
- A two-layer scour protection system is installed around the gravity base.

Structures such as gravity caissons in the case of Nysted in Denmark (Danish Energy Authority 2006) have been installed in various conditions from relatively shallow water and low wave/current forcing to slightly deeper water, such as 20 m in the case of Thornton Bank, Belgium, and moderate wave/tidal conditions (Peire *et al.* 2009). Structures tend to be large and heavy, reaching thousands of tonnes. Jacket foundations are typically three- or four-sided steel structures that are typically 30 m wide and 60 m high connected to the seabed using micropiles and with a tower extending to the surface. Suction buckets are still under development, but have been

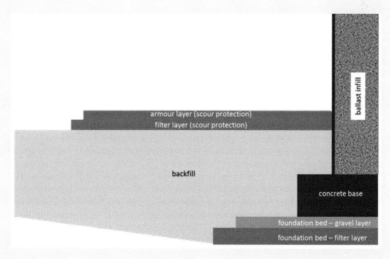

Figure 2.2 Gravity-base foundation installation at Thornton Bank, Belgium. (Modified from Peire *et al.* 2009)

used for meteorological masts such as at Dogger Bank (Nielson 2016). They consist of an inverted bucket that is either pushed into the seabed or installed by negative pressure. Typical dimensions for the bucket are up to 10 m diameter and 15 m high, with some foundations consisting of more than one bucket. Potential impacts of gravity-base structures are often classified as the realistic worst case scenario due to the potentially large scour pit and the degree of mitigation required (scour protection).

At the time of writing, only one offshore floating wind farm has been constructed, the Hywind Scotland Pilot Park in the Buchan Deep offshore of Peterhead on the east coast of Scotland, although another, the Kincardine OWF project offshore of nearby Aberdeen, is in planning. A number of prototypes have also been built and tested including Ideol, VoltunUS, Blue-H, WindFloat, and Sakiyama in several countries. Floating prototypes have employed a range of designs including semisubmersible platforms, spar-buoys and tension-leg platforms. All designs consist of a submerged, stable structure that is anchored to the seabed by a series of tethers. The tether anchor points could be either micropiles or gravity anchors, depending on the seabed conditions, or kedge anchors on catenary moorings. Potential impacts relate to the type of anchoring systems used, such as a kedge anchor or suction bucket, and the abrasion on the seabed caused by the catenary mooring line. In terms of coastal processes, impacts are expected to be less than with standard monopile designs but no evidence is available to demonstrate this.

Scope

This chapter focuses on experiences from the north-west European continental shelf, where the first rounds of OWFs have been concentrated. Specifically, experiences from the UK are explored, where the first wind farms in 10–30 m water depth were installed in regions with moderate- to high-magnitude wave and tidal exposure. Results have been extracted from the peer-reviewed literature, grey literature, in-depth review of Environmental Impact Statements and review of Monitoring Reports from licence conditions. These have been synthesised into national reports for the Convention for the Protection of the Marine Environment of the North-East Atlantic (OSPAR) and other international reporting mechanisms (OSPAR Commission 2008). Since the early 2000s, the design of structures, installation techniques and mitigation methodologies have advanced significantly from simple monopiles in shallow water (10–20 m) with post-installation scour protection, such as at Scroby Sands OWF, to complex jacket structures in relatively deep water (30–50 m) at the East Anglia One OWF. In parallel, understanding of the scale of the impacts on coastal processes has also improved, with targeted licence conditions and research programmes funded by the UK Government, The Crown Estate and the Natural Environment Research Council (NERC).

Changes to hydrodynamic and coastal processes resulting from OWFs include any change in water flow around the devices, any influence on the seabed sediments, such as scouring and suspension of seabed sediments into the water column, any influence on sediment transport patterns and any influence that devices may have on coastlines. The last factor also includes changes to hydrodynamic conditions such as tidal flow, tidal height and wave heights that may have an effect on either pelagic or benthic habitats, or both.

The most common foundation type is the monopile, but with increasing numbers of jacket-style foundations in deeper waters (Box 2.1). There is a reasonable body of evidence of the effects associated with these structures on coastal processes and the effects of related scour protection and cabling and cable protection. The *Themes* section discusses the impacts of OWF structures on the key physical processes on the continental shelf before describing the use of chemicals in the OWF sector and exploring mitigation methods.

Themes

Waves

All structures in the marine environment will be impacted by the local wave regime and, conversely, the structures will modify the wave regime. This is normally demonstrated by a reduction in wave height in the lee of the structure. Both observations using X-band radar (Rees 2006) and numerical modelling studies have identified this process at near field (ABPmer 2005). At far field, modelling studies (SMart Wind 2013; van der Molen *et al.* 2014) have shown that the footprint can extend beyond the licence area. No direct observations using wave-buoys have yet been undertaken with the aim of measuring the magnitude and shape of any footprint. Satellite-based observations using synthetic aperture radar appear to show wakes behind structures, but the frequency, magnitude and size of these wakes have not yet been established.

Research to quantify the effect of OWF structures on wind-driven waves, which propagate towards the coast, was undertaken by Alari and Raudsepp (2012). The main aim was to find out how much wave energy these structures will absorb and scatter. For this, data from two OWFs located in Estonia were used in the Simulating Waves Nearshore (SWAN) model to assess their potential effects upon wave heights. SWAN is a third generation phase-averaged spectral wave model developed at the Delft University of Technology in the Netherlands, used extensively in research and by consultancies. The impact of OWFs was assessed by calculating the ratio of difference of significant wave height, with and without wind turbines, to significant wave height without turbines. The authors concluded that the reduction of significant wave height near the coast below 10 m isobaths does not exceed 1%. While the potential effects of the reduction of significant wave heights by this order of magnitude on benthic communities have not been assessed, it seems unlikely that these would be significant.

Tidal currents

Tidal currents can be modified by the presence of the wind-farm structure by either decreasing the flows in the wake of the structure or accelerating the flows around the edges of the farm. In the far field, significant modifications of the flow regime are neither predicted nor observed. Computer modelling has also shown that the influence on hydrodynamics from a typical OWF layout is localised to individual structures and at stages of peak flow (ABPmer 2005). The ABPmer (2005) study also showed the combined consequences of the modified flow regime and the presence of a large number of small physical obstacles to a sediment pathway. These factors appear to have minimal influence on the net deposition patterns predicted for sand transport for different grades of sediment, and the structures seem to have little effect on waves.

The interactions between OWF structures, especially turbine foundations and the adjacent substrata, were examined by Wilhelmsson and Malm (2008). Such structures have the potential to alter local hydrodynamic patterns, which can cause changes in benthic biomass and diversity as well as scouring and sediment build-up (e.g. Guichard *et al.* 2001; Shyue & Yang 2002; Düzbastilar *et al.* 2006).

Sediment resuspension and deposition

The construction of OWFs can potentially lead to sediment resuspension (Bergström *et al.* 2013). The requirement of gravity-base and caisson foundations for a level platform on which to sit is created by dredging, which results in the mobilisation of seabed sediments into the water column, temporarily increasing turbidity and, as the sediments settle out of suspension, will deposit a new layer of fine sediment. Similarly, driven or drilled monopiles will also cause sediments to be remobilised, creating a suspended sediment plume. Cable-laying operations will also disturb sediments from the seabed. The extent of any sediment plumes from these activities depends on sediment type, grain-size distribution and the hydrodynamic regime, and thus can vary greatly between sites. Observations of sediment plumes from satellite and aerial platforms have identified their orientation and spatiotemporal extent (Figure 2.3) (Vanhellemont & Ruddick 2014; Forster 2017). However, the impact of these plumes on local habitats has not been determined.

Far-field direct and secondary impacts on sediment transport patterns and pathways are uncertain. Clark *et al.* (2014) describe that while most research suggests that changes to sediment dynamics will be local, the *combined* effects of multiple large-scale wind farms could alter net sediment transport and deposition, thereby affecting shorelines and bathymetry. Van der Molen *et al.* (2013) suggest that wind-farm arrays in areas with strong currents may lead to stronger far-field effects, and that if this were the case, and if multiple arrays were introduced, inter-array interaction might occur.

Scour

The flow of water around turbine bases can create scour pits in erodible sediments. The acceleration of the flow around the structure increases the erosion rate in the close proximity of the structure, which allows sediment to be transported away from it. A combination of local waves and currents and substrate type control the formation of scour pits. The Scroby Sands OWF off the East Norfolk coast of the UK is located in very dynamic waters where the extreme conditions have generated some of the largest scour pits around turbine foundations of up to 8 m (Rees 2006). Whitehouse *et al.* (2011) indicate that the maximum depth of scour (S) in relation to the foundation diameter (D) at a wind-farm foundation is $S/D=1.38$. The overall extent of scour in a sandbed is typically 4–5D where no protection has been placed.

Structures may also have the capability of mobilising sediment into the water column, creating a plume of suspended sediments, which may have a surface expression (Baeye & Fettweis 2015). Suspended sediment within plumes can be transported over longer distances depending on the particle size and the tidal excursion, which is typically up to 10 km. While in suspension, these additional sediments have the potential to impact on filter feeders and reduce the local light regime, which may impede primary production or the ability of higher predators to locate prey visually (Renouf 1980). Once deposited on the seabed, there is also the potential for sediment to smother benthic habitats, especially those of filter-feeding organisms (see Dannheim *et al.*, Chapter 4).

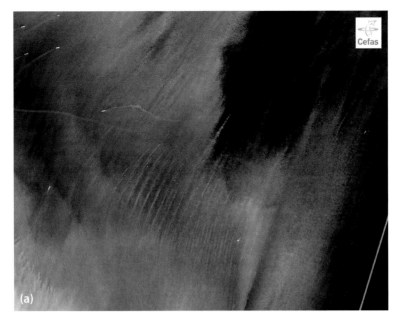

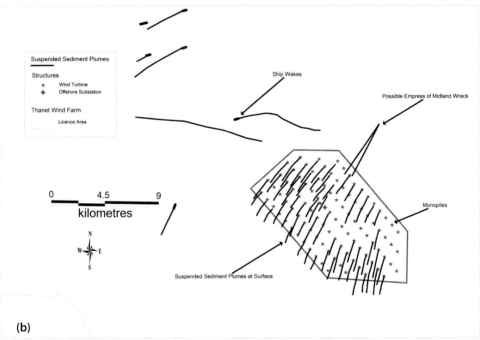

Figure 2.3 (a) Landsat 8 image of the Thanet offshore wind farm in the UK showing sediment plumes to the lee of the prevailing hydrodynamic energies; and (b) the interpretation of events.

As scour depths are controlled by a combination of structure size and shape, wave and tidal conditions and erodibility of local sediments, swath bathymetry surveys of the scour pits are often commonly included in UK licence conditions for OWFs. These show a wide range of scour depths from virtually nil, where combined wave and tidal bed shear stresses are insufficient to mobilise the local substrate or where scour protection

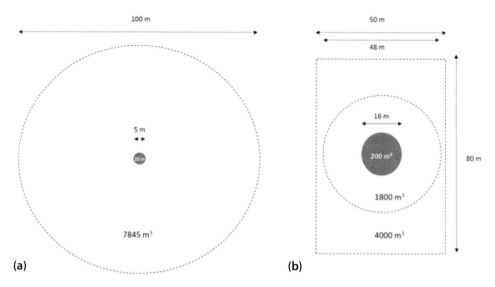

Figure 2.4 (a) Potential turbine footprint for a monopile in highly erodible sediment; (b) a typical gravity-based structure.

has been pre-laid to mitigate any changes in morphology, to over 5 m in areas where the potential for erosion is high and changes to metocean conditions can be significant. For instance, swath bathymetry surveys at the Scroby Sands OWF confirmed predictions made during the Environmental Impact Assessment (EIA) that the scour pits around individual monopile foundations would develop to 5 m depth and 100 m diameter (DECC & Defra 2008). A 5 m diameter monopile similar to that built at Scroby and without scour protection would only have a 20 m² footprint on the seabed. However, with scour to the scale seen at Scroby producing scour pits up to 100 m diameter, approximately 7,850 m² of seabed would be affected (Figure 2.4). It should be noted that Scroby Sands represents an extreme case and for most OWF sites the scale of scour effects will be much lower. In contrast, a 16 m diameter gravity-base foundation without scour protection will have a 200 m² footprint on the seabed, although the associated ground preparations would cover an area of approximately 1,800 m² for the gravel and filter layer and 4,000 m² for the armour and backfill (Figure 2.4).

The investigations at Scroby Sands OWF (DECC & Defra 2008) also showed that scour pits for monopiles at the site were independent of one another; that is, there was no connectivity between any two scour pits within a wind-farm array (Figure 2.5). To optimise utilisation of wind resources, turbines within a wind farm have often been spaced between approximately 500 m and 1,000 m apart, although larger designs of 7 MW capacity will increase this figure to 2,000 m or more. This suggests that with the relatively wide spacing of turbines required for efficient electricity generation, scouring impacts are likely to remain localised and not merge into continuous areas of disturbance, but instead remain as discrete and small areas separated by large areas of relatively undisturbed seabed.

Although the scour pits at Scroby Sands OWF have been shown to be independent of one another, Rees (2006) identified scour tails in the more exposed areas (Figure 2.5). Such effects are likely to be limited to the near field around the OWF array as the impacts on seabed sediments due to additional mobility and translocation of sediments are limited to within a few hundred metres of the wind-farm array (DECC & Defra 2008).

Figure 2.5 Scroby Sands offshore wind farm sediment transport features as produced by Rees (2006). The insert demonstrates where secondary scouring occurred around the scour protection, which led to further control measures being adopted (see Box 10.3).

Studies at Egmond aan Zee OWF in the Netherlands, with piles of 4.6 m pile diameter, showed that the residual stream from the scour pit downward of the scour protection had a maximum depth of 2.2 m.

Scour protection

In some cases, the creation of a scour pit can impact on the stability of a monopile and thus additional material is placed into the scour pit. Normally, this material is of such a size and density that it is not eroded by normal wave and tidal conditions. Types of scour protection include rocks of approximately 10 cm diameter, rock gabions, grout bags and shale.

At some sites, such as Burbo Bank in the UK (Carroll *et al.* 2010) and Egmond aan Zee OWF in the Netherlands (NoordzeeWind 2008), scour protection has taken the form of a layer of coarse material such as small pieces of slate, installed at each turbine location before the installation of the piles (DECC & Defra 2008). The purpose of this is to prevent scouring, rather than having to take remedial action to fill scour pits once they have formed. Whitehouse *et al.* (2011) further describe the approaches for the use of scour protection as: (1) placement of small-sized rock or gravel directly on the seabed as a preparatory layer, which acts as a filter layer on the seabed and can be placed before installation of the foundation or before scour has had time to develop; (2) placement of larger rock as an armour layer on top of the preparatory filter later; and (3) placement of widely graded rock in the scour pit formed around the structure.

These scour protection rocks constitute a colonising habitat for hard-substrate species, following a gradual development in species composition over the years, as shown by Bouma and Lengkeek (2012) for Egmond aan Zee OWF in the Netherlands (see Dannheim *et al.*, Chapter 4). Some proof is available that they also attract larger fish (Winter *et al.* 2010; van Hal *et al.* 2012; Gill & Wilhelmsson, Chapter 5). It has also been postulated that these hard substrates in soft or mobile substrate areas may provide the vector for the spread of certain invasive species (Adams *et al.* 2014).

Cabling and cable protection

The configuration of cables within an OWF includes the intra-array forming circuit and one or more cables forming the export to the onshore electricity grid. Depending on wind-farm design, one or more offshore substations and converters may also be included in the intra-array cable network.

The effects of cables on physical processes are wholly dependent on their dimensions and the methods used to install them. Cables are laid in trenches in soft sediments that will disturb surficial sediments and associated biota, or laid across the surface on hard substrates, which could cause some localised abrasion to the seabed. Cables laid on the surface may require protection with rock or concrete mattresses. The trench or any protection and the associated effects will be 2–5 m wide. The cables linking the individual wind turbines are several hundreds of metres in length (intra-array), whereas the cables to shore or export will be several kilometres or tens of kilometres in length. In Germany, the longest export cable to date is 125 km from Borwin Alpha to UW Diele. Cables are laid in an almost continuous length from specialised equipment towed by sea-going vessels. Cables are buried using a plough or trencher, or a jetting device which releases water under pressure to cut the trench. Typically, intra-array cables are ploughed for the majority of the route and a jetting technique is used for the last few metres to the J-tube on the monopile.

Where cables cross other cables or pipelines, or where potential seabed conditions allow exposure or the ploughing tool did not reach the designed burial depth, rock dumping or concrete mattress protection may be required to protect the integrity of the cable from trawling or anchoring for example. Rock dumping material and the associated effects can be very similar to scour protection around foundations and typically consists of material with a mean size of approximately 10 cm.

Chemicals and oil spills

Within the various designs of OWFs, chemicals are used for a variety of purposes in the turbine and foundation pieces as well as offshore substations (Table 2.1). Also included are chemicals used during construction by jack-up vessels, cable layers and crew transfer vessels, and during operation and maintenance activities, with decommissioning activities being envisioned as tending to be the reverse of construction. To the best of the authors' knowledge, only two OWFs have been decommissioned to date (2017), the earliest OWF built at Vindeby in Danish waters and the Swedish site at Yttre Stengrund. Two turbines have also been removed at Robin Rigg OWF on the west coast of the UK. The various designs of structures used by the OWF sector involve the use of chemicals in open and closed systems. Open systems can vent into the open sea, including biocides from open piles via J-tubes, whereas chemicals such as fuels are retained within closed systems.

Chemical contamination relates to the potential unlicensed release of chemicals into the marine, terrestrial or atmospheric environment. The likelihood of release and potential for mitigation are provided in Table 2.2. Note that drill tailings arising from the insertion of a monopile can be licensed for release at authorised disposal sites based on an EIA. Contamination effects are possible during the construction phase of OWFs, such as from accidental pollution incidents. At the time of writing, only two types of such incidents associated with OWF construction have been reported in the OSPAR maritime area, suggesting that such events are rare. The first type was two reported cases of failures in the seal between the turbine transition piece and the pile, resulting in a loss of approximately 30 tonnes of grout that was released under pressure. Regular monitoring of the equipment and instrumentation did not identify the problem, which only became apparent after

Table 2.1 Types of chemicals used in offshore wind farms

Chemical	Purpose
Lubricants	Gear boxes, bearings
Fuels	Used by vessels constructing or servicing installations as well as Commissioning equipment such as power generators
Biocides	Bio fouling management (internal and external to monopile and on jacket structures)
Corrosion inhibitors	External and internal anodes/cathodes to reduce degradation of the structure
Paints	Protective coatings to minimise corrosion
Grouts	Cementation of transition piece to foundation

completion when it was observed that only a small fraction of the grout had entered the transition piece. In both cases, diver inspections after the accident failed to identify grout on the seabed. The grout used was from the OSPAR Harmonised Mandatory Control System list of notified offshore chemicals (OSPAR Commission), so its ecotoxicological properties were known. Furthermore, RenewableUK Guidelines recommend the use of the Offshore Chemical Notification Scheme (Cefas) when hazardous substances are used offshore (RenewableUK 2014). In these two examples, as the incident was limited to a single turbine and the material was apparently rapidly dispersed, the impacts were concluded to be minimal.

Table 2.2 Likelihood, impact and mitigation measures for various chemical groups used in offshore wind farms

Chemical type	Likelihood	Impact	Mitigation
Lubricants	Low	Depending on density: surface plumes	Bunding within turbine or tower to capture total volume of lubricants Use of chemicals authorised in other marine sectors, e.g. oil and gas
Fuels	Low	Surface plumes	Bunding within turbine or tower to capture total volume of lubricants Use of chemicals authorised in other marine sectors, e.g. oil and gas
Biocides	Low	Diffusion into water column: impacts on pelagic environment	Use of chemicals authorised in other marine sectors, e.g. oil and gas
Paints	Low	Diffusion of active compounds; flaking on to seabed	Use of chemicals authorised in other marine sectors, e.g. oil and gas Sand blasting and repainting through industry standard practices
Grouts	Low	Low	Use of chemicals authorised in other marine sectors, e.g. oil and gas
Drill cuttings	Low/ medium	Low	Licensed disposal after risk assessment and impact statement

Figure 2.6 Offshore wind turbine tower and yellow paint spill on the surface of the water during paint application. (Martin Perrow)

The second type of pollution reported concerned two incidences where the protective paint on the monopiles failed. This was manifested by osmotic/electrolytic blistering in the splash zone due to encapsulated solvent following the application of a single thick coating of paint (Figure 2.6). This was contrary to the recommended and usual multilayered approach in which solvent is allowed to evaporate between applications.

These incidents highlight the need to use chemicals of known origin, to follow manufacturers' instructions on chemical use and, where such incidences could be anticipated, to ensure that each offshore development has a mandatory Marine Pollution Contingency Plan.

In-combination and cumulative effects

A range of activities have the potential to affect physical process, including the installation and presence of infrastructure, benthic trawling, navigation dredging, subsea cables, marine aggregate extraction, coastal protection and flood risk management. The individual effects on physical processes associated with these activities have the potential to interact with each other. The magnitude, duration and spatial extent of physical effects will be dictated by the prevailing physical environmental conditions with the tidal excursion of the development, including tidal speed, tidal direction, wave height, wind speed, vertical stratification, substrate type, substrate cohesiveness and seabed morphology. These local conditions also control which of these parameters will interact and how they will interact in combination. For example, the concentration of suspended sediments in a plume may increase if concurrent cable laying, foundation drilling, dredging and scouring are all occurring in the same tidal axis within the same tidal excursion; or habitat changes result from the combination of foundation installation, scouring, anchoring, dredging, placement of scour protection and cable installation.

As such, in-combination effects on physical processes may be split into two distinct typologies: (1) effects accumulating within the same local spatial and temporal footprint, such as suspended sediments; and (2) effects accumulating at a large number of discrete locations, such as scour pits, which, while they do not overlap, may contribute to an overall adverse effect. The suite of in-combination effects from one wind farm may then add to another in sufficiently close proximity as a cumulative effect across sites.

The magnitude, duration and spatial extent of chemical effects will also be dictated by the prevailing physical environmental conditions listed above but will also be dependent on the behaviour and reactivity of the constituent chemicals, including their persistence, bioaccumulation and toxicity.

Modelling undertaken as part of the EIAs for OWFs generally predicts that the risks of in-combination and cumulative effects on physical and chemical processes are low. However, the methods and monitoring do not yet have the power to measure accurately the full range of effects on marine ecosystems.

Table 2.3 Assessment of spatiotemporal impacts for physical process and chemicals (with confidence levels according to the Marine Climate Change Impacts Partnership approach)

Process	Near field	Far field	Cumulative
Waves	Known reduction in wave heights in lee of structures Confidence: high	Numerical modelling results demonstrate diminishing impact on wave heights Confidence: high	Processes and issues are understood in the near field, whereas in the far field non-linear processes and highly episodic events result in the deviation from the baseline of varying significance Confidence: low
Tides	Observed reduction in wake of turbines and acceleration around edges of farm Confidence: high	Changes in tidal currents rarely predicted or observed Confidence: high	Well defined in near field and insignificant in far field Confidence: high
Sediment transport	Processes and rates understood through observations Confidence: high	Direct and secondary impacts on sediment transport patterns and pathways uncertain Confidence: low	Far-field effects on coastlines and sand banks are ill defined owing to non-linear effects (see Waves) and highly variable nature of geomorphological change Confidence: low
Chemicals	Quantities and likelihoods are low but track record suggests minimal incidents. Any impacts are likely to be short term for single incidents Confidence: medium/ low owing to lack of data	Dispersion and dilution will result in negligible impacts Confidence: medium/ low owing to lack of data	Significance is directly proportional to the number and severity of incidents. Incidents are thought to be rare, but further evaluation of the significance of chemical release (e.g. a systematic release of biocides) is required Confidence: low

Concluding remarks

Understanding of the physical and chemical effects of monopile and gravity-base turbine installation and operation has increased substantially since the turn of the millennium. Effects from individual foundations and structures have been simulated in numerical models with increasing accuracy. However, understanding of the in-combination effects, especially from gravity base, is still being developed.

The spatiotemporal effects are summarised in Table 2.3. This uses a series of confidence ratings derived using the Marine Climate Change Impacts Partnership approach, based on a scoring system which combines the amount of evidence available and the level of scientific consensus based on several reviewers (MCCIP).

The majority of evidence (ABPmer 2005; Danish Energy Authority 2006; Rees 2006; Carroll 2010; Whitehouse *et al.* 2011; Alari & Raudsepp 2012; van der Molen *et al.* 2014) suggests that most effects on the physical environment are near field; however, waves and far-field effects on sediment transport patterns and pathways are uncertain.

In general, where OWFs are sited on fine substrates such as silts, these are likely to be low-energy, depositional environments where effects are more likely to be near field than far field owing to reduced seabed transport and remobilisation rates.

Acknowledgements

The authors wish to thank the UK Government and The Crown Estate, which have funded projects to inform the writing of this chapter.

References

ABPmer (2005) Assessment of potential impact of round 2 offshore wind farm developments on sediment transport. Report no. R1109.

Adams, T.P., Miller, R.G., Aleynik, D. & Burrows, M.T. (2014) Offshore marine renewable energy devices as stepping stones across biogeographic boundaries. *Journal of Applied Ecology* 51: 330–338.

Alari, V. & Raudsepp U. (2012) Simulation of wave damping near coast due to OWFs. *Journal of Coastal Research* 28: 143–148.

ASCOBANS (2014) Development of noise mitigation measures in offshore wind farm construction 2013. Information Document 3.2.2.b. 21st ASCOBANS Advisory Committee Meeting, Gothenburg, Sweden, 29 September–1 October 2014. Retrieved 21 September 2017 from http://www.ascobans.org/sites/default/files/document/AC21_Inf_3.2.2.b_NoiseMitigation_OffshoreConstruction.pdf.

Baeye, M. & Fettweis, M. (2015) In situ observations of suspended particulate matter plumes at an offshore wind farm, southern North Sea. *Geo-Marine Letters* 35: 247.

Bergström, L., Sundqvist, F. & Bergström, U. (2013) Effects of an offshore wind farm on temporal and spatial patterns in the demersal fish community. *Marine Ecology Progress Series* 485: 199–210.

Bouma, S. & Lengkeek, W. (2012) Benthic communities on hard substrates of the offshore wind farm Egmond aan Zee (OWEZ): including results of samples collected in scour holes. Bureau Waardenburg. Commissioned by Noordzee Wind. Retrieved from https://www.buwa.nl/fileadmin/buwa_upload/Bureau_Waardenburg_rapporten/BW_research_OWEZ_benthic_communities_-_compressed_file.pdf

Carroll, B., Cooper, B., Dewey, N., Whitehead, P., Dolphin, T., Rees, J. & Judd, A. (2010) A further review of sediment monitoring data. Retrieved 21 September 2017 from https://www.thecrownestate.co.uk/media/5873/a_further_review_of_sediment_monitoring_data.pdf

Centre for Environment Fisheries and Aquaculture Science (Cefas). Offshore Chemical Notification Scheme. Retrieved from https://www.cefas.co.uk/cefas-data-hub/offshore-chemical-notification-scheme/

Clark, S., Schroeder, F. & Baschek, B. (2014) The influence of large offshore wind farms on the North Sea and Baltic Sea – a comprehensive literature review. Geesthacht: Helmholtz-Zentrum Geesthacht Zentrum für Material- und Küstenforschung. Retrieved 21 September 2017 from https://tethys.pnnl.gov/sites/default/files/publications/Clark_et_al_2014.pdf

Danish Energy Authority (2006) Offshore wind farms and the environment: Danish experience from Horns Rev and Nysted. Report by the Danish Energy Agency. Retrieved 21 September 2017 from https://tethys.pnnl.gov/publications/offshore-wind-farms-and-environment-danish-experiences-horns-rev-and-nysted

Department for Energy & Climate Change (DECC) & Department of the Environment, Fisheries and Rural Affairs (Defra) (2008) Review of cabling techniques and environmental effects applicable to the offshore wind farm industry. Retrieved 21 September 2017 from http://webarchive.nationalarchives.gov.uk/+/http:/www.berr.gov.uk/files/file43527.pdf

Düzbastilar, F.O., Lök, A., Ulaş, A. & Metin, C. (2006) Recent developments on artificial reef applications in Turkey: hydraulic experiments. *Bulletin of Marine Science* 78: 195–202.

Forster, R.M. (2017) The effect of monopile-induced turbulence on local suspended sediment patterns around UK wind farms: field survey report. An IECS report to The Crown Estate.

Guichard, F., Bourget, E., & Robert, J.-L. (2001) Scaling the influence of topographic heterogeneity on intertidal benthic communities: alternate trajectories mediated by hydrodynamics and shading. *Marine Ecology Progress Series* 217: 27–41.

Malhotra, S. (2011) Selection, design and construction of offshore wind turbine foundations. In Al-Bahadly, I.H. (ed.) *Wind Turbines*. London: InTechOpen. pp. 231–264. Retrieved 21 September 2017 from http://www.intechopen.com/books/wind-turbines/selection-design-and-construction-of-offshore-wind-turbine-foundations

Marine Climate Change Impacts Partnership (MCCIP). Confidence assessments. Retrieved from http://www.mccip.org.uk/impacts-report-cards/full-report-cards/2013/confidence-assessments/

Natural Environment Research Council (NERC) Marine renewable energy. Retrieved from http://www.nerc.ac.uk/research/funded/programmes/mre/

Nielsen, S.A (2016) Innovations in foundations: the mono bucket. Oceanology International and Catch the Next Wave. Retrieved 1 November 2016 from http://www.oceanologyinternational.com/__novadocuments/230227?v=635950441263470000

NoordzeeWind (2008) Off shore windfarm Egmond aan Zee. General report OWEZ_R_141_20080215. Retrieved from http://www.noordzeewind.nl/wp-content/uploads/2012/02/OWEZ_R_141_20080215-General-Report.pdf

OSPAR Commission. Offshore chemicals. London: OSPAR. Retrieved from http://www.ospar.org/work-areas/oic/chemicals

OSPAR Commission (2008) Guidance on environmental considerations for offshore wind farm development. London: OSPAR. Retrieved from https://www.ospar.org/work-areas/eiha/offshore-renewables

Peire, K., Nonneman, H. & Bosschem, E. (2009) Gravity base foundations for the Thornton Bank offshore wind farm. *Terra et Aqua* 115: 19–29. Retrieved 21 September 2017 from https://www.iadc-dredging.com/ul/cms/terraetaqua/document/2/5/8/258/258/1/article-gravity-base-foundations-for-the-thornton-bank-offshore-wind-farm-terra115-3.pdf

Rees, J.M. (2006) Contract AE0262: Scroby Sands coastal processes monitoring. Final report. Retrieved 21 September 2017 from http://www.google.co.uk/url?sa=t&rct=j&q=&esrc=s&source=web&cd=1&ved=0ahUKEwjC77HsoLbWAhXpB8AKHU7zB9oQFggoMAA&url=http%3A%2F%2Frandd.defra.gov.uk%2FDocument.aspx%3FDocument%3DAE0262_3757_FRP.pdf&usg=AFQjCNGyUbGz8Y_oI1ko6yNZO-JKZ3yKoTg

RenewableUK (2014) Offshore wind and marine energy health and safety guidelines, Issue 2, Section C.9.1.2. Retrieved from http://www.renewableuk.com/resource/collection/AE19ECA8-5B2B-4AB5-96C7-ECF3F0462F75/Offshore_Marine_HealthSafety_Guidelines.pdf

Renouf, D. (1980) Fishing in captive harbour seals (*Phoca vitulina concolor*): a possible role for vibrissae. *Netherlands Journal of Zoology* 30: 504–509.

Shyue, S. & Yang, K. (2002) Investigating terrain changes around artificial reefs by using a multi-beam echosounder. *ICES Journal of Marine Science* 59 (Supplement 1): S338–S342.

SMart Wind (2013) Section 7.5.1.2. Annex 5.1.2: Wave modelling. London: SMart Wind. Retrieved 20 October 2017 from https://infrastructure.planninginspectorate.gov.uk/wp-content/ipc/uploads/projects/EN010033/EN010033-000556-7.5.1.2%20Wave%20Modelling.pdf

van der Molen, J., Smith, H.C.M., Lepper, P. Limpenny, S. & Rees, J. (2014) Predicting the large-scale consequences of offshore wind turbine array development on a North Sea ecosystem. *Continental Shelf Research* 85: 60–72.

van Hal, R., Couperus, A.S., Fassler, S.M.M., Gastauer, S., Griffioen, B., Hintzen, N.T., Teal, L.R., van Keeken, O.A. & Winter, H.V. (2012) Monitoring- and evaluation program near shore wind farm (MEP-NSW): fish community. IMARES Report C059/12. Ijmuiden: IMARES. Retrieved 13 December 2017 from https://www.wur.nl/en/Publication-details.htm?publicationId=publication-way-343337363933

Vanhellemont, Q. & Ruddick, K. (2014) Turbid wakes associated with offshore wind turbines observed with Landsat 8. *Remote Sensing of the Environment* 145: 105–115.

Whitehouse, R., Harris, J., Sutherland, J. & Rees, J. (2011) The nature of scour development and scour protection at offshore windfarm foundations. *Marine Pollution Bulletin* 62: 73–88.

Wilhelmsson, D. & Malm, T. (2008) Fouling assemblages on offshore wind power plants and adjacent substrata. *Estuarine, Coastal and Shelf Science* 79: 459–466.

Winter, H., Aarts, G. & van Keeken, O.A. (2010) Residence time and behaviour of sole and cod in the offshore wind farm Egmond aan Zee (OWEZ). Report no. OWEZ_R_265_T1_20100916. IMARES Report C038/10. Ijmuiden: IMARES. Retrieved 13 December 2017 from https://tethys.pnnl.gov/sites/default/files/.../Winter%20et%20al%202010.pdf

CHAPTER 3

Atmosphere and ocean dynamics

GÖRAN BROSTRÖM, ELKE LUDEWIG, ANJA SCHNEEHORST and
THOMAS POHLMANN

Summary

This chapter considers the effect of offshore wind farms (OWFs) on the atmosphere and ocean dynamics by first outlining the basic features of ocean dynamics that underpin the oceanic response. In simple terms, OWFs extract energy from the wind, thereby decreasing wind speed and increasing turbulence levels in the lower atmosphere. The impact on the dynamics of the upper ocean is not yet well understood and is based on theoretical predictions and a single case study at the Alpha Ventus wind-farm test site in German waters. Nevertheless, lower wind speed implies that the wind stress on the ocean surface becomes smaller. More importantly, the horizontal gradients in the wind stress will become artificially large, thereby creating a convergence/divergence in the wind-driven (Ekman) transport. Thus, this wind-wake effect will create both upwelling and downwelling, with the additional possibility of relatively strong horizontal currents. Exploratory analysis shows that a wind farm of 5×5 km^2 may induce an artificial upwelling in the order of 1 m/day over an area of 200 km^2 or around eight to ten times the area of the wind farm, which corresponds to a total upwelling of roughly 2,000 m^3/s or about 2×10^8 m^3/day. Upwelling is a known driver for ocean productivity and the prediction is that OWFs may increase productivity substantially. By simply extrapolating numbers on upwelling and estimated nutrient levels, it can be speculated that artificially induced upwelling may double the natural primary production in an area that is, say, ten to 20 times larger than the area of the wind farm. It should be noted, however, that such calculations have large uncertainties, and more studies are needed to quantify these numbers more precisely.

Introduction

Offshore wind farms (OWFs) have some potential for negative impacts upon some marine fauna such as marine mammals (see Nehls *et al.*, Chapter 6) and seabirds (Vanermen & Stienen, Chapter 8; King, Chapter 9), while generating positive reef and refuge effects benefiting seabed communities (Dannheim *et al.*, Chapter 4), fish (Gill & Wilhelmsson, Chapter 5), and perhaps some marine mammals (Nehls *et al.*, Chapter 6) and birds (Vanermen & Stienen, Chapter 8). Some localised effects on the wider environment through coastal processes are predicted (Rees & Judd, Chapter 2), but otherwise the generation of offshore wind power is generally considered to be environmentally benign. Since wind energy is extracted from the atmosphere, it follows that the wind energy in the atmosphere decreases; that is, the wind slows down. In addition, the disturbance of the wind field by turbines will create higher turbulence levels in the atmosphere. The effects of decreased wind speed and increase in the levels of turbulence are most visible at the hub level, but the wind deficit and high turbulence levels will spread both upwards and downwards as well as downwind from the turbine or the wind farm. This is an inevitable consequence of wind farms connected to the energy-extraction process. Indeed, it is well known that the wake from a wind turbine stretching downwind within a wind farm will influence the efficiency of any turbines downwind (Frandsen *et al.* 2006; Barthelmie & Jensen 2010; Barthelmie *et al.* 2010). The increased turbulence levels will also influence downwind turbines as the wind becomes gustier. Not only is this an important consideration for the construction of wind farms, in order to maximise their efficiency, but also it may interact with other marine activities such as shipping and fishing.

Although more work is needed before accurate statements on environmental impacts can be made, there is increasing evidence that wind farms do have a local influence on the environment (Baidya Roy & Pacala 2004; Kirk-Davidoff & Keith 2008; Baidya Roy & Traiteur 2010; Baidya Roy 2011; Walsh-Thomas *et al.* 2012; Zhou *et al.* 2012, 2013; Zhang *et al.* 2013; Takle 2017). Local influence is mainly derived from the increased turbulence levels due to the turbines. More turbulence means more mixing and the turbulent exchanges of heat and moisture are altered and, in general, the surface becomes drier. The exact response depends on the vertical stratification, but most often it is expected that in the presence of turbines the surface will become cooler in the daytime while the surface grows warmer during the night-time. It may also be noted that if projections for future wind-energy requirements are true, it is likely that wind farms will have an impact on the global climate (Keith *et al.* 2004; Kirk-Davidoff & Keith 2008; Santa Maria & Jacobson 2009; Wang & Prinn 2010; Vautard *et al.* 2014). The projections indicate that the global climate will not change much, but that there will be regional changes in temperature that exceed 1°C. This is mainly the result of the increased drag that the turbines place on the atmosphere. It is also likely that the changes in atmospheric temperature and wind directions, and perhaps also storm tracks, will have an impact on the water cycle (Takle 2017). This has not been studied to the same extent, but may be an important effect to consider in future studies.

In the same way that each wind turbine will have a particular wind wake that stretches downwind, and spreads upwards and downwards to the surface, the wind farm, as a collection of turbines, will leave a similar pattern behind itself. The wake behind an entire wind farm has been much less studied than the wake behind each turbine, for the simple reasons that the wake behind each turbine is (i) easier to study and (ii) has considerable economic value as it connects directly to energy production and wind-farm optimisation. Thus, the extent of the wind wake behind large wind farms is much less understood than the wake from each turbine. Nonetheless, it may be claimed with some certainty that the wind velocity is reduced by, say, 10–30% and possibly up to 50%, compared to the surrounding area (Christiansen & Hasager 2005; Fitch *et al.* 2012; Ludewig 2013, 2015) The size of the wake is difficult to study but appears to be rather long, possibly with a characteristic e-folding length scale that may be two to four times the characteristic size of the wind farm offshore. High wind speed in ocean areas results from low turbulence over the smooth ocean, and low turbulence levels imply that the return to normal atmospheric conditions beyond a wind farm will take longer distance over the ocean than over land. The exact properties of the wake will depend on the size and configuration of the wind farm and the wind conditions at the specific location.

The reduction in wind speed and increase in turbulence levels will impact on ocean dynamics through changes in wind stress and exchange of heat and moisture at the surface. In addition, the slowing down of the wind by the wind farm will cause air to rise above a wind farm, which will be likely to increase precipitation above it and bring drier air downwind (Rooijmans 2004; Takle 2017). Again, this effect has not been targeted in many studies and additional studies are required in order for it to be quantified.

Scope

The aim of this chapter is to outline the response of the upper ocean circulation on the wind wake downwind from a wind farm. There have not been many studies on this topic and much is based on theoretical predictions, simplified numerical simulations, and a single observational study at the Alpha Ventus wind farm in the German economic zone on the north-west European continental shelf (Box 3.1).

Box 3.1 Research at Alpha Ventus: a detailed investigation of the effects of a wind farm on ocean dynamics

Completed in 2009, Alpha Ventus was the first official offshore wind farm in the German exclusive economic zone (EEZ). The Research at Alpha Ventus (RAVE) initiative, consisting of more than 20 subprojects, accompanied the development, construction and operation of the site. The aim was to gain a better understanding of the influences of wind farms on the marine environment and to improve knowledge about the dynamics and forces that act on turbines. RAVE was sponsored by the Federal Ministry for Economic Affairs and Energy (BMWi) following a resolution by the German Federal Parliament.

The Alpha Ventus wind farm is located in a water depth of about 30 m in the southern part of the German Bight, 45 km north of the island of Borküm (Figure 3.1). The considerable distance of the site from the coast, with prevailing westerly winds, was to guarantee profitable amounts of renewable energy.

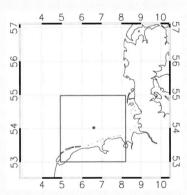

Figure 3.1 Position of the Alpha Ventus wind farm (red point) and model area (blue square), which covers 240×240 km² and has an ocean depth of 30 m. (From Ludewig 2015)

Figure 3.2 Panoramic view of the Alpha Ventus wind farm showing nine of the 12 turbines using two different turbine foundations. (© DOTI 2010/Alpha Ventus; Matthias Ibeler)

The wind farm spans an area of 4 km², or as big as 500 football fields, and comprises 12 5 MW wind turbines with two different foundation designs (Figure 3.2). Six wind turbines stand on jacket foundations installed after the supporting piles were driven into the seabed, whereas the other six turbines stand on tripod foundations with piles that were rammed into the sea floor after the tripods were placed on the seabed. These piles reach a depth of 35 m in the sea floor.

Four selected turbines were equipped with more than 1,200 sensors to collect environmental and technical data. The research focuses on load analysis, safety and logistic questions, construction and operational noise measurements, modelling of sound propagation, fatigue originators and environmental research including meteorological, ecological, geological and oceanographic questions. The Federal Maritime & Hydrographic Agency (BSH) collects the oceanographic and geological data and coordinates the measurements of the different scientific subprojects of the RAVE initiative. The research platform FINO1 is in close proximity to Alpha Ventus and has collected environmental data for about 10 years.

The first two themes outline the theoretical basis and start with a brief description of the dynamics of the upper ocean and how earth rotation impacts the ocean response. This is considered as general knowledge given at BSc or MSc courses in oceanography, but is outlined here to provide a basis for the theoretical predictions. The third and fourth themes describe some model predictions and empirical evidence of the ocean response to a wind farm. This is partly based on general theoretical predictions and model simulations. Some observations, and results from model simulations, taken at the Alpha Ventus wind farm are also presented. This was presented in a PhD thesis by one of the authors, Elke Ludewig, but is condensed into a more manageable text in this chapter. The fifth theme is devoted to possible increases in primary production as induced by the artificial upwelling.

This study is based on the few earlier studies, mainly in form of relevant peer-reviewed literature, of which the authors are aware. Some grey literature on the topic may exist, probably in the form of MSc theses, but these have not been systematically investigated in this study. There have not been many studies on the subject and there is a need for more studies before reliable predictions on the impact on large wind farms on the ocean dynamics and thus the environmental impact of wind farms can be made. However, the underlying dynamics described here are based on well-known theoretical frameworks that have long been established in physical oceanography.

Themes

Basic ocean dynamics

Atmospheric low-pressure systems have a typical size in the order of 1,000 km, while the corresponding features in the ocean have characteristic scales in the order of 5–30 km, and are called eddies. In nature, open ocean eddies draw their energy from the large-scale ocean flow through various instability mechanisms and not directly from the atmosphere. The reason for this is that direct forcing of ocean eddies from the atmosphere must derive energy from the horizontal shear in the wind field, not the wind field itself. Given that atmospheric low-pressure systems are much larger than ocean eddies, it follows that the horizontal wind shear is weak at the scale of ocean eddies, and the direct wind forcing on the ocean eddy field is also weak. Thus, atmospheric low-pressure systems and ocean eddies are decoupled despite their dynamic similarity. However, by placing a wind farm in the ocean, a wind field that has much smaller spatial scales will be created, implying the creation of an unnatural strong direct forcing on the oceanic eddy field. In other words, a persistent forcing field is created that acts directly on the natural scales of the ocean. In simple terms, wind farms generate wind forcing with an impact on the resonance scales of the ocean and hence generate a surprisingly strong impact (Broström 2008; Ludewig 2013, 2015). This is sometimes called the wind-wake effect from wind farms. There is small-scale variability in wind at the same scale as the ocean eddies; however, this is not as persistent as the wind-farm wake. For more details see Box 3.2.

Box 3.2   Divergent Ekman transport

When wind blows over an ocean it forces an ocean current. Owing to the rotation of the earth, the upper ocean current is to the right of the wind direction in the northern hemisphere. At the surface, the current is deflected about 30° to the right and the current veers more to the right with depth, such that the total mass transport is deflected 90° to the right of the wind direction. An additional feature is that the ocean current will reach its final strength after about one inertial period, or after about 14 h at 60°N.

To illustrate the calculation, a wind speed of 10 m/s will create a surface current of about 0.2 m/s. The total volume transport is often called the Ekman transport (M) and for a wind in the x-direction it is in the negative y-direction according to:

$$M_y = \frac{\tau_x}{\rho f}$$

where τ_x is the wind stress on the surface (N/m²), ρ is the density of water (e.g. 1,000 kg/m³), and f is the Coriolis parameter, which varies with latitude but is about 1.2×10^{-4}/s at 60°N.

The wind stress at the surface can then be estimated as

$$\tau = \rho_A C_D U_{10}^2$$

where ρ_A is the density of air (e.g. 1.25 kg/m³), C_D (~1.5×10⁻³) is a drag coefficient, and U_{10} is the wind speed at 10 m height. As example, a wind stress of 10 m/s gives 0.19 N/m² and a wind speed of 20 m/s gives a wind stress of roughly 0.75 N/m².

Thus, for a wind speed of 10 m/s the transport is about 1.5 m²/s, that is 1.5 m³/s/m. If the wind speed varies horizontally, the transport in the upper layer will also vary in the horizontal. This is called a divergent flow field. The volume is essentially conserved and there must be a compensating vertical flow if the horizontal flow in the surface layer is divergent. An extreme example is a coastal area with wind that is directed along the coast. Depending on the direction of the wind, upwelling of deep water or downwelling of surface water will result. In many places the climatological wind is such that upwelling prevails, such as in Peru, California, Portugal and the western Sahara, all areas known for their rich marine life and fisheries fuelled by the nutrients in the upwelled water.

The fact that wind farms extract wind energy implies that the wind stress behind the wind farm is reduced, which induces a divergent Ekman transport in the upper ocean. The consequence is that offshore wind farms will generate upwelling and downwelling patterns in their vicinity. The exact response in the ocean to a divergent Ekman flow field depends on the size of the disturbance in the flow field and the natural response scales in the ocean. This is discussed in Box 3.3.

Theory behind wind-farm induced upwelling

The idea that the wind wake from wind farms may impact strongly on ocean circulation was first put forward by Broström (2008). He used theoretical predictions and some idealised numerical model simulations to show that wind farms may produce a strong response on the ocean. A key parameter that determines the ocean response is the size of the wind farm; that is, the scale of the wake as compared to the dynamic scale of the ocean, which is also known as the internal radius of deformation or the internal Rossby radius (Box 3.3). This length scale depends on stratification and for coastal areas it is expected to be in the order of 5–30 km, depending on stratification: becoming larger for stronger stratification and smaller for weaker stratification. If a wind farm is much smaller than the internal deformation radius, the consequent response from the wind farm is weak, while it becomes much larger as its size approaches the internal radius of deformation (Broström 2008). The large wind farms beginning to be installed today, and those in the near future, are probably equal in size to, or becoming larger than, the internal deformation radius. When they are of similar sizes, theory suggests that the signal in the upper ocean structure starts to become significant, but also that the ocean response will propagate out from the area directly influenced by the wind farm. Thus, the upwelling would be expected to be distributed over a much larger area than both the wind farm and its wake, covering an area perhaps ten times the area of the wind farm. For more details see Box 3.3.

Figure 3.3 shows a simulation based on an idealised wind wake from a wind farm of 5×5 km² under a wind speed of 7.5 m/s and where the wind wake is about twice as long as the wind farm (see also Broström 2008). The results show an upwelling/downwelling dipole that is 20–40 times as large as the wind farm itself, and where the upwelling/downwelling areas are of about equal size. The resulting vertical velocity is about 0.5–1 m/day and the

Box 3.3 Ocean dynamics and scales

The rotation of the earth exerts strong control over the large-scale ocean currents. The motion of the rotating earth implies that a moving object will tend to feel a force perpendicular to the motion, known as the Coriolis force. The fact that the earth rotates also implies that the potential vorticity, that is, the rotation of the water that on earth also includes the earth's own rotation, becomes important for how a fluid can move on the earth. For instance, if we try to force water to go over a subsea canyon, the water column will stretch and it will start to rotate owing to the conservation of angular momentum. This rotation will steer the flow in such a way that ocean currents tend to flow along isobaths.

On the rotating earth, it takes about the local rotation, that is $1/f$, before the currents turn as a result of rotation. Disturbances in the ocean travel with the speed of large-scale waves, implying that perturbations can only propagate a certain distance before they are redirected by the rotation. In mathematical terms, the natural length scale of ocean eddies is the distance that an internal wave travels over the inertial period. This is also known as the internal deformation radius or the internal Rossby radius. It is given by:

$$L = \frac{\sqrt{g'h}}{f}$$

where the square root is the wave propagation speed of the disturbance and $1/f$ is the timescale, h is the thickness of the upper layer and g' ($=g\Delta\rho/\rho_0$, where g is gravity and $\Delta\rho$ and ρ_0 are the densities of the upper and lower layers, respectively) is the reduced gravity of the upper layer.

L is the primary horizontal length scale of the upper ocean variability, and it is the fundamental length scale for the oceanic response to perturbations. Thus, if the disturbance is smaller than this scale the ocean response will be smeared out over the distance described by the length scale. The forcing on the potential vorticity is the curl of the wind stress, written as:

$$curl(\tau) = \frac{\partial \tau_y}{\partial x} - \frac{\partial \tau_x}{\partial y}$$

This is the horizontal gradient of the wind stress on the sea surface, as previously discussed. Notably, the upwelling is proportional to the curl of the wind stress on a rotating ocean, which in the above formula describes the divergence in the wind-driven (Ekman) current. To provide a numerical example, if the wind in a storm at 20 m/s or 0.75 N/m² (see Box 3.2) changes to zero over 500 km, this gives a curl(τ) that is 1.5×10^{-6} N/m. In a wind farm, 0.2 N/m² may be measured outside the wind farm from a 10 m/s wind and 0.1 N/m² inside the wind farm from a 7.5 m/s wind. If this is over 5 km, the result would be 2×10^{-5} N/m, or 13 times the value from a storm. The corresponding induced vertical velocities are about 1 and 14 m/day, respectively, and the latter magnitude represents both the upwelling and downwelling dipole of the wind farm. However, upwelling and downwelling of this magnitude would not be expected as the signal will be redistributed over a larger area, perhaps ten times as large, thus giving an upwelling or a downwelling in the order of 1.4 m/day. In the case of upwelling/downwelling of 1 m/day, with a width of 10 km wide and a length of 20 km approximating to eight times the area of the wind farm, the upwelling would be $2{\times}10^8$ m^3/day or roughly 2,000 m^3/s.

Given that the wind farms of today are comparable in size to the internal radius of deformation, it follows that the oceanic response is distributed over an area that is larger than the wake area. However, at the same time the upwelling velocities become smaller, since horizontal gradients are reduced. The upwelling and downwelling will also create density gradients in the upper layer that create horizontal ocean currents. In combination with the existing background velocity field, this density driven flow will, in turn, advect the density field, creating a response pattern that could eventually cover a large area.

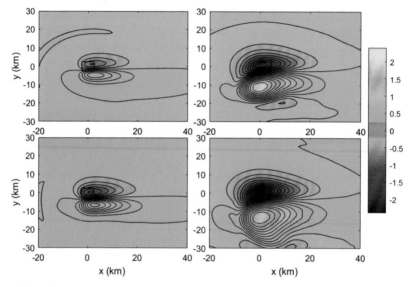

Figure 3.3 Disturbance of the thermocline after 1 day (left panels) and after 5 days (right panels) for a background wind speed of roughly 5 m/s (upper panels) and 7.5 m/s (lower panels) (see also Broström 2008). The red squares represent the position of the 5×5 km^2 wind farm in the simulations. Colours are measured in metres.

total upwelling is thus in the order of $2{\times}10^8$ m^3/day. If the upwelling brings nutrient-rich water to the surface it will lead to increased primary production (Box 3.4). It is noted that the wind wake used by Broström (2008) is based on very crude assumptions; it seems likely that the assumed size of the wind wake is actually too small, and increasing the length of the wake will increase the areas of upwelling and downwelling. The long-term evolution of the upwelling/downwelling pattern also needs further investigation. Nevertheless, both Paskyabi and Fer (2012) and Ludewig (2013, 2015) theoretically confirmed the general predictions of Broström (2008). In particular, Paskyabi and Fer (2012) showed that the impact on the ocean may become stronger if the role of surface waves for the air–sea exchange of momentum is accounted for. Further details of the study by Ludewig (2013, 2015) are provided below, in *Empirical evidence of upwelling* and *Effects on temperature*.

Moreover, in addition to the upwelling process itself, disturbances in the thermocline depth will also generate horizontal velocities. For the cases studied by Broström (2008),

Box 3.4 'Back of an envelope' calculation on upwelling and production

Wind farms will induce artificial upwelling, implying that nutrient levels will increase in the productive zone and thereby ultimately increasing primary production. A more comprehensive analysis requires a coupled physical–biogeochemical model, and probably also expertise on how ecosystem regimes may change in response to increased nutrient levels. However, it is an interesting exercise to provide some possible numbers on the induced productivity. As a reference, it is assumed that the North Sea has an area of 750,000 km^2 and annual production is roughly 30–50 g carbon (C)/m^2 per year (Richardson & Pedersen 1998; Richardson *et al.* 2000; Heath & Beare 2008), or thus roughly 20–40×10^6 t C/year for the entire North Sea. The North Sea may not be the best example since the southern North Sea is well mixed and nutrients originate from rivers rather than from upwelling. However, this area is used here as there are many plans for offshore wind farms. The calculations will be relevant for the northern rather than the southern North Sea. Readers should note that these calculations use new annual production; this is different from total annual net production, which includes recirculation of nutrients. Total annual net production is frequently a factor of two larger than new production.

   The calculations made here are based on some simplifications that may influence the results significantly. It is therefore not a best estimate but rather a simple and rough calculation to highlight the possible impact of the artificially induced upwelling. The assumptions are:

- All the divergence in the wind field from a wind farm is used to drive upwelling.
- The upwelling is taken from deep water; here with a nitrate level of 10 mmol NO$_3$/m^3 or about 10 μmol NO$_3$/kg.
- All nutrients are used up in the surface layer, and it is assumed that the new production follows the Redfield ratio; that is, organic matter contains nitrogen and carbon atoms in a 16:106 ratio.
- It is assumed that there is no shift in ecosystem functioning due to the upwelling.

Using the assumptions above gives the following rough results that for a single 5×5 km^2 wind farm in open water:

- Total volume of upwelling: As a starting point, a wind-driven upwelling of about 2,000 m^3/s is assumed (see *Potential ecosystem effects*, Boxes 3.2 and 3.3), although the exact number is unknown for a realistic wind farm.
- Upwelling of nutrients: If the nitrate in the deeper layer is 10 mmol NO$_3$/m^3, this could fuel a new primary production of ~2×10^6 mol N/day from (2,000 m^3/s) (86,400 s/day) (10×10^{-3} mol NO$_3$/m^3).
- Conversion to carbon: 2×10^6 mol N/day (106/16 C:N) (12 g C/mol) ~160 t C/day.
- Annual production of carbon: If the wind blows on 100 days a year in the productive season, this results in 16,000 t C per year in artificially induced primary production. If the area of the upwelling is 20 times as large as the wind farm, this corresponds to about 33 g C/m^2 per year in new annual production, which is the same as the lower estimate from the literature outlined above. The estimate of the increased annual productivity from a single wind farm is about

0.5‰ of the total production in the entire North Sea, thus 20 wind farms would correspond to, say, a 2% increase in productivity for the entire North Sea.

- Total production of carbon in herring: With a trophic efficiency of 10%, the above estimate of the artificially induced primary production corresponds to an increase of $16×10^3$ t herring C per year.

- Total herring production: If the weight of wet fish is about ten times the weight of C in fish, this implies that an additional 160,000 t of Atlantic Herring *Clupea harengus* may be produced per year. If 1 kg of herring is worth €1, the total value is €160 million per year.

- Total production of salmon/cod: If 10% is transferred from Atlantic Herring to Atlantic Salmon *Salmo salar* or Atlantic Cod *Gadus morhua* according to trophic efficiency, then 16,000 t salmon/cod per year may be produced.

Accordingly, the artificially induced new production around the wind farm would appear to be about the same size as the natural new production in an area that is about ten to 20 times as large as the wind farm. However, the ocean response to a wind farm under realistic conditions needs to be verified and quantified in much more detail. In addition, the ecosystem response to the increased nutrient upwelling may not be linear as assumed here. For example, this increased production can also add to eutrophication and lower oxygen concentrations in deep water, and, hence, have an inverse negative effect on fish productivity. Nevertheless, the calculations indicate that wind farms may induce highly productive areas surrounding the wind farm, which may benefit the local ecosystem and human interests.

the density-induced ocean currents were in the order of 0.3–0.4 m/s. The maximum velocities were located at the wind farm or upwind from the wind farm; currents against the wind and waves are known to generate high short-crested waves that may pose a problem for shipping and maintenance of the wind farm. However, the wake may also induce a self-advective pattern similar to a smoke ring. This implies that the upwelling and downwelling patterns evolve with time and may spread over large areas, and they may also interact with local currents. As an example, ocean eddies are known to move westwards in an open ocean, although the implication on the dipole eddies found here is unknown. Further studies are needed to quantify how the velocity field will behave and evolve in more realistic situations, and over longer periods.

Empirical evidence of upwelling

Following on from Broström (2008), Ludewig (2013, 2015) considered a much broader and more realistic set of conditions and found that although the findings of Broström (2008) were realistic overall, they were, as to be expected, too simplified. Both the wind wake and the ocean response are more complex in a realistic situation. Ludwig's study was based on simulations of a coupled atmosphere–ocean model, the Hamburg Shelf Ocean Model (HAMSOM), forced with meteorological data from the three-dimensional MEsoscale TRAnsport and Stream (METRAS) model. HAMSOM is a three-dimensional baroclinic ocean model, which was developed for shelf seas. Simulations were performed in an idealised test box with a horizontal domain size of 240×240 km² using the assumption of four open boundaries. Moreover, a constant wind forcing with a geostrophic south-westerly wind of 8 m/s was prescribed over 1 day. Owing to the limitations of this idealised case, tidal effects have been neglected. In agreement with the real conditions at Alpha Ventus (see Box 3.1), 12 wind turbines were placed in the centre of the model domain and the water depth was set uniformly to 30 m. The model resolution was set to 2.5′×1.5′ in the horizontal, approximating to 3 km, and 2 m in the vertical. For more details, see Ludewig (2013, 2015).

Beside these model simulations, some much needed observational evidence that the wind wake from a wind farm may produce upwelling and downwelling was derived from measurements conducted during cruise 141 with the research vessel VWFS Wega between 10 and 13 May 2013 around Alpha Ventus, where 39 conductivity, temperature and depth (CTD) profiles were taken. The positions of the CTD profiles are shown in Figure 3.4.

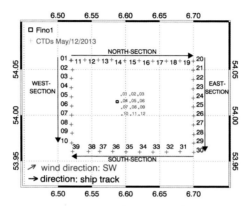

Figure 3.4 Map of the investigation area around test wind farm Alpha Ventus. Wind turbines are marked in blue; conductivity, temperature and depth (CTD) measurements are marked by red crosses, and the Fino1 platform is marked with a black square. Black arrows indicate the direction of the ship track during CTD measurements along sections west, north, east and south. (After Ludewig 2015)

The change in the wind field simulated by METRAS is illustrated in Figure 3.5. In front of the OWF placed in the centre of the model area, there is a surge area with slightly decreased wind speeds. Behind the wind farm a wind wake develops, with a maximum wind deficit of around 70%. Two areas with wind speeds increased by about 10% can be found at the flanks of the wind wake. The affected area is approximately 100 times bigger

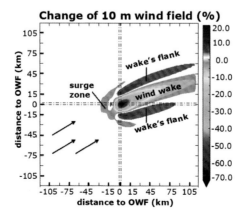

Figure 3.5 Offshore wind-farm (OWF) induced change in the 10 m wind field after 4 h operating wind turbines, simulated with the MEsoscale TRAnsport and Stream (METRAS) model, based on a south-westerly geostrophic wind of 8 m/s. (After Ludewig 2015)

than the wind farm itself, with a wind wake length of more than 100 km. This wind wake is much longer than considered in Broström's (2008) study, but is consistent with Fitch *et al.* (2012), for example. In addition, the wind deficit in the wake is stronger than suggested by Broström (2008), but here results from different studies do not converge (Christiansen & Hasager 2005; Fitch *et al.* 2012; Ludewig 2013, 2015). Christiansen and Hasager (2005) and Fitch *et al.* (2012) find and predict that the deficit in the wind stress is about 20–30%, while Ludewig (2015) finds a deficit in the wind stress that is up to 70%. Variation in results will influence the predictions of the wind-wake effects from wind farms, and their consequences on upper ocean dynamics and marine life. More research on this subject is clearly needed.

Whatever the scale of variation, modelled changes in the wind pattern will have a significant impact on the Ekman transport in the wind-driven upper ocean layer. A numerical model analysis presented by Ludewig (2015) demonstrates that as a result of the disturbance in the wind field, a convergence–divergence dipole of Ekman transport is created (Figure 3.6), which is consistent with the theoretical predictions by Broström (2008). The main effects resulting from this dipole are the formation of a surface elevation dipole structure and the related development of upwelling and downwelling cells, which, in turn, induce an excursion of the thermocline in the vicinity of the wind farm.

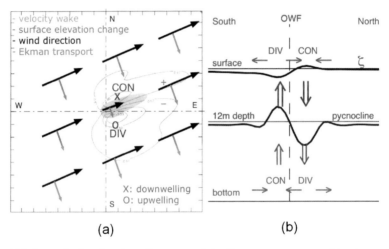

(a) (b)

Figure 3.6 (a) Schematic showing the adjustment of the Ekman transport resulting in upwelling and downwelling. Ekman transport (green arrows) is theoretically deflected by 90° to the right (in the northern hemisphere) of the wind direction (black arrows). As a result of wind wake (light blue area) the Ekman transport is reduced, which causes convergence (CON)/divergence (DIV) of water masses and a related change in surface elevation depicted by light grey isolines, in addition to downwelling (X) and upwelling (O). (b) Schematic of conditions on a vertical south–north cross-section through an offshore wind-farm (OWF) affected ocean system. A decline in surface elevation and shallowing of the pycnocline are related to downwelling, with the opposite behaviour generated by upwelling. (After Ludewig 2015)

Model simulations under representative conditions have shown that resulting changes in the surface elevation are in order of ±0.01 m. Associated with this, vertical cells with an average horizontal dimension of 15–30 km are created, with maximum vertical velocities of ±3.5 m/day, leading to an excursion of the pycnocline of about 10 m (Ludewig 2015).

Effects on temperature

The VWFS Wega measuring campaign at Alpha Ventus described above (see *Empirical evidence of upwelling*, Figure 3.3) allowed the measurement of temperature and salinity profiles to detect any OWF-induced upwelling and downwelling zones affecting the depth of the thermocline. Figure 3.4 shows the four sections (transects) of interest around the 12-turbine wind farm, with a length varying between 10.9 and 12.8 km and a distance between 5.5 and 6.4 km from the centre of the OWF. This analysis focuses on the temperature as this showed the clearest signal. Along these sections, Figure 3.7 contrasts the measured temperature profiles with the corresponding simulated profiles for the observed and simulated south-westerly wind conditions. The modelled and observed vertical temperature structures, accentuated by the black solid and dashed lines, show remarkably good agreement. In particular, the locations of simulated upwelling and downwelling zones downwind of the wind farm, that is on the northern and eastern sections, were also detected under real conditions. This substantiates the conclusions from previous theoretical and numerical considerations. However, there are also clear differences between the observations and model results, in particular on the upwind side, which can be explained by the partly idealised model set-up comprising constant topography and a constant wind field.

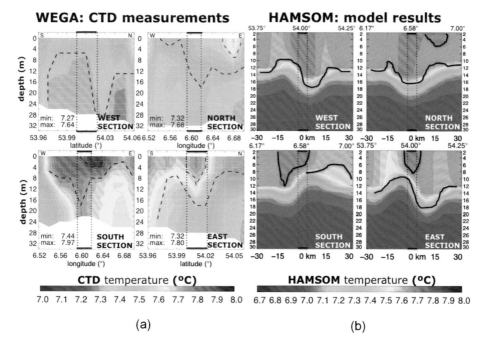

Figure 3.7 Comparison between (a) measured and (b) simulated temperature distributions around the Alpha Ventus offshore wind farm. (a) Measured conductivity, temperature and depth (CTD)– temperature profiles around Alpha Ventus in the four different sections; (b) results from the Hamburg Shelf Ocean Model (HAMSOM). Black dotted vertical lines mark the borders of the wind farm and black dashed lines accentuate the vertical temperature structure for comparison with the HAMSOM results. Results are presented for 12 May, corresponding to 3 days of simulation with constant wind forcing. (After Ludewig 2015)

Potential ecosystem effects

General theoretical arguments suggest that the response of ocean dynamics will increase quickly with the size of the wind wake, which in turn depends on the size of the wind farm (Broström 2008). In other words, size matters when evaluating the impact on the ecosystem, although more precise estimates of the response for different wind speeds and turning of wind direction are required before accurate predictions on the ecosystem can be made. Possible changes in species composition and ecosystem functioning may also be crucial for evaluating the likely ecosystem response. Nonetheless, some estimates can be provided from the current knowledge base. An important part of evaluating the impact from a wind farm is determining the effect upon productivity in the natural system. Taking the northern North Sea as an example, new production is roughly 30–50 g carbon (C)/m^2 per year (Richardson & Pedersen 1998; Richardson *et al.* 2000; Heath & Beare 2008). The system has a strong annual cycle with a pronounced spring bloom, while the summer usually has lower productivity owing to low nutrient levels (Burchard *et al.* 2006).

For a wind farm that is 5×5 km^2, there may be an upwelling in the order of 2,000 m^3/s or about 2×10^8 m^3 per day (Box 3.3 and Box 3.4). If this water is supplied from nutrient-rich layers near the sea floor with, say, a nitrate concentration of 10 mmol NO$_3$/m^3, this may fuel new production in the order of 30 g C/m^2 per year, which is similar to the natural annual production in the North Sea. This could be in an area around 20 times that of the wind farm. However, making a detailed prediction on the response in the production is beyond the scope of what can be achieved here, and this number is highly uncertain. In addition, the size of the affected area probably depends on the ecosystem's ability to utilise the increased nutrient levels and it is questionable whether all upwelled nutrients would drive new production. There are some questionable assumptions in the 'back of an envelope' calculations presented in Box 3.4, but, nevertheless, it points to a likely scenario that upwelling induced by wind farms may have a strong effect on the productivity around the wind farms.

Increased productivity will lead to an increase in fish production that would be of benefit to the fishing industry as well as attracting and supporting various birds, larger fish and marine mammals. Thus, wind farms may produce areas with rich and diverse marine life (see Dannheim *et al.*, Chapter 4; Gill & Wilhelmsson, Chapter 5) exploited by top predators (Nehls *et al.*, Chapter 6; Vanermen & Stienen, Chapter 8). However, a highly productive wind-farm area may have some disbenefits, such as the attraction of seabirds to the site (Vanermen & Stienen, Chapter 8) and the potential for more bird–blade collisions (King, Chapter 9). In addition, increased primary production could carry a price in the form of eutrophication and the potential for toxic algal blooms and increased areas with low oxygen levels near the seabed, although this would depend heavily on natural water circulation and tidal currents. Furthermore, as already mentioned, the natural cycle of the North Sea is characterised by a spring bloom and lower summer productivity; the latter may be affected by relatively constant upwelling, with important ecological consequences.

Natural pathways of transport, both locally and at some distance from the wind farm, may also be disturbed. The fact that a dipole structure is created may imply that the disturbance will move away from the area where it is generated and be manifested as a far-field effect, which has not yet been demonstrated for coastal processes (see Rees & Judd, Chapter 2). In addition, if the dipole interacts with local and large-scale topography, some disturbance in the transport pathways may result, thereby affecting connectivity between different areas, with implications for the transport of invertebrate and fish larvae and the patterns of colonisation, for example. More studies are clearly required before accurate predictions of the positive and negative consequences on marine life can be made.

Besides the impact of the wind-wake effect on ocean dynamics, the presence of the turbine bases and monopoles will also have direct effects on the local current and wave fields (see Rees & Judd, Chapter 2) and indirect effects by operating as artificial reefs (Dannheim *et al.*, Chapter 4; Gill & Wilhelmsson, Chapter 5). In agreement with Rees and Judd (Chapter 2), the authors think it unlikely that turbines and thus wind farms will have a large impact on the large-scale ocean currents, although one should not rule out the potential for far-field effects. In contrast, near-field effects are likely, with increased turbulence levels leading to sediment resuspension, as clearly shown as visible tracks from turbine foundations on satellite images from the London Array and Thanet wind farms (Vanhellemont & Ruddick 2014; the former of which was until recently the largest OWF in the world (Jameson *et al.*, Chapter 1). Together with the altered current field created by the wake, it is likely that sedimentation in and around the wind-farm area will be altered (Rees & Judd, Chapter 2).

Concluding remarks

The purpose of wind farms is to extract energy from the wind. In this process, the wind speed will decrease and the turbulence levels will increase as unavoidable consequences of the energy-extraction process. The reduction in wind speed implies that the wind stress on the ocean surface will decrease. Moreover, given the relatively small scales of the wind farms compared to the size of low-pressure systems, the forcing of potential vorticity will increase greatly in the presence of a wind farm. The wind-driven upwelling and downwelling depend directly on the strength of this forcing, which implies that large wind farms may generate unnaturally strong vertical motions in the ocean. More studies are needed, but the predictions are based on solid and well-known features of ocean dynamics. Fine-resolution measurements around an operating wind farm confirmed that the dynamics of the North Sea may be significantly affected by OWFs in general. The observations follow theoretical predictions validating the general theoretical considerations behind earlier studies. Following this first investigation, further cruises around Germany's wind farms are planned to gain a more comprehensive view of the full impact of wind farms on the marine system.

The ocean eddies generated by the wind wake are likely to generate stronger variability in the currents. The long-term development of these eddies has not yet been considered. Interactions with topography and naturally occurring ocean currents and eddies are not yet understood, which is also an area where possible large impacts on the ocean dynamics and the ecosystem are still to be discovered. Synthetic aperture radar images at Horns Rev (Christiansen & Hasager 2005) showing the bottom structures may be related to an ocean current travelling around closed depth contours corresponding to a general theoretical prediction.

Some 'back of an envelope' calculations on the possible impacts on ecosystems have been presented in this chapter. For example, 20 large wind farms in the North Sea may increase the total primary production by a few per cent. The local impact closer to wind farms is much stronger, with an estimated increase in primary production of 50–80%. However, more realistic studies to evaluate these numbers are badly needed.

As a final remark, the upper ocean may respond strongly to the presence of large OWFs as predicted by a solid theoretical foundation. Moreover, this chapter also presents some observational evidence that reinforces the point that theoretical predictions are reliable, although more observations on the wind wake and the upper ocean response are needed to confirm those theoretical predictions. Furthermore, the response to more realistic wind-farm configurations needs to be assessed, along with more detailed analyses on the effects of increased upwelling and downwelling on the marine ecosystem. The indication from limited study to date is that the response in the ocean covers a much larger area than the wind farm itself. Accordingly, if large wind farms become abundant, the interaction between wind farms will start to play an important role. The focus so far has been on the short-term ocean response, and evaluation of long-term change needs further study. There may be some surprises along the research road concerning impacts at a far distance from wind farms.

References

Baidya Roy, S. (2011) Simulating impacts of wind farms on local hydrometeorology. *Journal of Wind Engineering and Industrial Aerodynamics* 99: 491–498.

Baidya Roy, S. & Pacala, S.W. (2004) Can large wind farms affect local meteorology? *Journal of Geophysical Research* 109: D19101.

Baidya Roy, S. & Traiteur, J.J. (2010) Impacts of wind farms on surface air temperatures. *Proceedings of the National Academy of Sciences of the United States of America* 107: 17899–17904.

Barthelmie, R.J. & Jensen, L. (2010) Evaluation of wind farm efficiency and wind turbine wakes at the Nysted offshore wind farm. *Wind Energy* 13: 573–586.

Barthelmie, R.J., Pryor, S., Frandsen, S.T., Hansen, K.S., Schepers, J., Rados, K., Schlez, W., Neubert, A., Jensen, L. & Neckelmann, S. (2010) Quantifying the impact of wind turbine wakes on power output at offshore wind farms. *Journal of Atmospheric and Oceanic Technology* 27: 1302–1317.

Broström, G. (2008) On the influence of large wind farms on the upper ocean dynamics. *Journal of Marine Systems* 74: 585–591.

Burchard, H., Bolding, K., Kühn, W., Meister, A., Neumann, T. & Umlauf, L. (2006) Description of a flexible and extendable physical–biogeochemical model system for the water column. *Journal of Marine Systems* 61: 180–211.

Christiansen, M.B. & Hasager, C.B. (2005) Wake effects of large offshore wind farms identified from satellite SAR. *Remote Sensing of Environment* 98: 251–268.

Fitch, A.C., Olson, J.B., Lundquist, J.K., Dudhia, J., Gupta, A.K., Michalakes, J. & Barstad, I. (2012) Local and mesoscale impacts of wind farms as parameterized in a mesoscale NWP model. *Monthly Weather Review* 140: 3017–3038.

Frandsen, S., Barthelmie, R.J., Pryor, S.C., Rathmann, O., Larsen, S.E., Højstrup, J. & Thøgersen, M. (2006) Analytical modelling of wind speed deficit in large offshore wind forms. *Wind Energy* 9: 39–53.

Heath, M.R. & Beare, D.J. (2008) New primary production in northwest European shelf seas, 1960–2003. *Marine Ecology Progress Series* 363: 183–203.

Keith, D.W., DeCarolis, J.F., Denkenberger, D.C., Lenschow, D.H., Malyshev, S.L., Pacala, S. & Rasch, P.J. (2004) The influence of large-scale wind power on global climate. *Proceedings of the National Academy of Sciences of the United States of America* 101: 16115–16120.

Kirk-Davidoff, D.B. & Keith, D.W. (2008) On the climate impact of surface roughness anomalies. *Journal of Atmospheric Science* 65: 2215–2234.

Ludewig, E. (2013) *Influence of offshore wind farms in atmosphere and ocean dynamics*. PhD thesis, Max Planck Research School for Maritime Affairs, Hamburg.

Ludewig, E. (2015) *On the Effect of Offshore Wind Farms on the Atmosphere and Ocean Dynamics*. Hamburg Studies on Maritime Affairs. Volume 31. Cham: Springer.

Paskyabi, M.B. & Fer, I. (2012) Upper ocean response to large wind farm effect in the presence of surface gravity waves. *Energy Procedia* 24: 245–254.

Richardson, K. & Pedersen, F.B. (1998) Estimation of new production in the North Sea: consequences for temporal and spatial variability of phytoplankton. *ICES Journal of Marine Science* 55: 574–580.

Richardson, K., Visser, A. & Pedersen, F.B. (2000) Subsurface phytoplankton blooms fuel pelagic production in the North Sea. *Journal of Plankton Research* 22: 1663–1671.

Rooijmans, P. (2004) *Impact of a large-scale offshore wind farm on meteorology.* MSc thesis, Utrecht University.

Santa Maria, M.R.V. & Jacobson, M.Z. (2009) Investigating the effect of large wind farms on energy in the atmosphere. *Energies* 2: 816–838.

Takle, E.S. (2017) Climate. In Perrow, M.R. (ed.) *Wildlife and Wind Farms, Conflicts and Solutions. Volume 1. Onshore: Potential Effects.* Exeter: Pelagic Publishing. pp. 24–39.

Vanhellemont, Q. & Ruddick, K. (2014) Turbid wakes associated with offshore wind turbines observed with Landsat 8. *Remote Sensing of Environment* 145: 105–115.

Vautard, R., Thais, F., Tobin, I., Bréon, F.-M., de Lavergne, J.-G.D., Colette, A., Yiou, P. & Ruti, P.M. (2014) Regional climate model simulations indicate limited climatic impacts by operational and planned European wind farms. *Nature Communications* 5: 3196.

Walsh-Thomas, J.M., Cervone, G., Agouris, P. & Manca, G. (2012) Further evidence of impacts of large-scale wind farms on land surface temperature. *Renewable and Sustainable Energy Reviews* 16: 6432–6437.

Wang, C. & Prinn, R.G. (2010) Potential climatic impacts and reliability of very large-scale wind farms. *Atmospheric Chemistry and Physics* 10: 2053–2061.

Zhang, W., Markfort, C.D. & Porté-Agel, F. (2013) Experimental study of the impact of large-scale wind farms on land–atmosphere exchanges. *Environmental Research Letters* 8 (1): 015002.

Zhou, L., Tian, Y., Roy, S.B., Thorncroft, C., Bosart, L.F. & Hu, Y. (2012) Impacts of wind farms on land surface temperature. *Nature Climate Change* 2: 539–543.

Zhou, L., Tian, Y., Roy, S.B., Dai, Y. & Chen, H. (2013) Diurnal and seasonal variations of wind farm impacts on land surface temperature over western Texas. *Climate Dynamics* 41: 307–326.

CHAPTER 4

Seabed communities

J. DANNHEIM, S. DEGRAER, M. ELLIOTT, K. SMYTH
and J.C. WILSON

Summary

This review of published and unpublished information demonstrates that offshore wind farms (OWFs) have major effects on the benthos; that is, the seabed flora and fauna. By adding artificial hard substrata to the marine ecosystem, OWFs create new habitat for colonising benthic species, allowing attachment and attraction of hard-substratum species, in 'the artificial reef effect'. The general exclusion of fisheries further creates flourishing soft-sediment benthic communities. Although wind farms hardly extend the distribution range of hard-substratum species, they may be stepping stones for non-indigenous species. Such an increase in benthic diversity, however, is countered by the loss of, disturbance to and/or alteration of the natural seabed. Despite this, it may be concluded that OWFs create local hotspots of benthic diversity, directly influencing the local marine food web. During construction, the biomass of forage species decreases, affecting predatory and scavenging species negatively and positively, respectively. Mobile predatory species tend to leave the area during construction. Once installed, the flourishing benthic communities greatly increase in benthic foraging species and attract predators. The surrounding natural sediments are affected by the deposition of organic matter from the epibionts on the turbine monopoles and scour protection and by the altered predator community. Given that a new ecological equilibrium in the benthic system will develop over 20–30 years, it is arguable whether a return to the pre-construction state following full decommissioning would be feasible or desirable. In contrast, a 'renewables-to-reefs' decommissioning scheme involving only partial removal of the wind farm could ensure protection for ecologically valuable sites. While many data already exist, it is difficult to detect significant effects because these are proportional to the degree of change and the changes may take place at different spatial scales. This should be taken into account in OWF monitoring.

Introduction

Benthic communities are of significant ecological and socio-economic importance at a global level. This includes acting as habitat for numerous species at all life-cycle stages, and as a feeding ground for a range of predators. From a socio-economic perspective, these predators can include species of commercial importance. Benthic communities operate both directly and indirectly as food resources for such species. Therefore, the study of the benthic environment around any activity in the marine environment, including offshore renewable energy, is vital to identify potential effects and their significance.

All human activities in the marine environment have the potential, by their very nature, to affect its natural structure and functioning. Because of both the direct effects on the seabed and the intimate links between the water column and benthos (Gray & Elliott 2009), the seabed will always be directly or indirectly affected. The effects of offshore wind farms (OWFs) on seabed communities comprise one of the most important elements when considering the potential impacts of such developments, due to the inevitability of effects arising, especially from monopiles bored into the substratum or gravity bases supported on the seabed, their surrounding erosion protection layer and the installation of cable routes (Wilson *et al.* 2010) (Figure 4.1). Even developments in floating wind technology still require anchor points and the connection of associated infrastructure, such as inter-array and export cables (e.g. Butterfield *et al.* 2005; Statoil ASA 2017; see Chapter 1 in this volume). Therefore, an understanding of the ways in which seabed communities are affected by OWFs is vital, in part, so that appropriate mitigation measures can be identified and deployed.

A set of key hypotheses have been generated for this chapter:

- Changes in seabed ecology as a result of installing a wind farm in the marine environment are viewed as neither positive or negative in ecological terms, but just different.

- The inherent variability of the seabed biota and hydrodynamic conditions may prevent the subtle effects of OWFs being detected, in particular in current wind-farm locations around the North Sea.

- Hard structures associated with OWFs are available as colonisation sites and 'stepping stones' for non-indigenous species.

- A focus on the structure of the benthos rather than its ecological functioning does not satisfactorily assess impact.

- The effect of a wind-farm structure on the seabed is mirrored by an effect of the seabed and its biota on the structure.

- Given the many human activities and pressures, there are in-combination synergistic and antagonistic effects of all aspects of the same development, and cumulative effects of different developments in the same area, which need to be disentangled.

- Location provides opportunities (for habitat creation) as well as threats (to the local biota and habitats), and both need to be considered together.

- Climate change will increase the variability of an already highly variable system, making it increasingly difficult to detect the effects of the wind farm and its structures.

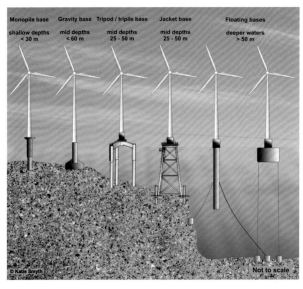

Figure 4.1 Different types of wind turbine foundation. From left to right: monopile base, gravity-base support structure, tripod, jacket base and floating structures. (Katie Smyth)

Scope

This chapter mainly draws on experiences from north-western European waters as this region is the focus for offshore wind, although theoretical considerations and other evidence is also given for sites elsewhere, such as in the USA. The review builds on previous work by the authors and their collaborators and uses both primary and secondary published and peer-reviewed literature as well as grey literature in the public domain. Unfortunately, there are many studies and much information held by the offshore wind industry that are not yet publicly available. In addition, the chapter covers both empirical results and theoretical considerations, especially as the number of publicly available studies is still relatively limited and the consequences of wind farms for individuals, populations and communities are still poorly understood.

By considering the construction, operation and decommissioning aspects, this chapter has been subdivided into four overarching themes relating to seabed communities and their interactions with OWFs. These are as follows:

- **Offshore wind farm artificial reefs create benthic diversity hotspots.** Adding new artificial hard substratum to the marine ecosystem, OWFs create new habitat for benthic species, some of which migrate from outside the area. The exclusion of fisheries further contributes to the development of the soft-sediment benthic communities. This promotion of biodiversity, however, is counteracted by the disturbance of the natural seabed by the installation of the wind farms. This theme evaluates the level at which OWFs truly create hotspots of benthic diversity.

- **Offshore wind farms alter the benthic food web.** Owing to the changes in benthic diversity, trophic interactions between benthic species and higher trophic levels will change as a consequence of OWFs. Predatory and scavenging species attraction and organic enrichment all contribute to the creation of a wind-farm food web that is very different from that occurring naturally.

- **Restoration potential following decommissioning.** As there are similarities in methodologies for construction and decommissioning, a consideration of decommissioning is appropriate for an overall understanding of the effects that may arise for the seabed and its communities.

- **Monitoring programmes, sampling design and analysis.** Numerous studies on OWF effects have been conducted in the framework of environmental impact assessments. Whether these investigations delivered the information needed to appropriately assess the impacts on the benthic ecosystem is evaluated.

Themes

Offshore wind farm artificial reefs create benthic diversity hotspots

The introduction of any new object into the marine environment has the capacity to cause a change in the volume or area, diversity and quality of habitats present in both the immediate and adjacent marine environment. With regard to offshore wind developments, this has been the subject of several studies (e.g. Wilson & Elliott 2009). Throughout the offshore wind industry, monopiles (Figure 4.1) are still the most commonly used form of turbine foundation, used for 75% of all turbines installed (EWEA 2014). However, there have been significant developments in other technologies, including floating structures supporting turbines and gravity bases, which will bring with them their own impacts from a benthic community perspective.

The key element to consider is not just the introduction of new habitat, which may be considered a positive development (Wilson & Elliott 2009), but the introduction of a new type of habitat, that is the introduction of hard material into an area where this generally did not exist. As a result, this must be considered an important habitat replacement, rather than merely habitat creation.

Box 4.1   Co-locating offshore wind farms and marine protected areas

Given that offshore wind farms (OWFs) have the potential to act as biodiversity hotspots for benthic communities and thereby for associated food chains all the way up to large mammal predators, the opportunity to build on this must also be considered, together with co-location with protected areas for the conservation of both habitats and species. Co-location, as a means of siting two or more marine activities together but without adverse environmental effects (Christie *et al.* 2014), may give benefits. Examples include linking OWFs with conservation areas, aquaculture or non-trawling methods of fishing. In turn, these could even be used for habitat creation and biodiversity enhancement, such as the renewables-to-reefs option in following the rigs-to-reefs approach in the USA (Coates *et al.* 2013; Smyth *et al.* 2015), or with ecological engineering principles used to design future wind farms, and other offshore structures, in a way that provides enhanced opportunities for colonisation by biota.

Figure 4.2 Scour protection around the base of a monopile: (a) its creation; (b) *in situ* view; (c) erosion monitoring points (red rods). (With permission of (a) Jan de Nul, Hofstade-Aalst, Belgium; (b) Seatower Wind Turbine Foundations, Glasgow, UK; (c) DHI, Hørsholm, Denmark)

Any structure in the marine environment has the potential to become an artificial reef through colonisation by marine biota (Wilson & Elliott 2009; Wilson *et al.* 2010), and will inevitably create different and/or additional habitat to that originally present. Although the footprint of a typical turbine foundation is estimated at less than 1% of the wind-farm area (Petersen & Malm 2006) and despite turbine foundations and scour protection removing the underlying infaunal habitat, the monopile and scour protection with which it is replaced can create 2.5 times the amount of area that was lost (Wilson & Elliott 2009). For example, in the German Bight, the surface of a hard foundation similar to a wind turbine base (1,280 m²) was covered by an average of 4.3 t marine organism biomass and contained 35 times more macrozoobenthic biomass than the same area of soft bottom in the German Exclusive Economic Zone (0.12 kg/m²) (Krone *et al.* 2013a). This introduction of a hard structure with (potentially) more or different biomass than the surrounding habitat type is known as the reef effect. In addition, during the operational phase, OWFs are generally accompanied by the cessation of bed-modifying activities such as trawling: eliminating both physical disturbance and the removal of larger predators such as fish and macro-crustaceans. This combination of effects results in the creation of *de facto* protected areas, which can become hotspots for biodiversity (Box 4.1). Here, a biodiversity hotspot is defined as an area of high species and habitat richness that includes representative, rare or threatened features (Hiscock & Breckels 2007). Naturally, any further benefits still need to be quantified, such as the increased productivity of reefs overspilling into the wider areas or even increased fish feeding in the wider area with repercussions for fisheries.

Hard-substratum associated fauna

The most obvious habitat change coinciding with the construction of an OWF is undoubtedly the introduction of artificial hard substrata. The wind-turbine foundation, erosion or scour protection layer (Figure 4.2) and possible cable toppings all contribute to the introduction of habitat for hard-substratum epifauna typically in a predominantly soft-sediment

Figure 4.3 Artificial substratum in the North Sea providing new habitat for hard-substratum fauna in the mainly sedentary benthic environment. (WG Ecosystem Functions, Alfred Wegener Institute, Helmholtz Centre for Polar and Marine Research)

environment, representing up to 0.5 ha (monopile foundation) to 1.2 ha (gravity-base foundation) artificial hard substratum added to the marine environment per square kilometre of wind-farm area (Rumes *et al.* 2013).

Offshore wind turbine foundations and the monopile provide habitat throughout the full water column. Natural vertical hard substratum in offshore waters is rare in areas where OWFs are usually constructed, with the prevailing natural hard substratum being either coastal in the form of rocky shores or horizontal in the form of gravel beds. Most wind farms are directly fixed to soft substrata such as sandbanks and mixed sediment areas, and only floating or gravity-base structures could be on hard substrata. Unlike shipwrecks, which are widely distributed in the marine environment, the vertical structure provided by turbines includes intertidal hard substratum, which is rare in offshore waters. Oil and gas, and meteorological platforms provide similar habitat, but these structures are less numerous than offshore wind turbine monopiles.

This new artificial substratum allows for the attachment of epibionts and provides refuges for smaller species against predation (Gutierrez *et al.* 2003), enhancing the development and colonisation of other species and the growth of a stable benthic community, as demonstrated in Figure 4.3.

Several floral and faunal zones can hence be distinguished on wind-turbine foundations, from intertidal sessile taxa such as green, brown and red algae, barnacles (e.g. *Semiblananus* spp. and *Balanus* spp.) and Blue Mussels *Mytilus edulis* in north-western Europe (Figure 4.4) and *Mytilus galloprovincialis* in the Mediterranean, to subtidal communities dominated by mobile crustaceans (amphipods, cumaceans and isopods) and

Figure 4.4 Algae, barnacles and blue mussels dominate the intertidal zone of the wind turbine foundations. (Francis Kerckhof, Royal Belgian Institute of Natural Sciences)

anemones such as *Metridium* spp. (Kerckhof *et al.* 2010; Krone *et al.* 2013a; de Mesel *et al.* 2015). Given these changes, the introduction of new and different habitat types leads to an increased local biodiversity, which is one of the most commonly reported changes due to OWFs (Gutow *et al.* 2014; Lindeboom *et al.* 2015). At the German wind farm Alpha Ventus, the abundance of macroinvertebrates was 100 times higher on the foundations just 2 years after construction compared to the surrounding seabed, with over 2,000 Edible Crabs *Cancer pagurus* inhabiting single foundations compared to 29 in reference areas (Krone & Krägefsky 2013). At a wind-farm scale, sessile macrofaunal species richness, for example, may increase considerably, perhaps in the order of 3–14-fold (Rumes *et al.* 2013). The sessile macrofaunal community composition, however, seems impoverished from that expected on natural hard substrata in similar oceanographic conditions. In North Sea waters, the vertical structures especially tend to be dominated by only a few species, such as the amphipod *Jassa herdmani*, the plumose anemone *Metridium senile*, the hydroids *Tubularia* spp. and the Blue Mussel *M. edulis*, while *J. herdmani* proved to be less prevalent on the erosion protection layers (e.g. de Mesel *et al.* 2013, 2015).

Scour protection around the base of turbines and associated infrastructure is a key component of the design and engineering of an OWF (Whitehouse *et al.* 2011). With careful planning and design, the varying types and methods of scour protection have the capacity to enhance existing habitats, for example through placement of boulder protection or concrete mattressing (Wilson & Elliott 2009) (Box 4.2). These structures are attractive sheltered environments for species such as Edible Crab and European Lobster *Homarus gammarus*, which prefer a sheltered reef area adjacent to sandy areas for feeding. Mobile species, such as fish including Atlantic Cod *Gadus morhua* and Pouting *Trisopterus luscus*, may also be attracted by prey among the rich epifaunal growth (see Gill & Wilhelmsson, Chapter 5); and the presence of juveniles such as Atlantic Cod suggests the use of OWFs as nurseries for fish (e.g. Langhamer 2012; Reubens *et al.* 2013; Gill & Wilhelmsson, Chapter 5).

Box 4.2   Designing offshore wind farms to promote biodiversity

Ecological engineering is a widely used tool (Elliott *et al.* 2016). Construction of seawalls for habitat complexity instead of a uniform surface has been highly successful in increasing diversity and colonisation by creating tidepool-type habitats (CMA 2009; Browne & Chapman 2011). Pipeline construction in the Indian Ocean was designed to include habitat and nursery areas for fish (Pioch *et al.* 2011), and habitat restoration and creation schemes using reef balls have taken place in several locations, with positive impacts for both diversity and sea-defence purposes (Barber 2012). The creation of refuges on offshore structure foundations has significantly increased stocks of the Edible Crab *Cancer pagurus* (Langhamer *et al.* 2009). Therefore, using ecological engineering principles at the construction stage of future wind farms would allow the possibility for a valuable habitat to develop, providing conservation aspects, niches and prey for predators which could remain at the end of the wind farm's life. However, care must be taken to provide the right initial artificial substratum, as benthic communities have been noted to be less diverse on foundations made of steel than on concrete foundations, even though abundances and biomasses were not necessarily different (Qvarfordt *et al* 2006; Wilhelmsson & Malm 2008).

Providing hard-substratum habitat in a predominantly soft-sediment environment, OWFs extend the natural habitat range of several hard-substratum species. However, most, if not all subtidal species are also known from natural and other artificial subtidal hard substrata, such as rock outcrops, gravel beds and shipwrecks. Wind farms do not appear to add to the regional species pool (Rumes *et al.* 2013). Wind farms thus contribute little to the range of such species, although extension may occur for obligate intertidal hard substrata species, such as the Common Limpet *Patella vulgata*, for which offshore habitat is rare to non-existent. As such, wind farms may allow non-indigenous introduced species and range-expanding species to expand their population size and hence to colonise formerly unsuitable areas in the 'stepping stone hypothesis' advanced by Kerckhof *et al.* (2011). Indeed, it should be noted that, in some instances, there may be negative effects associated with the local changing biodiversity around OWFs leading to unwanted effects. For example, anthropogenic hard substratum such as OWFs in the southern Baltic Sea has the potential to increase the abundance of the Moon Jellyfish *Aurelia aurita*, conflicting directly with the energy and tourism sectors along the Danish, Polish and German coastlines (Janßen *et al.* 2013).

Soft-substratum associated fauna

Most OWFs are built on soft substrata such as sandy seabeds, muds and clays. These sites are generally preferred by wind-farm developers, first because of the relative ease of installing turbines in soft rather than harder substrata, and secondly because these habitats predominate in the relatively shallow waters usually selected for wind-farm development, such as the southern North Sea. In soft substrata, communities are generally dominated by burrowing organisms, which themselves play a key role in the structuring of the environment through biogenic transformations including bioconstruction, bioturbation, sediment stabilisation and sediment destabilisation (Reise 2002; Gray & Elliott 2009). Epibenthic organisms may act as harder substratum in a sedimentary environment, providing anchorage points for algal species, which in turn provide habitat and food for primary consumers (Gutierrez *et al.* 2003).

Soft substrata and their inhabiting communities are impacted by the introduction of artificial hard substratum at three scales: (1) the micro-scale, involving changes to the material and texture of the habitat such as sediment composition; (2) the meso-scale, considering, for example, the introduction of scour protection at a single turbine scale; and (3) the macro-scale, considering changes at the scale of the whole wind farm (Petersen & Malm 2006). Changes to the soft-sediment environment occur particularly at the micro- and meso-scales.

Two direct impacts on soft substrata can be distinguished. First, habitat loss occurs under the footprint of the foundation and scour protection, but also below toppings of inter-array and export cables. Although the total loss of communities is permanent, this impact may still be considered minor given that the artificial hard substrata typically only covers less than 1% of the total wind-farm surface area (Petersen & Malm 2006). Secondly, the original habitat may be smothered by the dredging, disposal and displacement of sediments in the preparation of the construction area (Gray & Elliott 2009), ultimately covering a larger proportion of the wind-farm surface area than the footprint of the turbines (van den Eynde *et al.* 2013). This is particularly true for gravity-base foundations for which much sediment is moved, but is less of a problem for the more commonly used monopile foundations. While the impact of smothering of the soft-substrata communities may be more important in spatial extent, its duration appears short lived (Coates *et al.*

2013) and the original sediment type may be regained if the introduced fine sediment is winnowed out of the sediment by the prevailing hydrographic regime, including the tidal, wind-driven and residual currents.

Indirect effects are likely to be most prominent on soft-sediment communities. For example, turbine foundations change local currents (Leonhard *et al.* 2006; Rees & Judd, Chapter 2), which influence the prevailing currents, leading to small-scale alterations of the spatial distribution of erosion and deposition and hence sediment composition (Gill 2005; Leonhard *et al.* 2006; Vaissière *et al.* 2014). Such changes alter the soft-sediment community composition and distribution (Coates *et al.* 2014; de Backer *et al.* 2014), with typical fine-sandy communities found close to the turbine monopile and adjacent to the scour protection. In particular, it is expected that with any structure in place, the local environment will adjust to the new hydrographic conditions and so the substrata and biota will adjust accordingly. This, therefore, creates a new equilibrium. With the removal of that structure, it is unknown whether the original equilibrium will be re-created even if marine systems have an ability to regain that structure (Elliott *et al.* 2007). The latter effect, however, is difficult to distinguish from the effect of biodeposition of organic matter from the wind-turbine epifauna, which locally enhances soft-sediment macrofaunal abundance and biomass (Coates *et al.* 2014).

The cessation of fisheries activities inside OWF arrays further contributes to the development of the soft-sediment communities surrounding the wind turbines in the absence of trawling disturbance. One example of the fisheries' cessation effects is the *Sabellaria spinulosa* reefs associated with the Thanet OWF in eastern England. During construction and turbine installation, precision mapping and micro-siting were employed to ensure minimal damage to the biogenic reefs within the wind-farm footprint of 35 km², and following construction, the reefs recovered and expanded in extent (Pearce *et al.* 2014). Such reefs are listed under Annex I of the European Union (EU) Habitats Directive as a marine habitat to be protected by designation of Special Areas of Conservation (SACs). The natural construction of intertwined dwelling tubes by *Sabellaria* and similar reef-building species creates stable habitat for associated species, allowing epibenthic and crevice faunal communities to become established (Hendrick & Foster-Smith 2006) and thereby increasing biodiversity between the wind-farm structures.

Offshore wind farms alter the benthic food web

In addition to changes in environmental conditions, trophic interactions are the main forcing factors for the occurrence, distribution and behaviour of organisms (e.g. MacArthur 1955; Pimm 1982; McCann 2000). Feeding relationships constitute the biological base of ecological functioning, as the species is the evolutionarily derived basic entity of ecological functioning. Consequently, alterations in the species inventory directly affect the food web of the benthic system (May 1983; Chapin *et al.* 2000).

During the construction and decommissioning of OWFs, forage species are smothered by mobilised sediment. The recovery rate of benthic animals depends on the sediment type and on the species' ability to migrate to the surface (Newell *et al.* 1998; Powilleit *et al.* 2009). It is axiomatic that species in mobile sandbanks are adapted to rapid changes of sediment, such as from storms, and so are readily able to recover from introduced sediment and smothering (Gray & Elliott 2009). Despite this, these species may be unavailable for predators for a short to medium period, although scavenging species may be able to prey or scavenge on species damaged during construction works as an additional food source, as is known to occur with bottom trawling (Gray & Elliott 2009).

Construction works also lead to changes in the sediment biogeochemistry and increased turbidity in the water column (Watling *et al.* 2001). Resuspended organic matter from deeper sediment layers may thereby become accessible to deposit and suspension feeders in the wider surrounding area by the spread of prevailing currents, defined as far-field effects with the dispersion of fine material (see Rees & Judd, Chapter 2).

There is increasing evidence of the effects of noise and vibration on fish and the benthos (Roberts *et al.* 2015, 2016a,b; Roberts & Elliott 2017). For example, the noise generated during the construction of an OWF can cause fish to avoid the area and thus release the benthos from predation pressure on a short timescale (Andersson 2011; Roberts *et al.* 2016a,b). Andersson (2011) showed a behavioural response of Atlantic Cod and Common Sole *Solea solea* in that the fish showed directional swimming away from the pile-driving noise with an increased swimming speed, thereby decreasing feeding pressure on the benthic forage species in the direct vicinity of the construction activities.

During the operational phase, three major interactions potentially lead to long-term effects on the benthic food web: (1) the addition of hard-substratum species as additional food sources; (2) organic enrichment around turbines; that is, increased organic material available for deposit and suspension feeders; and (3) the cessation of active trawling, with its consequences for the food web in the wind-farm area.

The introduction of monopiles and erosion protection, which are quickly colonised by an epifouling community, add additional benthic producers and consumers to the marine ecosystem (e.g. Wolfson *et al.* 1979; Lindeboom *et al.* 2011; Krone *et al.* 2013a). These species thus serve as additional prey items for higher trophic levels such as benthic and benthopelagic fish, including Pouting and Atlantic Cod (Maar *et al.* 2009; Reubens *et al.* 2011; Bergström *et al.* 2013), or larger invertebrates such as crabs, which accumulate around the artificial structures (Krone *et al.* 2013b) (Box 4.3). At the same time, the accumulation of larger benthic predators can increase the predation pressure on the surrounding soft-bottom communities by selectively removing those species with biological traits which make them susceptible to predation, for example mobile crustaceans which leave the sediment if disturbed (Bergström *et al.* 2013).

Organic enrichment has been demonstrated for soft-bottom sediments adjacent to turbine monopiles (Joschko 2007; Coates *et al.* 2014), favouring the occurrence of deposit and suspension feeders and a shift in the trophic structure of the surrounding community. For example, Coates *et al.* (2014) demonstrated an increase of the polychaete worms *Lanice conchilega* and *Spiophanes bombyx* up to 25 m from the platform. On the one hand, these tube-building polychaetes have the potential to enhance the changes in hydrodynamic flow, grain-size distribution and food availability (Coates *et al.* 2014) by collecting organic matter, stabilising the sediment and increasing habitat complexity for other species as a result of their tubes. On the other hand, they constitute a considerable food resource for higher trophic levels (Dannheim *et al.* 2014). The increase in total organic carbon content results from different organic components such as faeces, pseudofaeces or the decaying tissue of dead animals released from the epifouling community on the artificial structures (Wolfson *et al.* 1979; Ambrose & Anderson 1990; Widdows *et al.* 1998; Joschko *et al.* 2008). Organic matter is likely to be patchily distributed as a result of changes caused by the presence of the turbine (Leonhard *et al.* 2006), with accumulation of organic material by small gyres. Scour could also create seabed depressions which act as sinks for organic material. In both cases, deposit feeders or interface feeders that profit from the additional organic material may accumulate at a very local scale.

Finally, safety issues and the legal requirement in some countries to create safety buffer zones around individual monopiles are likely to prevent active demersal, benthic and pelagic trawling within the wind-farm array (e.g. Lindeboom *et al.* 2011). Given that

Box 4.3 Predatory fish are differentially affected by offshore wind farms

On a local scale within the Belgian part of the North Sea, during a four-year study, specific age groups of Atlantic Cod *Gadus morhua* and Pouting *Trisopterus luscus* were shown to be seasonally attracted towards offshore wind farms, showing high site fidelity (Reubens *et al.* 2013; Reubens *et al.*, Box 5.1, Chapter 5). The fish fed upon the dominant epifaunal prey species present (Figure 4.5) and growth was observed throughout the period for which the fish were present, indicating extra biological production on a local scale (Reubens *et al.* 2013). However, not all fish species may benefit from these hotspots; for instance, stomach content analysis shows that hard-substratum associated organisms were an insignificant food component of pelagic Atlantic Mackerel *Scomber scombrus* and Horse Mackerel *Trachurus*

Figure 4.5 Large predators such as large mobile crabs, fish and starfish profit from the shelter and the additional food source, such as these Blue Mussels *Mytilus edulis*, provided by offshore wind-farm structures. (WG Ecosystem Functions, Alfred Wegener Institute, Helmholtz Centre for Polar and Marine Research)

trachurus caught within the German Alpha Ventus wind farm (Figure 4.6), with these fish showing a significantly reduced degree of stomach fullness compared with those caught outside the wind farm (Krone & Krägefsky 2013).

Figure 4.6 Turbine bases typically attract a range of predatory fish species that feed on invertebrates and smaller fish associated with epifaunal growth, although dietary studies suggest that the attraction of pelagic species such as these Horse Mackerel *Trachurus trachurus* is not linked to improved foraging conditions. (WG Ecosystem Functions, Alfred Wegener Institute, Helmholtz Centre for Polar and Marine Research)

seabed trawling is recognised as a major stressor on the seabed (Hall 1999; Kaiser *et al.* 2006), the cessation of trawling causes two major changes to the benthic food-web structure: (1) the absence of physical sediment disturbance caused by fishing gear; and (2) the absence of fisheries-generated additional food sources, such as gear-induced mortality and discards (Dannheim 2007). The absence of physical disturbance changes the biogeochemistry of the sediment, which particularly affects the occurrence of deposit feeders and increases the settlement of small tube-building deposit or interface feeders such as spionid worms like *Spiophanes bombyx* (Kröncke *et al.* 2004; Wieking & Kröncke 2005; Dannheim *et al.* 2014). Large mobile generalists, including starfish such as *Asterias rubens* or swimming crabs such as *Liocarcinus holsatus*, generally profit from the comparably high trophic level food generated by fisheries in the form of gear-induced mortality and discards. Trawling primarily causes mortality of benthic organisms, but many survivors such as mobile predators and scavengers profit from the gear-generated food items (Groenewold & Fonds 2000; Rumohr & Kujawski 2000). Such species may thus actively migrate from non-trawled to trawled areas (Dannheim *et al.* 2014). At the same time, the missing predation may increase the abundance within the wind-farm area of large mobile species, such as predatory and scavenging decapods, which could subsequently increase the predation pressure on the soft- and hard-bottom species. In some wind farms, such as Sheringham Shoal in the UK, there are intensive pot fisheries of crabs and lobsters, the occurrence of which is favoured by the installation of rock scour protection, which may again reduce the number of large mobile predators.

Although marine mammals are a subject of another chapter within this publication (Nehls *et al.*, Chapter 6), tracking of seals has shown individuals visiting individual turbines within an OWF, as well as following the routes of subsea pipelines (Russell *et al.* 2014), suggesting the use of wind farms for foraging. The presence of top-level predators such as these requires established and supporting benthic communities and suggests that wind farms are acting as hotspots where species such as seals can find enhanced foraging opportunities.

Restoration potential following decommissioning

A wind farm reaches the end of its operational life (20–30 years) when it can no longer function properly or satisfy the expectations of its user (Ortegon *et al.* 2013). At this point, it can be either repowered, whereby existing foundations are repurposed to hold a longer monopile and more powerful turbine and blade system following technological advances in the intervening time, or decommissioned, whereby some or all of the physical structures are removed and the environment is returned to a more natural state.

At the time of writing in 2018, owing to offshore wind being a relatively new industry, only the earliest built Dong wind farm at Vindeby off the Danish island of Lolland and the Vattenfall wind farm at Yttre Stengrund, Sweden, have been decommissioned, in 2017 and 2016, respectively. Hence, cases from the offshore oil and gas industries provide a more studied, yet similar, starting point for assessing the potential consequences of decommissioning (Smyth *et al.* 2015). Here, the removal of the upper structure is required but the bed structures such as pipelines and mattress seabed protection can be left in place to prevent greater damage to the seabed (Kaiser 2006; Kaiser & Pulsipher 2005). As an example, in the north-east Atlantic, Recommendation 98/3 of the Oslo and Paris Convention (OSPAR) requires that Member States remove the upper structures such as the turbine towers (monopiles), or at least ensure that there is 55 m clearance to allow navigation, whereas the seabed structures can be left in place with the permission and agreement of the national regulatory bodies.

Box 4.4   Guidelines for decommissioning offshore wind farms

International Maritime Organization (IMO) Guidelines and Standards for the Removal of Offshore Installations and Structures on the Continental Shelf and in the Exclusive Economic Zone [IMO Resolution A.672 (paragraph 3.5)] states that 'where entire removal would involve an unacceptable risk to the marine environment, the coastal State may determine that it need not be fully removed.' And the Convention for the Protection of the Marine Environment of the North-East Atlantic (OSPAR) Windfarm Guidance 2008-3 (paragraph 93) correspondingly states that if the 'competent national authority decides that a component of the windfarm should remain at site (e.g. parts of the piles in the seabed, scour protection materials), it should be ensured that they have no adverse impact on the environment, the safety of navigation and other uses of the sea.' This guidance especially links to a provision in the OSPAR document Decision 98/3 on the Disposal of Disused Offshore Installations, which states in paragraph 8d that 'The assessment of the disposal options shall take into account…impacts on the marine environment, including exposure of biota to contaminants associated with the installation, other biological impacts arising from physical effects, conflicts with the conservation of species, with the protection of their habitats, or with mariculture' which could allow a partial removal of the decommissioned wind farm if it would pose a risk to the biota that had developed at the site during the operational life. Any site for which a partial decommissioning is considered would need a full assessment of its viability. However, despite the navigational safety considerations, the energy costs, labour costs and safety issues during removal of such structures may mean that it is more cost-beneficial to leave parts in place, especially where the aim is to protect and enhance the habitat at the decommissioned site.

Analogous to structures such as oil and gas rigs, the installation of a wind farm induces changes to the local ecology and environment, eventually reaching a new equilibrium. Studies on physical disturbance in the marine environment, such as laying of pipelines, aspects of construction and bed-dredging activities, have concluded that even after such disturbances have ceased, many marine systems do not revert to the pre-stressor state (Elliott *et al.* 2007; Mazik & Smyth 2013; Duarte *et al.* 2015). Hence, it is arguable whether there would be a return to the pre-construction state following full decommissioning of a wind farm. For instance, in sandy sediments, upon which many of the earliest wind farms were built, scour can be as deep as 1.38 times the monopile diameter (Whitehouse *et al.* 2011) and so an extensive amount of scour protection may have been installed to protect the foundations. Hence, its removal is likely to generate adverse change to the new ecological equilibrium that has developed over the 20–30 years of operation.

The approach of leaving in place an artificial reef, with benefits for commercial and recreational fishing plus the reduced costs for developers, needs to be weighed against operational challenges of leaving parts in place, where these challenges relate to safety of navigation, ongoing maintenance costs, issues in relation to liability of the reef and potential for the spread of non-indigenous species. This would need to be assessed on a site-specific basis with all of these factors being considered (Smyth *et al.* 2015).

Decommissioning guidance for OWFs in the UK sea area generally recommends a full removal of all components together with the associated disturbances that this would cause to the benthos. However, legislation at both EU and national level could allow for a partial removal of components in some cases (Smyth *et al.* 2015) (Box 4.4). Provisions exist at a national level; for example, Danish law allows for partial decommissioning in cases where full decommissioning would lead to an 'environmental hazard' (CCC 2010) and Dutch law only requires monopiles to be cut to 4 m below sea level (CCC 2010), suggesting that leaving benthic components of wind farms in place may be possible in both Danish and Dutch waters. In UK waters, the recommendations (DECC 2011) suggest removal of the foundations and monopile, but leaving the scour protection *in situ*, and several site decommissioning plans cite this as their option, often mentioning artificial reefs as a reason (see a review of these plans in Smyth *et al.* 2015). Finally, in Belgian waters, a full restoration of the area to its original state is required.

If implemented appropriately and with consideration of the ten-tenets framework for achieving sustainable management (Elliott 2013; Barnard & Elliott 2015), a 'renewables-to-reefs' (Smyth *et al.* 2015) decommissioning scheme involving only partial removal of the wind farm, hence leaving in places benthic aspects such as foundations, scour protection and possibly the lower part of the monopile (or jacket/tripod in the case of alternative models), could ensure protection for ecologically valuable sites and allow for the regeneration of the disturbed marine environment, at the same time conferring societal benefits. The ten tenets require that ecological, economic, technological, legal, political, administrative, ethical, societal, cultural and communication aspects are included in the decision making. Co-locating other activities such as recreation or conservation at the decommissioned site is also possible (Christie *et al.* 2014) and has the added benefit of easing demands on limited space for development in the marine environment. These measures could be achieved with an integrated marine management framework (Elliott 2014; Elliott *et al.* 2017). Such an integrated framework is required to include the causes of the impacts from the activities and pressures in the sea, the consequences of those impacts for the natural and societal systems, including effects on ecosystem services and societal goods and benefits, the main stakeholders and the regulatory regimes, and the management responses. Importantly, it requires either modifying the ecosystem to enhance the ecology or at least providing the appropriate physical conditions to allow recovery, that is, ecological engineering (Elliott *et al.* 2016).

Monitoring programmes, sampling design and analysis

Environmental Impact Assessments (EIAs) are required to cover all stages of OWF development of the area: construction, operation and decommissioning. The EIA should be very precise in determining the impact of an activity, and even the removal of an activity, at a particular place and carried out at a certain time, constructed or removed in a certain way and with an agreed level of compensation and mitigation, and communicated widely (Gray & Elliott 2009). Accordingly, there are many OWF-related EIAs covering the benthos, of which the benthos–substratum interactions are perhaps the most important aspects of the impacts from the monopile and cable routing (Franco *et al.* 2015).

Despite this, as shown by Franco *et al.* (2015), this activity and its components show that subtle changes due to the OWF are difficult to detect against the background variability and that often the wind-farm operators are not obliged by the regulators to put in sufficient effort to detect subtle changes against that background variability. Furthermore, Franco *et al.* (2015) indicate that OWF operators may be asked unrealistically to detect changes that cannot be detected. For example, if the operators are requested by the regulators to detect a given change in the quality or quality of the benthos such as species richness, diversity, abundance and biomass, then the effort in terms of calculating the number of sample replicates should be designed, such as by using power analysis, to detect that required

Table 4.1 A synthesis of the dominant effects of offshore wind farms (OWFs) on the benthos

Original hypothesis	Commentary
The seabed ecology created by placing an OWF in the marine environment is neither improved nor degraded ecologically, just different	A wind farm merely replicates some of the habitats present in the natural system and benthic features similar to those of other natural and anthropogenic structures. The epithets 'improved' or 'degraded' should therefore focus on the human uses of the marine system where ecosystem services and societal benefits required by humans are lost or gained, rather than the ecological status
The inherent variability of the seabed biota and hydrodynamic conditions may prevent the subtle effects of OWFs being detected, in particular in current wind-farm locations	The seabed is a mosaic of habitats, and wind farms generally only replicate those naturally present, although they may increase the relative frequency of different habitats and increase habitat patchiness in the system. As yet, any changes to the benthos appear to be restricted to relatively small areas directly affected by the turbine tower, foundations, cabling route and scour protection
The hard structures associated with OWFs are available as colonisation sites and 'stepping stones' for non-indigenous species	Wind-farm structures are colonisation sites for epibenthic, sessile and hard-substratum species, and if any such species are introduced to the area then they will find an available niche. Given that there is already a plethora of equivalent structures such as shipwrecks and rock outcrops, it will not be possible to prove that wind farms will exacerbate the transfer of non-indigenous species. However, this is conceivable, albeit as yet unproven for the unique intertidal habitat in clear offshore water provided by wind farms
The focus on the structure of the benthos rather than this together with its functioning is not satisfactory for assessing impact	The species diversity, richness and abundance of the colonising benthos provide some indication of the number of niches created, the competition between species and the success of recruitment influenced by the presence of the structures. However, some quantification of ecological function is needed to show the impact on system dynamics, feeding relationships and colonisation potential from the wind farm to the wider environment
There are effects of the structure on the seabed, but also effects of the seabed (and biota) on the structure	Environmental managers will be predominantly interested in the effects of the structure on the environment, such as a loss or gain in particular habitats. However, the wind-farm operators will be interested in the effect of the attached organisms as stressors on the structure in relation to corrosion, as well as the increase in weight and thus hydrodynamic drag

Table 4.1 – *continued*

Original hypothesis	Commentary
Given the many human activities and pressures, there are in-combination effects (the synergistic and antagonistic effects of all aspects of the same development) and cumulative effects (the effects of different developments in the same area), which need to be disentangled	The effects of each component of the turbine, including the tower, foundation, scour protection and cabling, as well as any products of electrical generation, such as electromagnetic radiation and vibration, on the benthos need to be considered together with the effects of other seabed activities in the area, such as aggregate (sand and gravel) extraction, trawling, and oil and gas activities
Location provides opportunities as well as threats, and both need to be considered together	While some habitats such as sandy seabed may be lost or reduced, others are created, thus increasing the potential for colonisation. Similarly, habitat complexity is increased
Increasingly, these effects will be difficult to detect against a highly variable system whose variability will increase with climate change	The marine benthos varies with habitat, place, season, and natural and anthropogenic influences at short-, medium- and long-term spatial and temporal scales, all of which are influenced by climate change. All of these aspects contribute to the inherent 'noise' (variability) in the system against which the 'signal' of change due to the wind farm needs to be assessed

change. Most monitoring schemes around OWF sites do not have the statistical power to detect the required degree of change (Franco *et al.* 2015; Wilding *et al.* 2017).

The level of benthic monitoring required to detect significant effects is proportional to the degree of change expected (Gray & Elliott 2009), and the changes may be at the micro-scale (individual monopile), meso-scale (between monopiles within a wind farm), macro-scale (at the whole wind-farm scale) and mega-scale (between wind farms within a region). The monitoring should include the substratum type (e.g. particle-size analysis), organic content of sediment, habitat mapping, faunal abundance, species richness (number of species), diversity (e.g. using many diversity indices) and biomass (as an indication of fish prey present) (Gray & Elliott 2009). The monitoring may also be required to show an analysis of the biological traits of the benthos as an indication of the links between the fauna and its habitat.

The monitoring of the seabed follows well-defined and accepted methods (Gray & Elliott 2009; Eleftheriou 2013) but should be subjected to analytical quality control and quality assurance (Elliott 1993). This will ensure that the data produced in any EIA are legally defendable and can be used as a baseline against which future changes are judged.

Concluding remarks

This chapter has aimed to summarise the salient points relating to the influence of OWFs on marine benthos. While it is not possible to cover every study in an expanding field, the assessment may be summarised in the light of the eight underpinning hypotheses of this chapter (Table 4.1). This shows that there are many effects of OWFs on the benthos of which we already have a good understanding, but also suggests that often we do not take a sufficiently wide-ranging and holistic view.

As shown here, the construction of wind farms gives the potential for habitat management and enhancement of the seabed and, if managed correctly, there is the potential for the introduction of, and change to, habitats to produce positive effects (Wilson & Elliott 2009; Wilson *et al.* 2010). An increased local biodiversity is one of the most commonly reported impacts of OWFs (Gutow *et al.* 2014; Lindeboom *et al.* 2015). However, should OWFs be installed in areas with poor management and consideration for the receiving environment, then the introduction of hard substratum may result in habitat loss.

This review has indicated that benthic communities will develop on the installed hard structures and could redevelop, perhaps with an entirely different community structure, if those hard structures were removed at the end of the life of the wind farm. These trajectories may reflect similar ones for the benthos around and on other anthropogenic structures for which there is a large body of information (Elliott *et al.* 2007; Gray & Elliott 2009; Borja *et al.* 2010; Duarte *et al.* 2015). However, only a few studies have determined whether a stable community has been formed and how long it takes to form such a community.

As demonstrated in this chapter, there are positive and negative effects of wind farms on the marine benthos, and these require a risk and opportunity analysis and management framework (Cormier *et al.* 2013). Effects and benefits at the micro-, meso- and macro-scales require upscaling to the mega-scale to determine larger influences. This requires not only international collaboration, but also linkage to other activities in the same sea region (Boyes *et al.* 2016) and wider conservation initiatives, such as Marine Protected Area designation and Maritime Spatial Planning, as well as the implementation of marine statutes such as EU Directives aimed at protecting natural functioning and delivering ecosystem services and societal benefits (e.g. Borja *et al.* 2017).

Acknowledgements

The authors acknowledge several offshore wind operators and study groups with which and with whom they have worked; for example, the ICES Working Group on Marine Benthal and Renewable Energy Developments (WGMBRED). Despite this, any views here are those of the authors personally and not of any organisation.

References

Ambrose, R.F. & Anderson, T.W. (1990) Influence of an artificial reef on the surrounding infaunal community. *Marine Biology* 107: 41–52.

Andersson, M.H. (2011) Offshore windfarms – ecological effects of noise and habitat alteration on fish. Dissertation, Stockholm University. Retrieved 15 May 2018 from https://www.researchgate.net/profile/Mathias_Andersson/publication/267817047_Offshore_wind_farms_-_ecological_effects_of_noise_and_habitat_alteration_on_fish/ links/54dc5f8a0cf2a7769d95eec4/Offshore-wind-farms-ecological-effects-of-noise-and-habitat-alteration-on-fish.pdf?origin=publication_detail

Barber, T. (2012) The Reef Ball Foundation-designed artificial reefs. Athens, GA: Reef Ball Foundation. Retrieved 15 May 2018 from www.reefball.org

Bergström, L., Sundqvist, F. & Bergström, U. (2013) Effects of an offshore windfarm on temporal and spatial patterns in the demersal fish community. *Marine Ecology Progress Series* 485: 199–210.

Borja, Á., Dauer, D.M., Elliott, M. & Simenstad, C.A. (2010) Medium- and long-term recovery of estuarine and coastal ecosystems: patterns, rates and restoration effectiveness. *Estuaries and Coasts* 33: 1249–1260.

Borja, Á., Elliott, M., Uyarra, M.C., Carstensen, J. & Mea, M. (eds) (2017) *Bridging the Gap Between Policy and Science in Assessing the Health Status of Marine Ecosystems*, 2nd edn. Lausanne: Frontiers Media. Retrieved 15 May 2018 from http://www.frontiersin.org/books/Bridging_the_Gap_Between_Policy_and_Science_in_Assessing_the_Health_Status_of_Marine_Ecosystems_2nd/1151

Boyes, S.J., Elliott, M., Murillas-Maza, A., Papadopoulou, N. & Uyarra, M.C. (2016) Is existing legislation fit-for-purpose to achieve Good Environmental Status in European seas? *Marine Pollution Bulletin* 111: 18–32.

Browne, M. & Chapman, M. (2011) Ecologically informed engineering reduces loss of intertidal biodiversity on artificial shorelines. *Environmental Science & Technology* 45: 8204–8207.

Butterfield, S., Musial, W., Jonkman, J. & Sclavounos, P. (2005) Engineering challenges for floating offshore wind turbines. Golden, CO: National Renewable Energy Laboratory. Retrieved 15 May 2018 from https://www.nrel.gov/docs/fy07osti/38776.pdf

Chapin III, F.S., Zavaleta, E.S., Eviner, V.T., Naylor, R.L., Vitousek, P.M., Reynolds, H.L., Hooper, D.U., Lavorel, S., Sala, O.E., Hobbie, S.E., Mack, M.C. & Diaz, S. (2000) Consequences of changing biodiversity. *Nature* 405: 234–242.

Christie, N., Smyth, K., Barnes, R. & Elliott, M. (2014) Co-location of activities and designations: a means of solving or creating problems in marine spatial planning? *Marine Policy* 43: 254–261.

Climate Change Capital (CCC) (2010) Offshore renewable energy installation decommissioning study. Independent Report by Climate Change Capital (CCC) for the Department of Energy and Climate Change, United Kingdom Government. 17 November 2010. Retrieved from https://www.gov.uk/government/publications/offshorerenewable-energy-installation-decommissioning-study

Catchment Management Authority (CMA) (2009) Environmentally friendly seawalls. A guide to improving the environmental value of seawalls and seawall-lined foreshores in estuaries. Sydney Metropolitan Catchment Authority. New South Wales Government, Department of Environment and Climate Change NSW, Australia. Retrieved 15 May 2018 from http://www.hornsby.nsw.gov.au/__data/assets/pdf_file/0017/41291/Environmentally-Friendly-Seawalls.pdf

Coates, D., van Hoey, G., Reubens, J., Vanden Eede, S., de Maersschalck, V., Vincx, M. & Vanaverbeke, J. (2013) The macrobenthic community around an offshore wind farm. In Degraer, S., Brabant, R. & Rumes, B. (eds) *Environmental Impacts of Offshore Windfarms in the Belgian Part of the North Sea: Learning from the past to optimize future monitoring programmes*. Brussels: Royal Belgian Institute of Natural Sciences, Operational Directorate Natural Environment, Marine Ecology and Management Section. pp. 87–97. Retrieved 15 May 2018 from https://tethys.pnnl.gov/sites/default/files/publications/Degraer-et-al-2013.pdf

Coates, D.A., Deschutter, Y., Vincx, M. & Vanaverbeke, J. (2014) Enrichment and shifts in macrobenthic assemblages in an offshore windfarm area in the Belgian part of the North Sea. *Marine Environmental Research* 95: 1–12.

Cormier, R., Kannen, A., Elliott, M., Hall, P. & Davies, I.M. (eds) (2013) *Marine and Coastal Ecosystem-based Risk Management Handbook*. ICES Cooperative Research Report, No. 317, March 2013. Copenhagen: International Council for the Exploration of the Sea.

Dannheim, J. (2007) Macrozoobenthic response to fishery – trophic interactions in highly dynamic coastal ecosystems. PhD dissertation, University of Bremen. Retrieved 15 May 2018 from http://elib.suub.uni-bremen.de/diss/docs/00010865.pdf

Dannheim, J., Brey, T., Schröder, A., Mintenbeck, K., Knust, R. & Arntz, W.E. (2014) Trophic look at soft-bottom communities – short-term effects of trawling cessation on benthos. *Journal of Sea Research* 85: 18–28.

de Backer, A., van Hoey, G., Coates, D., Vanaverbeke, J. & Hostens, K. (2014) Similar diversity disturbance responses to different physical impacts: three cases of small-scale biodiversity increase in the Belgian part of the North Sea. *Marine Pollution Bulletin* 84: 251–262.

de Mesel, I., Kerckhof, F., Rumes, B., Norro, A., Houziaux, J.-S. & Degraer, S. (2013) Fouling community on the foundations of wind turbines and the surrounding scour protection. In Degraer, S., Brabant, R. & Rumes, B. (eds) *Environmental Impacts of Offshore Windfarms in the Belgian Part of the North Sea: Learning from the past to optimise monitoring programmes*. Royal Belgian Institute of Natural Sciences, Operational

Directorate Natural Environment, Marine Ecology and Management Section. pp. 123–137. Retrieved 12 June 2018 from https://odnature.naturalsciences.be/downloads/mumm/windfarms/winmon_be_2013.pdf

de Mesel, I., Kerckhof, F., Norro, A., Rumes, B. & Degraer, S. (2015) Succession and seasonal dynamics of the epifauna community on offshore windfarm foundations and their role as stepping stones for non-indigenous species. *Hydrobiologia* 756: 37–50.

DECC (2011) Decommissioning of offshore renewable energy installations under the Energy Act 2004: guidance notes for industry. Department of Energy and Climate Change (DECC), Department of Trade and Industry, p. 70.

Duarte, C.M., Borja, Á., Carstensen, J., Elliott, M., Krause-Jensen, D. & Marbà, N. (2015) Paradigms in the recovery of estuarine and coastal ecosystems. *Estuaries and Coasts* 38: 1202–1212.

Eleftheriou, A. (ed.) (2013) *Methods for the Study of Marine Benthos*, 4th edn. Oxford: Wiley-Blackwell.

Elliott, M. (1993) The quality of macrobiological data. *Marine Pollution Bulletin* 25: 6–7.

Elliott, M. (2013) The 10-tenets for integrated, successful and sustainable marine management. *Marine Pollution Bulletin* 74: 1–5.

Elliott, M. (2014) Integrated marine science and management: wading through the morass. *Marine Pollution Bulletin* 86: 1–4.

Elliott, M., Burdon, D., Hemingway, K.L. & Apitz, S. (2007) Estuarine, coastal and marine ecosystem restoration: confusing management and science – a revision of concepts. *Estuarine, Coastal & Shelf Science* 74: 349–366.

Elliott, M., Mander, L., Mazik, K., Simenstad, C., Valesini, F., Whitfield, A. & Wolanski, E. (2016) Ecoengineering with ecohydrology: successes and failures in estuarine restoration. *Estuarine, Coastal and Shelf Science* 176: 12–35.

Elliott, M., Burdon, D., Atkins, J.P., Borja, Á., Cormier, R., de Jonge, V.N. & Turner, R.K. (2017) 'And DPSIR begat DAPSI(W)R(M)!' – a unifying framework for marine environmental management. *Marine Pollution Bulletin* 118: 27–40.

European Wind Energy Association (EWEA) (2014) The European offshore wind industry – key trends and statistics 2013. Brussels: European Wind Energy Association. Retrieved 15 May 2018 from http://www.ewea.org/fileadmin/ files/library/publications/statistics/European_offshore_statistics_2013.pdf

Franco, A., Quintino, V. & Elliott, M. (2015) Benthic monitoring and sampling design and effort to detect spatial changes: a case study using data from offshore windfarm sites. *Ecological Indicators* 57: 298–304.

Gill, A.B. (2005) Offshore renewable energy: ecological implications of generating electricity in the coastal zone. *Journal of Applied Ecology* 42: 605–615.

Gray, J.S. & Elliott, M. (2009) *Ecology of Marine Sediments: Science to management.* Oxford: Oxford University Press.

Groenewold, S. & Fonds, M. (2000) Effects on benthic scavengers of discards and damaged benthos produced by the beam-trawl fishery in the southern North Sea. *ICES Journal of Marine Science* 57: 1395–1406.

Gutierrez, J.L., Jones, C.G., Strayer, D.L. & Iribarne, O.O. (2003) Mollusks as ecosystem engineers: the role of shell production in aquatic habitats. *Oikos* 101: 79–90.

Gutow, L., Teschke, K., Schmidt, A., Dannheim, J., Krone, R. & Gusky, M. (2014) Rapid increase of benthic structural and functional diversity at the *alpha ventus* offshore test site. In BSH & BMU (eds) *Ecological Research at the Offshore Windfarm Alpha Ventus – Challenges, results and perspectives.* Wiesbaden: Federal Maritime and Hydrographic Agency (BSH), Federal Ministry of the Environment, Nature Conservation and Nuclear Safety (BMU). pp. 67–81.

Hall, S.J. (ed.) (1999) *The Effects of Fishing on Marine Ecosystems and Communities.* Oxford: Blackwell Science.

Hendrick, V. & Foster-Smith, R. (2006) *Sabellaria spinulosa* reef: a scoring system for evaluating 'reefiness' in the context of the Habitats Directive. *Journal of the Marine Biological Association* 86: 665–677.

Hiscock, K. & Breckles, M. (2007) Marine biodiversity hotspots in the UK: a report identifying and protecting areas for marine biodiversity. WWF-UK. Online at: https://www.marlin.ac.uk/assets/pdf/marinehotspots.pdf

Janßen, H., Augustin, C.D., Hinrichsen, H.H. & Kube, S. (2013) Impact of secondary hard substrate on the distribution and abundance of *Aurelia aurita* in the western Baltic Sea. *Marine Pollution Bulletin* 75: 224–234.

Joschko, T.J. (2007) Influence of artificial hard substrates on recruitment success of the zoobenthos in the German Bight. Dissertation, Carl von Ossietzky University, Oldenburg.

Joschko, T.J., Buck, B.H., Gutow, L. & Schröder, A. (2008) Colonization of an artificial hard substrate by *Mytilus edulis* in the German Bight. *Marine Biology Research* 4: 350–360.

Kaiser, M.J. (2006) The Louisiana artificial reef program. *Marine Policy* 30: 605–623.

Kaiser, M.J. & Pulsipher, A.G. (2005) Rigs-to-reef programs in the Gulf of Mexico. *Ocean Development & International Law* 36: 119–134.

Kaiser, M.J., Clarke, K.R., Hinz, H., Austen, M.C.V., Somerfield, P.J. & Karakassis, I. (2006) Global analysis of response and recovery of benthic biota to fishing. *Marine Ecology Progress Series* 311: 1–14.

Kerckhof, F., Rumes, B., Norro, A., Jacques, T.G. & Degraer, S. (2010) Seasonal variation and vertical zonation of marine biofouling on a concrete offshore windmill foundation on the Thornton Bank (southern North Sea). In Degraer, S., Brabant, R. & Rumes, B. (eds) *Environmental Impacts of Offshore Windfarms in the Belgian Part of the North Sea: Early environmental impact assessment and spatio-temporal variability.* Brussels: Royal Belgian Institute of Natural Sciences, Operational Directorate Natural Environment, Marine Ecology and Management Section. pp. 53–68. Retrieved 15 May 2018 from https://odnature.naturalsciences.be/downloads/mumm/ windfarms/mumm_report_mon_win2010.pdf

Kerckhof, F., Degraer, S., Norro, A. & Rumes, B. (2011) Offshore intertidal hard substrata: a new habitat promoting non-indigenous species in the Southern North Sea: an exploratory study: In: Degraer, S., Brabant, R. & Rumes, B. (eds) *Offshore Windfarms in the Belgian Part of the North Sea: Selected findings from the baseline and targeted monitoring.* Brussels: Royal Belgian Institute of Natural Sciences, Operational Directorate Natural Environment, Marine Ecology and Management Section. pp. 27–37. Retrieved 15 May 2018 from https://odnature.naturalsciences.be/downloads/mumm/windfarms/ monwin_report_2011_final.pdf

Kröncke, I., Stoeck, T., Wieking, G. & Palojärvi, A. (2004) Relationship between structural and functional aspects of microbial and macrofaunal communities in different areas of the North Sea. *Marine Ecology Progress Series* 282: 13–31.

Krone, K. & Krägefsky, S. (2013) Effects of offshore wind turbine foundations on mobile demersal megafauna and pelagic fish – research at the offshore windfarm Alpha Ventus. Conference Presentation, StUKplus Conference, 30–31 October 2013, Berlin. Abstract. Retrieved 15 May 2018 from http://www.bsh.de/de/Meeresnutzung/Wirtschaft/Windparks/Windparks/ StUKplus/konferenz/praesentationen/praesentationen_13.pdf

Krone, R., Gutow, L., Joschko, T.J. & Schroder, A. (2013a) Epifauna dynamics at an offshore foundation – implications of future wind power farming in the North Sea. *Marine Environmental Research* 85: 1–12.

Krone, R., Gutow, L., Brey, T., Dannheim, J. & Schroder, A. (2013b) Mobile demersal megafauna at artificial structures in the German Bight – likely effects of offshore windfarm development. *Estuarine Coastal and Shelf Science* 125: 1–9.

Langhamer, O. (2012) Artificial reef effect in relation to offshore renewable energy conversion: state of the art. *The Scientific World Journal* 2012: 386713.

Langhamer, O., Wilhelmsson, D. & Engström, J. (2009) Artificial reef effect and fouling impacts on offshore wave power foundations and buoys – a pilot study. *Estuarine, Coastal and Shelf Science* 82: 426–432.

Leonhard, S., Birklund, O. & Birklund, J. (2006) Change in diversity and higher biomass. In Andersen, S. (ed.) *Danish Offshore Wind – Key environmental issues.* Holbæk: PrinfoHolbæk–Hedehusene. pp. 44–63.

Lindeboom, H.J., Kouwenhoven, H.J., Bergman, M.J.N., Bouma, S., Brasseur, S., Daan, R., Fijn, R.C., de Haan, D., Dirksen, S., van Hal, R., Lambers, R.H.R., Ter Hofstede, R., Krigsveld, K.L., Leopold, M. & Scheidat, M. (2011) Short-term ecological effects of an offshore wind farm in the Dutch Coastal Zone: a compilation. *Environmental Research Letters* 6 (2011) 035101.

Lindeboom, H., Degraer, S., Dannheim, J., Gill, A. & Wilhelmsson, D. (2015) Offshore wind park monitoring programmes, lessons learned and recommendations for the future. *Hydrobiologia* 756: 169–180.

Maar, M., Bolding, K., Petersen, J.K., Hansen, J.L.S. & Timmermann, K. (2009) Local effects of blue mussels around turbine foundations in an ecosystem model of Nysted offshore windfarm, Denmark. *Journal of Sea Research* 62: 159–174.

MacArthur, R. (1955) Fluctuations of animal populations and a measure of community stability. *Ecology* 36: 533–536.

May, R.M. (1983) The structure of food webs. *Nature* 301: 566–568.

Mazik, K. & Smyth, K. (2013) Is 'minimising the foot-print' an effective intervention to maximise the recovery of intertidal sediments from distur-bance? Phase 1: Literature review. Natural England Commissioned Reports, Number 110. Retrieved 15 May 2018 from http://publications. naturalengland.org.uk/publication/5091106

McCann, K.S. (2000) The diversity–stability debate. *Nature* 405: 228–233.

Newell, R.C., Seiderer, L.J. & Hitchcock, D.R. (1998) The impact of dredging works in coastal waters: a review of the sensitivity to disturbance and subsequent recovery of biological resources on the sea bed. *Oceanography and Marine Biology: An Annual Review* 36: 127–178.

Ortegon, K., Loring, F. & Sutherland, J. (2013) Preparing for end of service life of wind turbines. *Journal of Cleaner Production* 39: 191–199.

Pearce, B., Hill, J., Wilson, C., Griffin, R., Earnshaw, S. & Pitts, J. (2013) *Sabellaria spinulosa Reef Ecology and Ecosystem Services. The Reef Ecosystem Ecology Report.* The Crown Estate, ISBN: 978-1-906410-27-8. pp. 1–120.

Petersen, J. & Malm, T. (2006) Offshore windmill farms: threats to or possibilities for the marine environment. *Ambio* 35: 75–80.

Pimm, S.L. (ed.) (1982) *Food Webs.* London: Chapman and Hall.

Pioch, S., Saussola, P., Kilfoyleb, K. & Spieler, R. (2011) Ecological design of marine construction for socio-economic benefits: ecosystem inte-gration of a pipeline in coral reef area. *Procedia Environmental Sciences* 9: 148–152.

Powilleit, M., Graf, G., Kleine, J., Riethmüller, R., Stockmann, K., Wetzel, M.A. & Koop, J.H.E. (2009) Experiments on the survival of six brackish macro-invertebrates from the Baltic Sea after dredged spoil coverage and its impli-cations for the field. *Journal of Marine Systems* 75: 441–451.

Qvarfordt, S., Kautsky, H. & Malm, T. (2006) Devel-opment of fouling communities on vertical structures in the Baltic Sea. *Estuarine, Coastal and Shelf Science* 67: 618–628.

Reise, K. (2002) Sediment mediated species interac-tions in coastal waters. *Journal of Sea Research* 48: 127–141.

Reubens, J.T., Degraer, S. & Vincx, M. (2011) Aggregation and feeding behaviour of pouting (*Trisopterus luscus*) at wind turbines in the Belgian part of the North Sea. *Fisheries Research* 108: 223–227.

Reubens, J.T., Degraer, S. & Vincx, M. (2013) The ecology of benthopelagic fishes at offshore windfarms: a synthesis of 4 years of research. *Hydrobiologia* 727: 121–136.

Roberts, L. & Elliott, M. (2017) Good or bad vibra-tions? Impacts of anthropogenic vibration on the marine epibenthos. *Science of the Total Environ-ment* 595: 255–268.

Roberts, L., Cheesman, S., Breithaupt, T. & Elliott, M. (2015) Sensitivity of the mussel *Mytilus edulis* to substrate-borne vibration in relation to anthro-pogenically-generated noise. *Marine Ecology Progress Series* 538: 185–195.

Roberts, L., Cheesman, S., Elliott, M. & Breithaupt, T. (2016a) Sensitivity of *Pagurus bernhardus* (L.) to substrate-borne vibration and anthropogenic noise. *Journal of Experimental Biology and Ecology* 474: 185–194.

Roberts, L., Pérez-Domínguez, R. & Elliott, M. (2016b) Use of baited remote underwater video (BRUV) and motion analysis for studying the impacts of underwater noise upon free ranging fish and implications for marine energy manage-ment. *Marine Pollution Bulletin* 112: 75–85.

Rumes, B., Coates, D., de Mesel, I., Derweduwen, J., Kerckhof, F., Reubens, J. & Vandendriessche, S. (2013) Does it really matter? Changes in species richness and biomass at different spatial scales. In Degraer, S., Brabant, R. & Rumes, B. (eds) *Environmental Impacts of Offshore Windfarms in the Belgian Part of the North Sea: Learning from the past to optimize future monitoring programmes.* Brus-sels: Royal Belgian Institute of Natural Sciences, Operational Directorate Natural Environment, Marine Ecology and Management Section. pp. 183–192. Retrieved 12 June 2018 from https:// odnature.naturalsciences.be/downloads/ mumm/windfarms/winmon_be_2013.pdf

Rumohr, H. & Kujawski, T. (2000) The impact of trawl fishery on the epifauna of the southern North Sea. *ICES Journal of Marine Science* 57: 1389–1394.

Russell, D.J.F., Brasseur, S.M.J.M. Thompson, D., Hastie, G.D., Janik, V.M., McClintock, B.T., Matthiopoulos, J., Moss, S.E.W. & McConnell, B. (2014) Marine mammals trace anthropogenic structures at sea. *Current Biology* 24: R638–R639.

Smyth, K., Christie, N., Burdon, D., Atkins, J.P., Barnes, R. & Elliott, M. (2015) Renewables-to-reefs? – Decommissioning options for the offshore wind power industry. *Marine Pollution Bulletin* 90: 247–258.

Statoil ASA (2017) Hywind – the world's leading floating offshore wind solution. Retrieved 15

December 2017 from https://www.statoil.com/en/what-we-do/hywind-where-the-wind-takes-us.html

Vaissière, A.C., Levrel, H., Pioch, S. & Carlier, A. (2014) Biodiversity offsets for offshore windfarm projects: the current situation in Europe. *Marine Policy* 48: 172–183.

van den Eynde, D., Baeye, M., Brabant, R., Fettweis, M., Francken, F., Haerens, P., Mathys, M., Sas, M. & van Lancker, V. (2013) All quiet on the sea bottom front? Lessons from the morphodynamic monitoring. In Degraer, S., Brabant, R. & Rumes, B. (eds) *Environmental Impacts of Offshore Windfarms in the Belgian Part of the North Sea: Learning from the past to optimize future monitoring programmes.* Brussels: Royal Belgian Institute of Natural Sciences, Operational Directorate Natural Environment, Marine Ecology and Management Section. pp. 35–47. Retrieved 12 June 2018 from https://odnature.naturalsciences.be/downloads/mumm/windfarms/winmon_be_2013.pdf

Watling, L., Findlay, R.H., Mayer, L.M. & Schick, D.F. (2001) Impact of a scallop drag on the sediment chemistry, microbiota, and faunal assemblages of a shallow subtidal marine benthic community. *Journal of Sea Research* 46: 309–324.

Whitehouse, R., Harris, J., Sutherland, J. & Rees, J. (2011) The nature of scour development and scour protection at offshore windfarm foundations. *Marine Pollution Bulletin* 62: 73–88.

Wilding, T.A., Gill, A.B., Boon, A., Sheehan, E., Dauvin, J.C., Pezy, J.P., O'Beirn, F., Janas, U.,

Rostin, L. & de Mesel, I. (2017) Turning off the DRIP ('Data-rich, information-poor') – rationalising monitoring with a focus on marine renewable energy developments and the benthos. *Renewable and Sustainable Energy Reviews* 74: 848–859.

Widdows, J., Brinsley, M.D., Salkeld, P.N. & Elliott, M. (1998) Use of annular flumes to determine the influence of current velocity and bivalves on material fluxes at the sediment–water interface. *Estuaries* 21(4A): 552–559.

Wieking, G. & Kröncke, I. (2005) Is benthic trophic structure affected by food quality? The Dogger Bank example. *Marine Biology* 146: 387–400.

Wilhelmsson, D. & Malm, T. (2008) Fouling assemblages on offshore wind power plants and adjacent substrata. *Estuarine, Coastal and Shelf Science* 79: 459–466.

Wilson, J. & Elliott, M. (2009) The habitat-creation potential of offshore windfarms. *Wind Energy* 12: 203–212.

Wilson, J., Elliott, M., Cutts, N., Mander, L., Mendao, V., Perez-Dominguez, R. & Phelps, A. (2010) Coastal and offshore wind energy generation: is it environmentally benign? *Energies* 3: 1383–1422.

Wolfson, C., Van Blaricom, N., Davis, N. & Lewbel, G.S. (1979) The marine life of an offshore oil platform. *Marine Ecology Progress Series* 1: 81–89.

CHAPTER 5

Fish

ANDREW B. GILL and DAN WILHELMSSON

Summary

The current distribution of, and future plans for, offshore wind farms are likely to influence fish assemblages in coastal waters. In this chapter, the evidence for predicted changes in the fish community as a result of interaction with wind farms is considered, keeping a balance in the analysis that the changes are neither better nor worse ecologically, but will be different owing to the presence of artificial substrate forming habitat. The available evidence focuses on the effects on species, whereas it is also necessary for the species assemblages to be taken into account and, if possible, to translate these into ecological impacts. The effects on fish are linked to changes or responses of biological/ecological relevance and they are impacts only if they are deemed significant in this context. Specific attention is paid to whether specific species or taxa are likely to be sensitive to the environmental emissions associated with wind-farm development phases, from pre-construction and construction through operation to decommissioning; however, our understanding is patchy. Existing pressures on fish populations such as fishing or habitat degradation are highlighted to provide the appropriate contextual basis, but only as a backdrop against which to interpret the changes associated with wind farms. Overall, the evidence shows that fish are associated with wind farms and there may be important changes to the fish community structure and the trophic interactions within the local marine ecosystem. With the gaps in understanding, the full consequences of wind-farm deployment to fish remain unclear. Targeted research to address these gaps, particularly relating to the cause-and-effect pathways of change and trophic relationships, as well as better understanding of temporal and spatial use of wind farms for different life stages, is required to understand the changes that occur to the fish community.

Introduction

Fish assemblages are already being demonstrably altered by past and present human activity. The single most significant negative impact is undoubtedly linked to fishing. In whatever form, commercial fishing has directly reduced target fish species' abundance and biomass (Myers *et al.* 1996) and altered population dynamics and fish assemblage composition (Greenstreet & Hall 1996) for a large number of species (Frank *et al.* 2006). Some of the targets of fishing have suffered from overharvesting while other species have been decimated as an unfortunate consequence of indiscriminate fishing in the form of bycatch. However, other anthropogenic activities have altered the nature of habitats considerably, such that they have become limited for some species at certain stages of life, the most common examples being the degradation of coral reefs and other hard-bottom areas, as well as the widespread loss of coastal mangrove forests (Mumby *et al.* 2006). Less direct consequences of human activity can be associated with global changes such as the widespread acidification of the oceans (Orr *et al.* 2005). The impact on fish development is yet to be fully determined but there is little doubt that this is another significant pressure on the fish assemblages (Munday *et al.* 2010).

The development of offshore wind farms (OWFs) and the associated power-cable network infrastructure represent one of the single most significant changes to be experienced in areas of coastal and offshore marine environments in recent times (Inger *et al.* 2009; Boehlert & Gill 2010). Fish are one the major groups of organisms, comprising thousands of species around the world sharing these coastal and offshore environments with existing or planned OWFs. A number of fish species also have importance in terms of their commercial value as fisheries. The questions then arise: How do OWFs come into the picture? Are the changes associated with them significant in the wider context and, if so, are these changes positive or negative for the fish?

In simple terms, any changes to habitat, routes of movement, feeding opportunities, predator abundance and distribution of fish and/or alteration of the ambient environment could each individually have some significance for fish, or perhaps more importantly, they may occur simultaneously in a particular area or at a particular time so as to act in combination (Willsteed *et al.* 2017). The prediction of how fish may be affected by OWFs will depend on their stage of life and how long the changes will last. Change may be over a short, defined period during which environmental conditions vary, for example construction over a few months, or a long-term feature, such as the turbine structures, present throughout the life of the fish. The effects also may be considered as direct or indirect in nature. Those that are direct, an example being construction noise causing fish to swim away from an area, are the easiest for researchers to perceive and study, whereas the indirect effects, such as food-web related changes, are more complex and likely to occur over longer periods, making them more difficult to understand while also complicating the determination of effects (Gill 2005).

The expectation from the literature is that OWFs will influence local fish presence and abundance, with consequences for species interactions such as predator–prey dynamics, food-web connectivity and the fish community structure in terms of species abundance, body size of individuals and population demographics (Gill 2005; Raoux *et al.* 2017). How the scale of these effects, both direct and indirect, may influence the ecological changes associated with OWFs in isolation and cumulatively is regarded as an important but as yet unanswered topic (Boehlert & Gill 2010; Wilding *et al.* 2017; Willsteed *et al.* 2017). Furthermore, the significance of existing pressures on fish, such as fishing, habitat degradation and ocean acidification, needs to be acknowledged to enable appropriate interpretation of the changes that are associated with OWFs. In essence, the key questions relate to whether OWFs cause population-level effects, perhaps for some species cumulatively, that can be compared to any degree with changes resulting from other major effectors of change on the fish community.

Scope

The basis for understanding knowledge on the effects of OWFs on fish comes from two primary sources: academic journals and texts; and environmental reports, associated with either monitoring for the Environmental Impact Assessment (EIA) process or outputs from commissioned studies on particular topics. The recognition of both peer-review and grey-literature sources is important, as relying on just one or the other results in poor evidence coverage in the case of journals, or limitations in the scope of understanding of the effects of OWFs on fish in relation to environmental monitoring reports or commissioned reports. The present authors undertook a semi-systematic review of the topic status. Both authors have written about and reviewed the topic previously; hence, this knowledge was updated through online search engines, principally Scopus and Google Scholar, for publications and reports since 2010 covering the following terms: fish and offshore wind farms, fisheries and offshore wind farms, and derivative terms, such as OWFs. Furthermore, the interest in how OWFs interact with the environment is global but most of the development has been in northern European waters; hence, there have been a number of reports speculating on the possible effects on fish. However, the evidence base is restricted to where the wind turbines have actually been deployed and this indicates that knowledge on how geographic location may play a role is lacking.

Any assessment of environmental effects of OWFs on fish needs to acknowledge clearly the existing status of and pressures on the fish assemblages, and has to be objective in nature, recognising that the fish community interacting with the OWF development may well be altered, but is neither better nor worse ecologically unless the evidence suggests otherwise.

There is opinion that there are positive and/or negative effects, and hence the view here is that 'positive' and 'negative' impacts are terms that depend on how they are defined and who is interpreting them. The evidence for effects on fish was considered as changes or responses of biological or ecological relevance, whereas impacts on fish were defined as biologically or ecologically significant changes, responses or effects, following the definition of Boehlert and Gill (2010). It was anticipated that the focus when interpreting whether impacts occurred would be on the biologically significant effects on species, although the consequences for species assemblages and the wider community also needed to be taken into account when attempting to translate impacts to the local marine ecosystem. Therefore, the evidence for effects on species was considered and, where possible, the species assemblages as well, as this provided the opportunity to present the case for or against, better or worse, and also to address future direction of research and topic knowledge gaps, particularly where the effects may translate further into ecological impacts.

Themes

Why, conceptually, fish may be affected by offshore wind farms

To consider why and how fish may be affected by OWFs, it is necessary to identify and determine the knowledge of ecological effects of OWFs on fish. As such, the starting point is to look at the biological attributes that may make the fish more or less likely to be affected, before considering the evidence with respect to OWFs.

Life history

The majority of fish species that inhabit the waters where OWFs have been developed or are planned have life histories that will bring them into contact with the developments. It may be a single life stage, such as adult spawning aggregations for reproduction, or multiple life stages, for example the egg and juvenile developmental stages. The key considerations are the life-cycle stage of the fish, the duration of exposure and the spatial overlap of the OWF with the fish's functional habitat linked to a particular life stage (Gill 2005; Inger *et al.* 2009).

The life stages that are most vulnerable to change are those where resource limitation is the main controlling factor, examples being extent of refuge habitat, food availability and spawning habitat. Species that have reproductive behaviour that causes adults to congregate in a single area will increase the potential overlap and hence the likely significance of effects. Those species that have a high dependency on benthic sources of food may experience changes to the type and abundance of prey items (see Dannheim *et al.*, Chapter 4). Species with eggs that drift with the prevailing currents may also increase their chances of encountering OWFs downstream.

Spatial behaviour and migratory patterns

One particular attribute that a number of species of fish have is their propensity to undertake significant movements that have direct consequences for the species distribution. These may be short-term but regular movements, for example to and from feeding grounds, or they may be longer term, seasonal movements whereby some life stages undertake significant migrations across large distances, such as European Eel *Anguilla anguilla* moving over thousands of kilometres during both the larval stage, on their way to rivers, and the adult reproductive phase, when returning to central spawning grounds (van Ginneken & Maes 2005). Such movements will have consequences for the local and regional distribution, species abundance and community composition. Changes could also occur if movement routes are impeded or if local habitats are altered such that the migratory fish cannot move around normally; that is, the connectivity of functional habitats is blocked.

Sensory biology

Within any life stage, fish rely on their sensory systems, namely, the acoustic, visual, gustatory, chemosensory, mechanosensory and, in some cases, magnetosensory or electrosensory systems, to interact with the environment in which they are immersed.

Of all the environmental effects associated with human activities in the sea, such as OWF development, a significant amount of attention has been paid to how acoustic emissions (noise), which propagate over many kilometres in the low-frequency range, may affect fish (Gill *et al.* 2012a; Hawkins & Popper 2016). Fish live in a highly acoustic medium and most species are readily able to hear and respond to underwater noise (Slabbekorn *et al.* 2010). Some taxa are highly sensitive; for example, clupeids (herring *Clupea* spp. and their allies) use sound to coordinate the movement of schools (Popper 2000) and gadoids (cod *Gadus* spp. and their allies) use sound in reproductive behaviour (Hawkins & Popper 2016).

Energy emissions into the coastal and offshore environment occur not only in the form of underwater sound, but also in the much less understood form of electromagnetic fields (EMFs), thermal emission in the form of heat, and light and radiation. The last two types are not considered relevant energy emissions from OWFs; hence, they are not referred to any further in this text.

The earth is surrounded by natural EMFs from external sources such as the sun and interstellar space and internal sources from the earth's core and mantle. Furthermore, seawater is a highly effective electrical conductor and movement through the natural magnetic sources creates localised EMFs that can propagate over tens of metres (Gill *et al.* 2014). Fish from many taxonomic classes have the ability to respond to these EMFs and utilise them in relation to diurnal or seasonal migration, such as in the case of anguillid eels and salmonids, or, like elasmobranchs, to use them to orientate within their local environment or to find prey (Collin & Whitehead 2004; Gill *et al.* 2014).

In terms of heat emitted, it is predicted that fish may associate with a heat source, although the transmission through the seabed and water column is expected to be a matter of centimetres. How fish may use thermal cues in their environment remains a highly uncertain topic in relation to OWFs and their biological effects on fish.

Influence of habitat

Fish contribute significantly to the functioning of marine ecosystems, inhabiting the sea from the pelagic zone (open water) to the seabed at all latitudes, and have a major influence on energy flow through the food web via their feeding on invertebrates, algae, detritus and each other at various trophic levels, which can lead to enormous biological productivity for a number of species that is reflected in species biomass.

Habitat features influence the distribution and community composition of fish, and this is most pronounced when it comes to bottom habitats. A range of factors may influence the fish community structure in association with different types of bottom habitats, such as structural factors including complexity, surface structure, void space and number of interior spaces, and habitat area; degree of isolation in relation to equivalent habitats; composition of prey items; predators; and the surrounding seabed (Grove *et al.* 1991; Kim *et al.* 1994; Lan & Hsui 2006). The importance of different factors varies between species, trophic groups and spatial use patterns of fish (Moffit *et al.* 1989; Grove *et al.* 1991; Kim *et al.* 1994). Many reef-associated fish are limited by the availability of shelter sites, and often the amount of suitable habitat is suggested to be limiting during certain life stages, for example the early benthic phase or spawning (Chojnacki 2000; Hunter & Sayer 2009).

Evidence of effects of offshore wind farms on fish and their ecosystem

OWFs have defined periods during which the changes to the marine environment that may affect fish will differ, namely the pre-construction surveying and construction, operation and decommissioning phases (level 1 in the environmental effects framework in Figure 5.1). The pre-construction/construction and the decommissioning phases are expected to have the strongest influence on fish in terms of acute, short-term effects (Figure 5.1), as they represent the greatest amount of structural and habitat change, and are the busiest periods in terms of boat traffic and the noisiest times over periods of several months. However, more interest is emerging in the long-term operational phase (20–30 years), where the structures represent new habitat and environmental attributes which are likely to cause some response in the fish over time through chronic, long-term effects (Figure 5.1). Furthermore, in this phase the wind energy is harnessed and converted into electrical energy and some of the harnessed energy is emitted acoustically, electromagnetically and

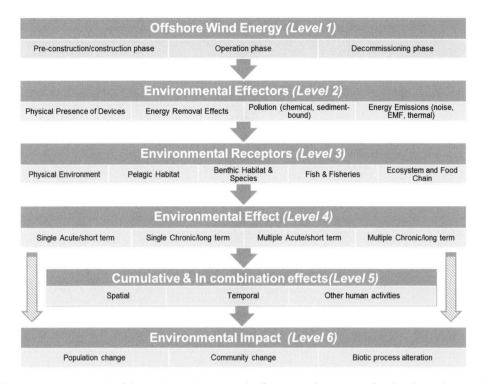

Figure 5.1 Framework of the main environmental effectors and receptors for the three phases of offshore wind-farm development. Each level feeds into the next and shows potential outcomes in the form of demonstrable effects over different scales and how these effects can manifest as biologically and ecologically relevant impacts, either directly (hatched arrows) or via cumulative and/or in-combination effects. EMF: electromagnetic field. (Modified from Boehlert & Gill 2010)

thermally into the surrounding environment, which leads to specific effector–receptor–effect pathways and predicted outcomes (Figure 5.1, levels 2–4) (Boehlert & Gill 2010). The effects will be evidenced by responses in fish and are determined as impacts (Figure 5.1, level 6) if they significantly alter fish population or community attributes either directly (Figure 5.1, levels 4–6, hatched arrows) or through cumulative or in-combination effects (Figure 5.1, levels 4–6).

Energy emissions into the environment

The acoustic emissions of wind turbines are acknowledged to have an influence on fish, particularly during the construction phase of development. The noise produced has two components: sound pressure and particle motion, described by particle acceleration or velocity (Hawkins & Popper 2016). Noise pressure associated with piling has been shown to have a very high peak intensity (Gill *et al.* 2012a) and also a low-frequency range coverage (1 Hz to 10 kHz), comprising the sound waves that travel the farthest and in the range of detection of fish (Hawkins & Popper 2016). Hence, underwater noise is known to propagate over tens of kilometres at levels exceeding those where a significant response in fish has been shown to occur. Anecdotally, there are reports of fish mortality seen during the early OWF constructions. Limited experimental studies on the effects of pile driving on caged fish have shown physiological and behavioural effects with small-scale piling, such as that used in port building (Nedwell *et al.* 2006). The OWF industry tends to employ noise-mitigation strategies, which significantly change the characteristics of the noise emitted, such as bubble curtains to reduce higher frequencies (Thomsen *et al.* 2015; Thomsen & Verfuß, Volume 4, Chapter 7), and will go some way towards reducing any direct mortality, although this has not been quantified. Acoustic standards to reduce the potential for negative effects on marine fauna have been developed, but these are poorly defined and not consistent across adjacent territorial state waters, such as in the North Sea, Europe, where the majority of OWF construction is taking place (Thomsen *et al.* 2015; Thomsen & Verfuß, Volume 4, Chapter 7).

There is no doubt that fish respond to the noise emitted by piling activity, and how the fish respond during OWF construction will depend on how far away they are from the noise source and for how long they are exposed (Figure 5.2) (Gill *et al.* 2012a). As sound propagates through seawater it loses energy; this happens more quickly in the higher frequencies but the low frequencies can still be detected tens of kilometres away. Hence, the effect of the noise upon fish is expected to be associated with the distance at which they received the sound, with injury occurring close to the noise source and behavioural response at a distance from it (Figure 5.2). The type of species will also determine how the noise is received, with species such as Atlantic Herring *Clupea harengus*, which are highly sensitive, detecting the pressure element of sound through an anatomical link between their swim bladders and their hearing apparatus (Popper 2000). In contrast, most species detect sound through particle motion (Hawkins & Popper 2016). Understanding of particle motion is poor, but acknowledging its importance highlights that researchers should not just concentrate on a few sensitive species, but consider OWF noise as a major factor in potentially influencing many different fish species and different life stages, thereby affecting assemblage attributes through time (Slabberkorn *et al.* 2010; Hawkins & Popper 2016). There is also a need to consider the extended exposure of fish during the

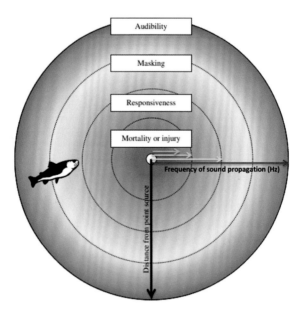

Figure 5.2 Theoretical zones of acoustic influence of a sound source (such as pile driving) on fish with distance. The source is depicted at the centre and the reduction in multifrequency propagation of the sound is represented by the reduction in coloured arrows with distance. Higher frequencies (>1,000 Hz) have limited emission range whereas low-frequency sounds (<100 Hz) dominate at a distance. (Modified from Gill *et al.* 2012a)

several decades of the OWF's operational phase, during which they will experience lower levels of noise but over a much longer period. If these noises are above ambient they may cause interference with daily functioning and perhaps cause effects over the longer term (Slabbekorn *et al.* 2010), although whether this has any biologically significant impact is undetermined (Hawkins & Popper 2016).

EMFs are emitted into the local environment mainly from subsea cables as a result of wind turbines generating electricity (Gill *et al.* 2012b; Thomsen *et al.* 2015). There are two main types of cable: high-voltage alternating current (HVAC) and high-voltage direct current (HVDC). Both types emit magnetic fields into the surrounding environment and HVAC also directly induces electric fields within the adjacent seawater (Figure 5.3). Furthermore, movement through the magnetic fields, either by an animal swimming or via tidal stream or water currents, induces electric fields (Figure 5.3). Both components of the EMF (i.e. magnetic and electric fields) propagate outwards at a decreasing rate with distance from the cable source, thereby creating an EMF zone in the water around the cable of several tens to hundreds of metres, which scales up as the cable current and voltage increase (Thomsen *et al.* 2015; Copping *et al.* 2016). The EMF zone will also extend along the length of a cable and will be associated with all cables within the array and in the export cables to shore (Gill *et al.* 2014). Thus, the emitted EMF may constitute a potential extended barrier to movement for sensitive migratory species in shallow coastal waters, such as migratory eels (Westerberg & Lagenfelt 2008).

Electromagnetic-sensitive species come from across many taxa, although there is a paucity of knowledge, on a restricted number of species, on how they respond to anthropogenic electric or magnetic fields compared with natural bioelectric or geomagnetic fields (Normandeau Associates *et al.* 2012). Of those that are potential receptors of OWF EMFs, magnetic-sensitive species are considered as those with a significant migratory phase in their life history, such as salmonids, and electrosensitive species with electroreceptor apparatus, including elasmobranchs (sharks and rays),

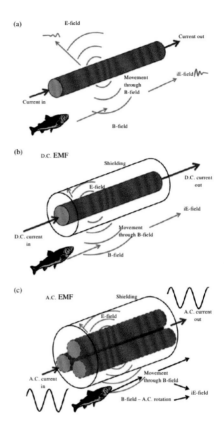

Figure 5.3 Schematic representation of the electromagnetic fields (EMF) associated with a subsea cable. Direct electric (E) fields, induced electric (iE) fields and magnetic (B) fields are shown, and wave magnitudes indicate the relative size of the EMFs with distance from the cable, in relation to: (a) EMFs are associated with an unshielded cable; (b) a high-voltage direct current (HVDC) cable including the shielding, which contains the E-field, whereas iE-fields occur as the fish (or seawater) moves through the EMFs; and (c) a high-voltage alternating current (HVAC) cable, with three cores that allow the AC to flow following a typical sine-wave pattern of transmission within each cable core. An asymmetric B-field is emitted from the interacting cores, causing a rotational B-field which induces its own iE-field. Not to scale. (Reproduced from Gill 2012b, with permission from Elsevier)

Acipenseridae (sturgeons) and Petromyzontiformes (lampreys). Only limited studies have been conducted on the responses of sensitive fish species to subsea cables. Nevertheless, there are demonstrable responses which appear to be variable, and hence no determination of any significant impact can be made without further research (Gill *et al.* 2014).

Heat is emitted during the transmission of electricity and if the cable is buried in sediment it is known to raise the temperature by a few degrees to tens of degrees Celsius, depending on the power rating in the cable and the length of time for which it is operating at maximum transmission (Worzyk 2009). There is no available evidence of any effects of thermal emissions on fish but, hypothetically, fish may respond at a very local level. Any impact would probably be indirect, via changes to the benthic infauna as a consequence of the sediment that they inhabit increasing in temperature. There is evidence of a greater abundance of invertebrates colonising subsea power cables (Worzyk 2009), which could alter the food availability for fish. Any trophic-related effect on fish may result in changes in distribution, movement and/or abundance, although there is no available evidence to support such changes.

Pollution

In the dynamic coastal environments where existing wind turbines are installed, questions may be raised about pollution, particularly if the area has a history of industrial activity (Gill 2005). The resuspension of old sediments, some of them with a legacy of industrial chemicals as a result of dredging, placement of construction vessel supports or cable laying, may release chemicals that become bioavailable. Alternatively, the disturbed sediments may simply smother spawning habitat or seabed-associated eggs or larvae. Potential pollutants also include liquids or materials used for the OWF, such as paints (see Rees & Judd, Chapter 2), but there is no evidence of contamination effects on fish. Furthermore, disentangling the likely consequences of past pollution and existing threats is inherently difficult.

Physical habitat change

The most commonly used wind-turbine foundations are monopiles; others have used tripod foundations with jacket structures and, to a much lesser extent, developments have used gravity foundations or suction caissons (see Jameson *et al.*, Chapter 1; Rees & Judd, Chapter 2; Dannheim *et al.*, Chapter 4). Deployments have been limited to coastal areas and offshore banks with sedimentary seabeds in water less than 50 m deep. In these exposed areas, the substrate is frequently disturbed and benthic invertebrate and algae communities are often dominated by opportunistic species that are adapted to the dynamic environment (see Dannheim *et al.*, Chapter 4). Nonetheless, offshore banks that are technically suitable for OWFs can provide a refuge opportunity for species that have been excluded by pollution, eutrophication or anthropogenic development farther inshore.

The single most significant habitat change associated with OWFs is the shift from an open-water (pelagic) environment to one with new, hard surfaces and a more complex three-dimensional structure (Boehlert & Gill 2010; Wilhelmsson *et al.* 2010; Dannheim *et al.*, Chapter 4). This shift is depicted in Figure 5.4, where the focus is on the physical habitat features that are likely to influence the presence or absence of fish within an area. As a consequence, one would expect significant changes in the trophic structure of an ecosystem where fish are a major component, in terms of both local biological production, that is, biomass, and trophic linkages within the ecosystem (Figure 5.5). This is discussed further in *Ecological change*, below.

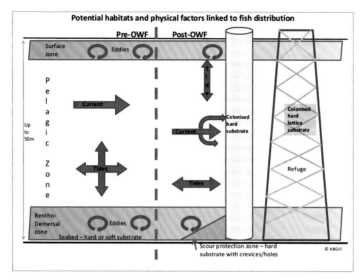

Figure 5.4 Depiction of the habitats available to fish before offshore wind farm (OWF) installation (left of dashed vertical line) and once the wind farm has been installed. Note the increase in hard surface and the potential changes to hydrodynamics.

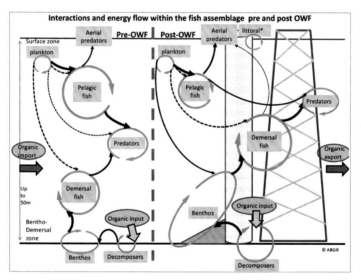

Figure 5.5 Depiction of the ecological interactions based on trophic linkage and energy flow with the habitats available to fish before offshore wind-farm (OWF) installation (left of dashed vertical line) and once the OWF has been installed. Note the increased number of links and the change in size of the grey circular representations of relative fish population.

The direct physical loss of habitat for soft-bottom dwelling species owing to the installation of wind turbines and scour protection typically represents a small percentage of the whole wind-farm area as a result of each turbine only claiming up to a few hundred square metres. Although the ecological footprint can extend beyond the physical structures, any changes in the significance of the original benthic community structure of the surrounding seabed may be relatively localised (Ashley *et al.* 2014). The key change is that portions of the seabed are replaced by hard-bottom substrata in the form of submerged foundations and predominantly rock scour protection of offshore wind turbine bases with a larger total surface area than the lost soft-bottom patches (Figure 5.4). It is anticipated that this availability of new substrata will influence the nature of the benthic community and thus the linked aspects of the composition, production and dynamics of the fish assemblage. The structures will also cause changes in the path of the prevailing water current accompanied by the potential for vertical zonation on the structures by tidal movements (Figure 5.4) and redistribution of sediment transported outside the wind-farm boundaries, all with ecological consequences for fish.

The artificial reef effect

By their very nature, wind-farm structures constitute small artificial reefs, providing new habitat for a number of sessile and motile colonising marine species, although they differ in substrate type, shape, time of submersion, and thereby also community composition, from natural reefs (Wilhelmsson *et al.* 2010; Ashley *et al.* 2014). Accumulated evidence suggests that artificial reefs generally hold greater fish densities and biomass, and provide higher catch rates, compared with the surrounding soft-bottom areas, and in many cases also in relation to adjacent natural reefs (for references see Wilhelmsson *et al.* 2010).

Worldwide, artificial reefs are constructed and deployed in coastal waters to manage fisheries, mitigate damage to the environment, protect and facilitate the rehabilitation of certain habitats or water bodies, or increase the recreational value of an area (e.g. Milon 1989; Wilhelmsson *et al.* 1998; Jensen 2002, Claudet & Pelletier 2004; Seaman 2007). Urban structures constructed primarily for other purposes in the sea, such as pier pilings, oil platforms, breakwaters and wind turbines, also serve as habitats for fish and invertebrate assemblages and were defined as secondary artificial reefs by Pickering *et al.* (1998).

Knowledge on the association between fish and wind turbines is slowly growing and highlights that fish are attracted to the structures (Figure 5.6), which leads to questions about how ubiquitous this association is for each marine wind turbine or wind farm and whether there are differences over time. The general grouping of fish and their position in the water column, such as pelagic and demersal species, and how trophic and predator–prey interactions may change in an area as a result of the installation of an OWF, are shown in Figure 5.5. Several benthic (bottom-dwelling) species, namely the European Eel, sculpins (Cottidae), and typical reef-associated species such as Goldsinny Wrasse *Ctenolabrus rupestris*, Eelpout *Zoarces viviparous* and Lumpsucker *Cyclopterus lumpus*, have been recorded in higher densities around turbines than in the surrounding waters and seabed in European waters (Wilhelmsson *et al.* 2006; Couperus *et al.* 2010; Bergström *et al.* 2013). A meta-analysis focusing on demersal and relatively stationary fish suggested that often only a few benthic and semi-pelagic hard-bottom associated species display

Figure 5.6 Underwater images showing large numbers of fish around wind turbine towers in Swedish waters. (Dan Wilhelmsson)

Figure 5.7 Atlantic Cod *Gadus morhua*: a large predatory species that is particularly attracted to the new habitats offered by wind-farm structures in the North Sea. (Martin Perrow)

large increases in local numbers, while densities of some soft-sediment associated species commonly remain unchanged or decrease (Ashley *et al.* 2014). However, more mobile pelagic species such as Horse Mackerel *Trachurus trachurus*, and demersal fish including Atlantic Cod *Gadus morhua* (Figure 5.7) and Pouting *Trisopterus luscus* have been recorded in large shoals closely associated with wind turbines (Couperus *et al.* 2010; Bergström *et al.* 2013). Catch rates of Pouting were noted as being nine to 100 times higher around wind turbines compared with surrounding sandy bottoms in a study by Reubens *et al.* (2013a).

The degree of aggregation of fish around wind turbines can also vary seasonally and diurnally (Reubens *et al.* 2013a), and the presence of wind-turbine foundations may increase residence time. For example, Reubens *et al.* (2011) recorded a school of approximately 22,000 Pouting, with an estimated total biomass of 2.5 t, residing around a single wind turbine for at least a year. Atlantic Cod have also shown relatively high site fidelity around wind turbines (de Troch *et al.* 2013) (Box 5.1). At Egmond aan Zee in the Netherlands, Winter *et al.* (2010) suggested that while Common Sole *Solea solea* did not spend more time in the wind farm compared with a control area, some of the juvenile Atlantic Cod population did reside for a longer time within the wind farm. However, the study was not designed to include any aggregations of fish in close association with the turbines, and parts of the fish assemblages may have been overlooked.

Little research has considered both spatial variation and temporal change, although the acoustic tracking study on Atlantic Cod by Reubens *et al.* (2014a) provides an intriguing insight into this topic (Box 5.1). Over a period of months, individual fish were found to be resident around particular turbines before they migrated away from the wind turbines to a nearby estuary in the winter (Reubens *et al.* 2014a). While the study was limited to a single year, the residency and distributional change by season is consistent with other Atlantic Cod movements recorded in the southern North Sea (Righton *et al.* 2007). Whether other species move between the offshore wind turbine habitat and other habitats is unknown, and whether the association with OWFs will have an influence on the population distribution of species is a fascinating area for future study. Moreover, with

Box 5.1 Acoustic telemetry to investigate the presence and movement behaviour of Atlantic Cod *Gadus morhua* in a Belgian offshore wind farm

Jan T. Reubens, Steven Degraer and Magda Vincx

In 2008, the construction of the C-Power offshore wind farm, the first in the Belgian part of the North Sea, was started. The first phase consisted of six 5 MW turbines with a gravity-base foundation. In a later phase, 48 6.15 MW turbines on jacket foundations were added and the wind farm became fully operational in 2013. The wind farm is situated on the Thorntonbank, a natural sandbank 27 km offshore, with water depths ranging between 18 and 24 m. The gravity-base foundations are surrounded by a scour protection layer with an average diameter of 45 m, comprised of pebbles and rocks (Reubens *et al.* 2014a).

To assess the impacts of the wind turbines on the presence and movement behaviour of Atlantic Cod *Gadus morhua*, an acoustic telemetry study was initiated. Acoustic telemetry allows the study of individual behaviour of undisturbed fish for a long period (Voegeli *et al.* 2001). This technique requires the use of two types of instruments: transmitters and receivers. Fish are tagged with a transmitter, which sends out an acoustic signal on a predefined delay. The receivers are deployed on the ocean floor and continuously monitor the presence of acoustic transmitters in their detection range. In this study, six receivers (type Vemco VR2W) were deployed around two turbines and 22 individual fish were tagged with V9 acoustic transmitters (Vemco Ltd) at those turbines (Figure 5.8). The study ran between May 2011 and July 2012. Further details on tagging and tracking procedures are provided in Reubens *et al.* (2013b). Here, the focus is on the results for the gravity-base foundations.

The fish showed a strong seasonal variation in presence at the turbines. During summer and autumn, many of the fish were highly resident as they were present for an extended period and detected on a daily basis. By the end of December, however, most of the fish had left the area and throughout the winter very few detections were encountered. In the spring of the following year, some of the fish reappeared (Figure 5.9). Based on triangulation, the receivers allowed the exact position of the fish when sending out an acoustic signal to be calculated.

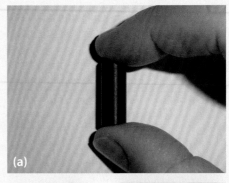

Figure 5.8 (a) Example of an acoustic tag; (b) tags being implanted into each of the 22 individual Atlantic Cod in the study. (Jan Reubens)

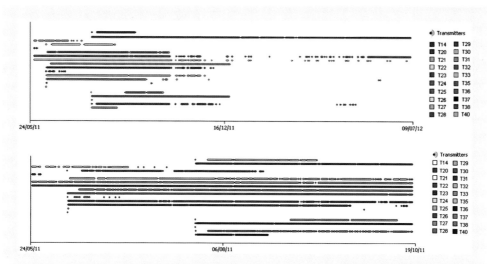

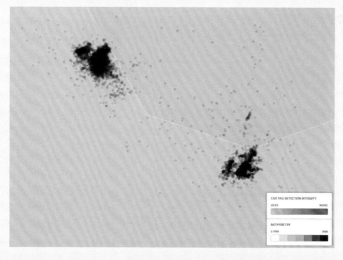

Figure 5.9 Detection plots of tagged Atlantic Cod over time: (upper) overview from the entire study; (lower) detail from 25 May 2011 to 19 October 2011. Each line represents the detections of one individual.

This further allowed the link between fish position and habitat type to be established. The turbines with scour protection form artificial hard substrates with rocks and pebbles, while the surrounding area consists of sandy sediments. Although soft sediments dominate the study area and only small patches of hard substrates are available, most of the detections from tagged fish were over or in close vicinity to the hard substrates (Figure 5.10). More than 95% of the calculated positions were situated within a 50 m range from the wind turbines, revealing strong aggregation behaviour of the study fish towards the hard substrates.

Figure 5.10 Detection intensities of tagged Atlantic Cod (13 individuals) around two turbines from C-Power in 2011, showing a preference for the hard-substrate associated areas in close proximity to the turbines.

Atlantic Cod are known to make extensive migrations between feeding (in summer and autumn) and spawning grounds (in winter), but during the feeding season they may largely reduce their foraging movements, sometimes to less than 1 km (Turner *et al.* 2002). This is consistent with the results of the current study. Reubens *et al.* (2014b) revealed that Atlantic Cod use the artificial hard structures of the wind turbines as feeding grounds in summer. During winter, the fish probably migrate towards spawning grounds and/or overwintering habitats. Unpublished ongoing research and recaptures by recreational fishers revealed that some of the individuals tagged at the wind farms moved to the Western Scheldt coastal area and estuary in the Netherlands. The specific reason why they moved to this area remains unresolved.

It can be concluded from this research that the offshore wind farms have a strong influence on the presence and movement behaviour of Atlantic Cod during summer and autumn, and play an important role in part of their life cycle.

earlier life stages of fish being attracted to wind-turbine towers and foundations, there is further potential for alteration in the spatial distribution of fish populations.

In general, the primary reasons suggested for the higher abundance and diversity of fish on and around artificial reefs include enhanced protection from predation and water movements, and increased availability of food as a result of the association of prey fish and invertebrates with the artificial structure, as reflected in Figure 5.5 (Bohnsack & Sutherland 1985; Jessee *et al.* 1985; Ambrose & Swarbrick 1989; Bohnsack 1989; Grove *et al.* 1991). Wilhelmsson *et al.* (2006) speculated that wind turbines may also function as 'larvae traps', triggering settlement of surface-oriented fish larvae, a portion of which otherwise may not have made it to suitable shallow coastal habitats. Gravity foundations and scour protection, in the form of rock rubble and boulders placed around the turbines, increase volume and structural complexity, which enhances their function as shelter and feeding grounds for fish. Wilhelmsson *et al.* (2006) also recorded high densities of several benthic and semi-pelagic fish around monopoles (Figure 5.6) that were simply driven into the seafloor without scour protection, indicating that these relatively structurally non-complex foundations may provide sufficient enhancement of conditions to aggregate fish.

The use of wind-turbine foundations as functional habitat varies with fish species. Small benthic species such as gobies (Gobiidae) and wrasses (Labriidae), or juveniles of other species, appear to benefit from the shelter afforded by the matrix of colonising and encrusting organisms, such as Blue Mussels *Mytilus edulis* and macroalgae (Bergström *et al.* 2013). Furthermore, a number of these species may gain from enhanced feeding conditions due to increased benthic productivity, as well as the higher flux of plankton close to the sea surface while sheltered by the turbines (Figure 5.5). Meanwhile, foundations are likely to function primarily as feeding habitats for larger mobile and pelagic species (Klima & Wickham 1971; Buckley & Hueckel 1985). De Troch *et al.* (2013) attempted to estimate the importance of food resources to explain the tendency of Atlantic Cod and Pouting to aggregate around turbines. Stomach analyses indicated that these fish primarily fed on the amphipod *Jassa herdmani* and the Porcelain Crab *Pisidia longicornis* located on the turbines, and energy profiling suggested that the turbines constituted patches of relatively favourable feeding grounds for the species.

Ecological change

Offshore wind turbines provide habitat for a number of species and can aggregate fish assemblages and alter the local fish community structure, as represented by the increase in biomass of the different trophic groups, changes to energy transfer and the increased complexity of the trophic linkages (Figure 5.5). As suggested by the increased connectivity in Figure 5.5, local biodiversity can therefore be expected to increase through the presence of typical reef-dwelling fish in areas otherwise largely devoid of hard-bottom habitat. The consequent effect on the production of fish biomass is, however, unclear. Although fish may reproduce and grow on and around the turbines, this may simply represent an aggregation of the bioproduction of that species, and the association with turbines may not lead to net added production over the wider area owing to the conditions being more favourable for survival, growth rates and/or reproduction compared with alternative and natural habitats. For some fish species that are clearly limited in terms of distribution and abundance by the availability of reef habitat for refuge, territory, food and behavioural requirements, and for heavily fished and vulnerable species, the habitat and protection provided by wind turbines may serve as total biomass enhancers. This is, however, considered to be of little relevance at the population level for most species (Bohnsack 1989; Wilhelmsson 2009), although the importance of this is likely to depend on the location, spatial scale and time schedule of OWF development.

OWFs may also influence local predator–prey dynamics. Below the surface in the surrounding benthodemersal zone, several species appear to be attracted to and benefit from the physical presence of turbines, whereas local abundances of other species may be negatively affected. It has been suggested that the densities of some benthic prey items can decrease with proximity to artificial structures owing to predation by fish attracted to and resident on the structures, of which some could be new to the area, and that deployment of these structures could even cause the elimination of prey species in adjacent areas (Davis *et al.* 1982; Kurz 1995; Jordan *et al.* 2005).

Reef-dwelling species can be limited in their distribution by too large distances between hard-bottom areas. Wind turbines could facilitate the establishment of the new species in the recipient region, including invasive non-native species, by providing 'beach heads' and 'stepping stones' for spread, as suggested by Wilhelmsson and Malm (2008) and Adams *et al.* (2014). This possibility has mostly been considered in relation to invertebrates with free-drifting larvae (Miller *et al.* 2014), but may also be applied to some fish species that are dependent on hard-bottom and/or shallow habitats during parts of their life cycles.

Within the water column and towards the surface, the change in fish abundance and distribution can also play a role in changes to predator–prey dynamics. Decreased numbers of fish prey have been shown to cause localised reductions in feeding by the seabird Little Tern *Sternula albifrons*, which was then linked to egg abandonment by adults and low levels of chick hatching (Perrow *et al.* 2011). Directed movement behaviour in Harbour Seal *Phoca vitulina* and Grey Seal *Halichoerus grypus* has been associated with the presence of prey around multiple turbines (Russell *et al.* 2014; Nehls *et al.*, Chapter 6). The few studies on predator–prey relationships indicate that fish associated with OWFs are having an influence on the distribution of piscivorous predators in the local area.

As indicated in Figure 5.5, the submerged turbine component may affect the trophic web functioning in the area (Raoux *et al.* 2017). Wind turbines seem to offer particularly favourable substrates for colonising invertebrate filterers such as Blue Mussel and cirriped barnacles, principally *Semibalanus balanoides* (de Mesel *et al.* 2015; Dannheim *et al.*, Chapter 4), with the resultant sessile biota being dominated either by these groups or by assemblages comprised of anemones, hydroids and solitary sea squirts (Wilhelmsson

& Malm 2008; Maar *et al.* 2009; Ashley *et al.* 2014). The matrices of sessile biota on the wind turbines also harbour macroinvertebrates that constitute a potential food resource for associated fish (Reubens *et al.* 2013a). The fish and sessile organisms associated with the turbines may, further, contribute to increased benthic productivity around the turbines through the deposition of organic material, such as faecal matter, organic litter and dead organisms, which may, in turn, attract benthos-feeding, soft-sediment associated fish (Figure 5.5) (Wilhelmsson *et al.* 2006). The associated productivity is taken into account in the cyclical arrows shown in grey in Figure 5.5.

Fisheries exclusion and conservation zones

The fact that OWFs exclude or largely restrict fishing trawls within their footprint may be of importance for certain fish species, in particular heavily fished and relatively stationary vulnerable species, as suggested by Ashley *et al.* (2014). OWF areas, including their safety zones, may resemble 'no-take zones' (NTZs), which could lead to average increases in the biomass of fish for the area as a whole, although there will always be a risk of redirecting fishing pressure to other areas. Meta-analyses of the average effects of NTZs globally suggest that fish density and biomass may be doubled and tripled, respectively, along with increased body size of individuals and species diversity (Halpern 2003). Being average estimates and site-specific predictions in relation to certain species or species assemblages, these analyses are inevitably uncertain but plausible. Fish densities and potential spillover effects to adjacent areas may also increase with the size of the NTZ (Claudet & Pellitier 2004); hence, the size of the OWF may mirror this prediction.

The empirical evidence for similar effects of fisheries exclusion on fish assemblages in wind-farm areas is weak. Results from surveys targeting fish assemblages within an OWF area as a whole, including the area between the turbines, in Belgium, Denmark, the Netherlands and Sweden indicate either increased abundance of some species, such as sandeels (Ammodytidae), Atlantic Cod, Whiting *Merlangius merlangus* and Common Sole, or no effects (Leonhard *et al.* 2011; Lindeboom *et al.* 2011; Bergström *et al.* 2013; Stenberg *et al.* 2015; Vandendriessche *et al.* 2015). One cannot exclude, however, that the interpretation of these effects is disguised by a concentration of fish around the artificial structures, which was not picked up by the surveying methods used. In addition, the investigated OWF had been in operation for only a few years. While significant increases in fish density and species richness within the NTZ have been recorded after 3 years of protection (Halpern & Warner 2002; Russ *et al.* 2005; Claudet *et al.* 2006), other studies have shown that it can take decades before results are effectively measurable (Micheli *et al.* 2004).

The effect of the limitation of fisheries activity on a fish population as a whole depends on overall fishing pressure, the proportion of the fish population that uses the area, which, in turn, is dependent on species-specific fish distribution patterns and wind-farm size, and the duration of residence of the fish in the OWF area. For fish that move over larger areas and spend only some of their time within the OWF area, the protection offered is of less significance, whereas more stationary fish assemblages would be likely to benefit more.

Notably, marine management strategies combining protection from exploitation with artificial structures such as those provided within OWFs are increasingly being recognised (Pitcher *et al.* 2002; Claudet & Pelletier 2004). At the same time, while OWFs that have been designed and sited with no primary focus on conservation or fisheries enhancement appear potentially to provide these functions to varying degrees, the habitats and species protected may not necessarily be of any significance in terms of conservation, restoration or fisheries management, but could be beneficial from an ecological perspective by maintaining or enhancing some ecosystem functions or processes.

Furthermore, larger scale effects of climate change and ocean acidification are likely to cause a shift in the distribution of fish species, hence bringing into question the idea of static protected areas of the sea. OWFs may play a role in combination with other factors, but seem unlikely to provide some of the significant key benefits that are often predicted, although better understanding and the increasing areal extent of OWFs in regional seas may increase this likelihood.

Cumulative and in-combination effects

The response of the fish community to OWFs will occur over different spatial scales through time, as determined by life-history factors, appropriate habitat development, resource availability and the presence of fish species in adjacent areas to take advantage of the new habitat. The changes are expected to be similar to those observed with structures in the offshore oil and gas sector (Box 5.2).

Viewing the fish community associated with an OWF as being defined by the presence of the wind turbines and the response of fish to OWFs themselves does, however, ignore the multiple other factors that determine the presence, or indeed the absence, of fish in the OWF area. The marine environment has many other human influences and some of these will act to enhance or hinder the likelihood of specific fish species colonising OWFs, with this also varying through time. Such effects may be assessed as being minor in the short term, but become biologically significant over the longer term (e.g. Slabbekorn *et al.* 2010); these changes through time and over different spatial scales are defined as in-combination effects in this context. The knowledge base is inadequate to assess either in-combination effects or the cumulative effects of the installation of multiple wind farms on the marine environment in particular areas (Willsteed *et al.* 2017), such as the North and Baltic Seas in north-western Europe. Determination of longer term and more ecologically relevant effects requires data to be collected on fish, their functional habitats and relationship with other species. This perspective over greater spatial and time scales is particularly relevant to understanding the relationship between OWFs and fish when set within the context of the extensive plans for OWFs in a variety of different environments, including subtropical and tropical as well as temperate seas around the globe over the coming decades (see Jameson *et al.*, Chapter 1).

Box 5.2 Significant fish community productivity associated with oil and gas platforms according to Claisse *et al.* (2014).

A number of studies of oil and gas platform structures in the marine environment have reported changes to the fish community in terms of abundance, species presence and distribution, with this scenario being repeated around the world, including in the Gulf of Mexico (increased fish abundance) (Stanley & Wilson 1996), North Sea (persistent residency) (Jørgensen *et al.* 2002), Adriatic Sea (higher species richness and diversity) (Fabi *et al.* 2004) and Australia (early life-stage recruitment) (Neira 2005).

Clear evidence of how these artificial hard substrates can significantly increase fish productivity has been shown during systematic and repeated surveys of oil and gas platforms off the Californian coast (Claisse *et al.* 2014). The study focused on the determination of secondary production, defined by the authors as "...new animal

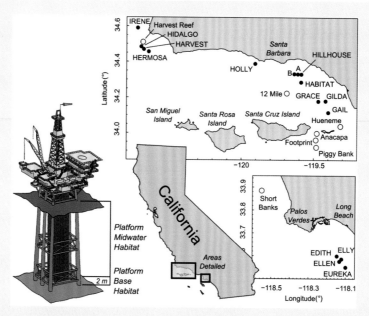

Figure 5.11 Platform diagram and map of the study area. The platform midwater habitat encompasses the hard substrate of the platform structure from the water surface to 2 m above the seafloor, whereas the platform base habitat is the bottom 2 m of the platform structure. The platform structure consists of outer vertical pilings and horizontal crossbeams as the platform jacket, with the vertical oil and gas conductors in the centre. Note that this is a general display diagram and the designs of these structures vary from platform to platform. The 16 platforms (filled circles; names in all capital letters) and seven natural reefs (open circles) used in the study were surveyed for at least 5 and up to 15 years between 1995 and 2011. (Reproduced from Claisse *et al.* 2014 with permission)

biomass from growth for all individuals in a given area during some period of time". The key reason for assessing secondary production, focusing on the fish community, was to take into account the multiple characteristics of the fish community based on density, body size, growth and survivorship in a single metric to represent ecological function. This enabled the authors to compare the annual secondary production of fish communities on the oil and gas platforms surveyed with those on southern Californian natural reefs (Figure 5.11) and with secondary production estimates from other marine ecosystems.

A productivity model was developed based on density and size structure data of fish recorded by annual observations between 5 and 15 years at each site. The total secondary production was made up of what the authors termed 'somatic' and 'recruitment' production, which predicted biomass in the subsequent year based on species growth and mortality and post-larval/juvenile growth and production, respectively (for details see Claisse *et al.* 2014). The productivity metric was then scaled per square metre of seabed for each entire platform surveyed.

While neither oil and gas platforms nor OWFs were designed with fish productivity in mind, Claisse *et al.* (2014) clearly show that the fish community associated with the platform significantly increased the local productivity, for all platform types, compared with the natural rocky reefs inhabited by the fish concerned (Figure 5.12). The principal reason for this huge boost in local productivity

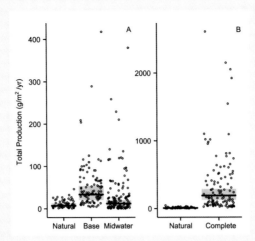

Figure 5.12 Annual total production: (a) annual production values are scaled to per square metre of habitat for natural reefs (*n*=56), platform habitat subtypes including base (*n*=111) and midwater (*n*=132); (b) annual production values are scaled to per square metre of seafloor for natural reefs (*n*=56) and complete platforms (*n*=111). Circles indicate individual data points and are jittered for visibility. Horizontal lines show the back-transformed estimated marginal means. The shaded box represents the 95% confidence intervals (CIs) of the mean. Differences were considered significant if the 95% CIs of their marginal means did not overlap. (Reproduced from Claisse *et al.* 2014 with permission)

is considered to be a result of the large increase in availability of structural habitat for different fish species and different life stages for a small number of dominant species. The authors suggest that their estimates may be conservative, as recruitment variability will play a large role in determining secondary production over time. They highlight that it is necessary to understand the local and regional oceanography responsible for larval fish delivery in relation to how the location of a structure will influence fish production.

Claisse *et al.* (2014) show that the complex hard substrate created by oil and gas platforms provides habitat for fish throughout the water column, leading to the highest estimate of secondary production per unit area of seabed of any comparable marine ecosystem studied so far. While oil and gas platforms are similar to offshore wind farms in terms of creating a hard, complex substrate, there is a difference in terms of the number of similar structures within a relatively small area of the sea. How this difference will affect fish community productivity remains an intriguing and as yet unanswered question.

Concluding remarks

To date, northern European waters have seen the greatest deployment of OWFs, and knowledge of the various disturbance effects on fish associated with OWFs is being increasingly substantiated owing to the realisation of several long-term monitoring programmes along with targeted studies and experiments. However, the majority of studies have been limited to single species, and tend to contribute primary data on the influence of OWFs on fish at the scale of a single installation or a small number of foundations. Thus, while the fish community evidently responds to the presence of OWFs, manifested as changing abundance, species assemblages, spatial and temporal distribution and movement or migration (see Perrow *et al.* 2011; Bergström *et al.* 2013; Reubens *et al.* 2013a) (Box 5.1), the scientific methods and approaches used have not yet generally been able to determine whether biological or ecologically significant changes have occurred, as previously noted for oil and gas platforms (Box 5.2).

Considerable knowledge gaps regarding the nature and longevity of impacts thus remain, particularly in relation to the scale of direct effects and the ecological outcomes of indirect effects on fish assemblages associated with OWFs. Continued, targeted and enhanced monitoring of species-specific attributes during construction and operation to recognise more reliably any 'positive' or 'negative' (adverse) impacts on the fish community is required to gauge properly the consequences of OWF developments (Lindeboom *et al.* 2015). The need for further research also requires answers to the question of how the anticipated positive or negative effects manifest themselves, such as increases or decreases in abundance, population size or species size structure and productivity. Without answering this question a precautionary approach to OWF may be seen as the most appropriate action, as the changes could put species assemblages of conservation interest at risk. However, it may well be that the local fish community could receive trophic benefits from increased biodiversity associated with OWFs, as suggested by Punt *et al.* (2009) and Inger *et al.* (2010). The potential ecological changes associated with OWFs as summarised in Figure 5.4 and Figure 5.5 are based on current understanding of the trophic relations between the primary production and organic sources, the primary consumers, different fish consumer groupings and predators within waters where OWFs are deployed. Ecological impact is expected to result from changes in the fish assemblages, an increase or decrease in the productivity of the fish groups (linked to individual fish size and abundance) and altered distribution associated with new habitat opportunities and refuges.

Critically, future research efforts need to be scaled from individual or a few turbines to entire OWFs and across several OWFs at a cumulative scale to measure ecosystem change, as any effects could be more relevant or pronounced than the summarised effects of individual turbines. An ecosystem-based perspective across fish distributional ranges is thought necessary to provide the knowledge at a scale relevant to both the management of marine systems and the fish themselves (Lindeboom *et al.* 2015). Ecosystem change is a difficult topic to address, but that should not be a reason to shy away from it. In fact, legislative existing environmental drivers, including the Marine Strategy Framework Directive in Europe, require the ecosystem approach to be applied for appropriate monitoring and management of the marine environment. Placing any changes observed within such an ecosystem context provides a clear picture of knowledge gaps, which can subsequently be filled by targeted research (Lindeboom *et al.* 2015). Furthermore, taking an ecosystem view conveys the need to take into account cause-and-effect pathways and the processes that cause or are predicted to cause changes to the ecosystem components. It is these effects which, if large enough and at a scale relevant to the species being studied, will be deemed to be significant and become impacts (*sensu* Wilding *et al.* 2017).

In conclusion, the overall question of whether OWFs are ecologically beneficial for fish remains open to debate. However, there is a requirement to consider this question because of existing local, regional and international legislation. But, rather than looking at the conflicts, perhaps a more productive, diversified and collaborative route should embrace the notion that OWFs are in general a good thing in the context of combating the global issue of fossil fuels contributing to climate change, while acknowledging that they will change the local environment and hence the fish community. This is with the caveat that we should try to improve our understanding of the ecological relevance of these changes and how they occur through targeted research and monitoring, and then consider alternatives. As OWFs have already been installed and new developments are being planned and constructed around the world, there is an opportunity for collaborative research efforts at the required ecosystem scale.

Acknowledgement

We thank the editor, Martin Perrow, for inviting us to write this chapter and for his comments, which helped to shape it.

References

Adams, T.P., Miller, R.G., Aleynik, D. & Burrows, M.T. (2014) Offshore marine renewable energy devices as stepping stones across biogeographic boundaries. *Journal of Applied Ecology* 51: 330–338.

Ambrose, R.F. & Swarbrick, S.L. (1989) Comparison of fish assemblages on artificial and natural reefs off the coast of southern California. *Bulletin of Marine Science* 44: 718–733.

Ashley, M.C., Mangi, S.C. & Rodwell, L.D. (2014) The potential for offshore windfarms to act as marine protected areas – a systematic review of current evidence. *Marine Policy* 45: 301–309.

Bergström, L., Sundqvist, F. & Bergström, U. (2013) Effects of an offshore wind farm on temporal and spatial patterns in the demersal fish community. *Marine Ecology Progress Series* 485: 199–210.

Boehlert, G.W. & Gill, A.B. (2010) Environmental and ecological effects of ocean renewable energy development – a current synthesis. *Oceanography* 23: 68–81.

Bohnsack, J.A. (1989) Are high densities of fishes on artificial reefs the result of habitat limitation or behavioural preference? *Bulletin of Marine Science* 44: 934–941.

Bohnsack, J.A. & Sutherland, D.L. (1985) Artificial reef research: a review with recommendations for future priorities. *Bulletin of Marine Science* 37: 11–39.

Buckley, R.M. & Hueckel, G.J. (1985) Biological processes and ecological development on an artificial reef in Puget Sound, Washington. *Bulletin of Marine Science* 37: 50–69.

Chojnacki, J. (2000) Environmental effects of artificial reefs in the southern Baltic (Pomeranian Bay). In Jensen, A.C., Collins, K.J. & Lockwood, A.P.M.L. (eds) *Artificial Reefs in European Seas.* Dordrecht: Kluwer, pp. 307–317.

Claisse, J.T., Pondella, D.J., Love, M., Zahn, L.A., Williams, C.M., Williams, J.P. & Bull, A.S. (2014) Oil platforms off California are among the most productive marine fish habitats globally. *Proceedings of the National Academy of Sciences of the United States of America* 111: 15462–15467.

Claudet, J. & Pelletier, D. (2004) Marine protected areas and artificial reefs: review of the interactions between management and science. *Aquatic Living Resources* 17: 129–138.

Claudet, J., Pelletier, D., Jouvenel, J.-Y., Bachet, F. & Galzin, R. (2006) Assessing the effects of marine protected area (MPA) on a reef fish assemblage in a northwestern Mediterranean case study: identifying community-based indicators. *Biological Conservation* 130: 349–369.

Collin, S.P. & Whitehead, D., (2004) The functional roles of passive electroreception in non-electric fishes. *Animal Biology* 54: 1–25.

Copping, A., Sather, N., Hanna, L., Whiting, J., Zydlewski, G., Staines, G., Gill, A., Hutchison, I., O'Hagan, A., Simas, T., Bald, J., Sparling, C., Wood, J. & Masden, E. (2016) *Annex IV 2016 State of the Science Report: Environmental Effects of Marine Renewable Energy Development Around the World.* Ocean Eneergy Systems. Retrieved 18 May 2018 from https://tethys.pnnl.gov/publications/state-of-the-science-2016

Couperus, B., Winter, E., van Keeken, O., van Koten, T., Tribuhl, S. & Burggraaf, D. (2010) Use of high resolution sonar for near-turbine fish observations (DIDSON). IMARES Wageningen UR. Report No. C138/10. Retrieved 18 May 2018 from https://tethys.pnnl.gov/sites/default/files/publications/Couperus-et-al-2010.pdf

Davis, N., VanBlaricom, G.R. & Dayton, P.K. (1982) Man-made structures on marine sediments: effects on adjacent benthic communities. *Marine Biology* 70: 295–303.

de Mesel, I., Kerckhof, F., Norro, A., Rumes, B. & Degraer, S. (2015) Succession and seasonal dynamics of the epifauna community on offshore wind farm foundations and their role as stepping stones for non-indigenous species. *Hydrobiologia* 756: 37–50.

de Troch, M., Reubens, J.T., Heirman, E., Degraer, S. & Vincx, M. (2013) Energy profiling of demersal fish: a case study in wind farm artificial reefs. *Marine Environmental Research* 92: 224–233.

Fabi, G., Grati, F., Puletti, M. & Scarcella, G. (2004) Effects on fish community induced by installation of two gas platforms in the Adriatic Sea. *Marine Environment Progress Series* 273: 187–197.

Frank, K.T., Petrie, B., Shackell, N.L. & Choi, J.S. (2006) Reconciling differences in trophic control in mid-latitude marine ecosystems. *Ecology Letters* 9: 1096–1105.

Gill, A.B. (2005) Offshore renewable energy – ecological implications of generating electricity in the coastal zone. *Journal of Applied Ecology* 42: 605–615.

Gill, A.B., Bartlett, M. & Thomsen, F. (2012a) Potential interactions between diadromous fishes of UK conservation importance and the electromagnetic fields and subsea noise from marine renewable energy developments. *Journal of Fish Biology* 81: 664–695.

Gill, A.B., Huang, Y., Spencer, J. & Gloyne-Phillips, I. (2012b) *Electromagnetic Fields Emitted by High Voltage Alternating Current Offshore Wind Power Cables and Interactions with Marine Organisms*. Liverpool: Institution of Engineering and Technology, IET Electromagnetics in Current and Emerging and Power Systems, University of Liverpool.

Gill, A.B., Gloyne-Phillips, I., Kimber, J.A. & Sigray, P. (2014) Marine renewable energy, electromagnetic fields and EM-sensitive animals. In Shields, M. & Payne, A. (eds) *Humanity and the Sea: Marine renewable energy and the interactions with the environment*. Dordrecht: Springer. pp. 61–79.

Greenstreet, S.P.R. & Hall, S. (1996) Fishing and the ground-fish assemblage structure in the northwestern North Sea: an analysis of long-term and spatial trends. *Journal of Animal Ecology* 65: 577–598.

Grove, R.S., Sonu, C.J. & Nakamura, M. (1991) Design and engineering of manufactured habitats for fisheries enhancement. In Seaman, W. & Sprague, L.M. (eds) *Artificial Habitats for Marine and Freshwater Fisheries*. San Diego, CA: Academic Press. pp. 109–149.

Halpern, B.S. (2003) The impact of marine reserves: do reserves work and does reserve size matter? *Ecological Applications* 13: 117–137.

Halpern, B.S. & Warner, R.R. (2002) Marine reserves have rapid and lasting effects. *Ecology Letters* 5: 361–366.

Harriman, J.A.E. & Noble, B.F. (2008) Characterizing project and strategic approaches to regional cumulative effects assessment in Canada. *Journal of Environmental Assessment Policy and Management* 10: 25–50.

Hawkins, A.D. & Popper, A.N. (2016) A sound approach to assessing the impact of underwater noise on marine fishes and invertebrates. *ICES Journal of Marine Science* 74: 635–651.

Hunter, W.R. & Sayer, M.D.J. (2009) The comparative effects of habitat complexity on faunal assemblages of northern temperate artificial and natural reefs. *ICES Journal of Marine Science* 66: 691–698.

Inger, R., Attrill, M.J., Bearhop, S., Broderick, A.C., James Grecian, W., Hodgson, D.J., Mills, C., Sheehan, E., Votier, S.C., Witt, M.J. & Godley, B.J. (2009) Marine renewable energy: potential benefits to biodiversity? An urgent call for research. *Journal of Applied Ecology* 46: 1145–1153.

Jensen, A. (2002) Artificial reefs of Europe: perspective and future. *ICES Journal of Marine Science* 59: 3–13.

Jessee, W.N., Carpenter, A.L. & Carter, J.W. (1985) Distribution patterns and density estimates of fishes on a Southern California artificial reef with comparisons to natural kelp-reef habitats. *Bulletin of Marine Science* 37: 214–226.

Jordan, K.K.B., Gilliam, D.S. & Spieler, R.E. (2005) Reef fish assemblage structure affected by small-scale spacing and size variations of artificial patch reefs. *Journal of Experimental Marine Biology and Ecology* 326: 170–186.

Jørgensen, T., Løkkeborg, S. & Soldal, A.V. (2002) Residence of fish in the vicinity of a decommissioned oil platform in the North Sea. *ICES Journal of Marine Science* 59 (Suppl): S288–S293.

Kim, C.G., Lee, J.W. & Park, J.S. (1994) Artificial reef designs for Korean coastal waters. *Bulletin of Marine Science* 55: 858–866.

Klima, E.F. & Wickham, D.A. (1971) Attraction of coastal pelagic fishes with artificial structures. *Transactions of the American Fisheries Society* 199: 86–89.

Kurz, R.C. (1995) Predator–prey interactions between gray triggerfish (*Balistes capriscus gmelin*) and a guild of sand dollars around artificial reefs in the northeastern Gulf-of-Mexico. *Bulletin of Marine Science* 56: 150–160.

Lan, C.-H. & Hsui, C.-Y. (2006) The development of artificial reef ecosystem: modelling, simulation and application. *Simulation Modelling Practice and Theory* 14: 663–675.

Leonhard, S., Stenberg, C. & Støttrup, J. (eds) (2011) Effect of the Horns Rev 1 offshore wind farm on fish communities follow-up seven years after construction. DTU Aqua, Orbicon, DHI, NaturFocus. Report commissioned by The Environmental Group through contract with Vattenfall Vindkraft A/S. Retrieved 18 May 2018 from http://www.aqua.dtu.dk/-/media/Institutter/Aqua/Publikationer/Forskningsrapporter_201_250/246_2011_effect_of_the_horns_rev_1_offshore_wind_farm_on_fish_communities.ashx?la=da

Lindeboom, H.J., Kouwenhoven, H.J., Bergman, M.J.N., Bouma, S., Brasseur, S., Daan, R., Fijn, R.C., de Haan, D., Dirksen, S., van Hal, R., Lambers, R.H.R., ter Hofstede, R., Krijgsveld, K.L., Leopold, M. & Scheidat, M. (2011) Short-term ecological effects of an offshore wind farm in the Dutch coastal zone; a compilation. *Environmental Research Letters* 6: 035101.

Lindeboom, H., Degraer, S., Dannheim, J., Gill, A.B. & Wilhelmsson, D. (2015) Offshore wind park monitoring programmes, lessons learned and recommendations for the future. *Hydrobiologia* 756: 169–180.

Maar, M., Bolding, K., Petersen, J.K., Hansen, J. & Timmerman, K. (2009) Local effects of blue mussels around turbine foundation in an ecosystem model of Nysted offshore wind farm Denmark. *Journal of Sea Research* 63: 159–174.

Micheli, F., Halpern, B.S., Botsford, L.W. & Warner, R.R. (2004) Trajectories and correlates of community change in no-take marine reserves. *Ecological Applications* 14: 1709–1723.

Miller, R.G., Hutchison, Z.L., Macleod, A.K., Burrows, M.T., Cook, E.J., Last, K.S. & Wilson, B. (2013) Marine renewable energy development: assessing the benthic footprint at multiple scales. *Frontiers in Ecology and the Environment* 11: 433–440.

Milon, J.W. (1989) Artificial marine habitat characteristics and participation behavior by sport anglers and divers. *Bulletin of Marine Science* 44: 853–862.

Moffit, R.B., Parrish, F.A. & Polovina, J.J. (1989) Community structure, biomass and productivity of deepwater artificial reefs in Hawaii. *Bulletin of Marine Science* 44: 616–630.

Mumby, P.J., Dahlgren, C.P., Harborne, A.R., Kappel, C.V., Micheli, F., Brumbaugh, D.R., Holmes, K.E., Mendes, J.M., Broad, K., Sanchirico, J.N., Buch, K., Box, S., Stoffle, R.W. & Gill, A.B. (2006) Fishing, trophic cascades, and the process of grazing on coral reefs. *Science* 311: 98–101.

Munday, P.L., Dixson, D.L., McCormick, M.I., Meekan, M., Ferrari, M.C. & Chivers, D.P. (2010) Replenishment of fish populations is threatened by ocean acidification. *Proceedings of the National Academy of Sciences of the United States of America* 107: 12930–12934.

Myers, R., Hutchings, J. & Barrowman, N. (1996) Hypotheses for the decline of cod in the North Atlantic. *Marine Ecology Progress Series* 138: 293–308.

Nedwell, J.R., Turnpenny, A.W., Lovell, J.M. & Edwards, B. (2006) An investigation into the effects of underwater piling noise on salmonids. *Journal of the Acoustical Society of America* 120: 2550–2554.

Neira, F.J. (2005) Summer and winter plankton fish assemblages around offshore oil and gas platforms in south-eastern Australia. *Estuarine, Coastal and Shelf Science* 63: 589–604.

Normandeau Associates, Exponent Inc., Tricas, T. & Gill, A. (2011) *Effects of EMFs from Undersea Power Cables on Elasmobranchs and Other Marine Species.* OCS Study BOEMRE 2011-09. Camarillo, CA: US Department of the Interior, Bureau of Ocean Energy Management, Regulation, and Enforcement, Pacific OCS Region.

Orr, J.C., Fabry, V.J., Aumont, O., Bopp, L., Doney, S.C., Feely, R.A., Gnanadesikan, A., Gruber, N., Ishida, A., Joos, F. & Key, R.M. (2005) Anthropogenic ocean acidification over the twenty-first century and its impact on calcifying organisms. *Nature* 437: 681–686.

Perrow, M.R., Gilroy, J.J., Skeate, E.R. & Tomlinson, M.L. (2011) Effects of the construction of Scroby Sands offshore wind farm on the prey base of little tern *Sternula albifrons* at its most important UK colony. *Marine Pollution Bulletin* 62: 1661–1670.

Pickering, H., Whitmarsh, D. & Jensen, A. (1998) Artificial reefs as a tool to aid rehabilitation of coastal ecosystems: investigating the potential. *Marine Pollution Bulletin* 37: 505–414.

Pitcher, T.J., Buchary, E.A. & Hutton, T. (2002) Forecasting the benefits of no-take human-made reefs using spatial ecosystem simulation. *ICES Journal of Marine Science* 59: 17–26.

Popper, A.N. (2000) Hair cell heterogeneity and ultrasonic hearing: recent advances in understanding fish hearing. *Philosophical Transactions of the Royal Society B: Biological Sciences* 355: 1277–1280.

Punt, M.J., Groeneveld, R.A., Van Ierland, E.C. & Stel, J.H. (2009) Spatial planning of offshore wind farms: a windfall to marine environmental protection? *Ecological Economics* 69: 93–103.

Raoux, A., Tecchio, S., Pezy, J.P., Lassalle, G., Degraer, S., Wilhelmsson, D., Cachera, M., Ernande, B., Le Guen, C., Haraldsson, M. & Grangeré, K. (2017) Benthic and fish aggregation inside an offshore wind farm: which effects on the trophic web functioning? *Ecological indicators* 72: 33–46.

Reubens, J.T., Degraer, S. & Vincx, M. (2011) Aggregation and feeding behaviour of pouting (*Trisopterus luscus*) at wind turbines in the Belgian part of the North Sea. *Fisheries Research* 108: 223–227.

Reubens, J.T., Vandendriessche, S., Zenner, A.N., Degraer, S. & Vincx, M. (2013a) Offshore wind farms as productive sites or ecological traps for gadoid fishes? Impact on growth, condition index and diet composition. *Marine Environmental Research* 90: 66–74.

Reubens, J.T., Pasotti, F., Degraer, S. & Vincx, M. (2013b) Residency, site fidelity and habitat use of Atlantic cod (*Gadus morhua*) at an offshore wind farm using acoustic telemetry. *Marine Environmental Research* 90: 128–135.

Reubens, J.T., Degraer, S. & Vincx, M. (2014a) The ecology of benthopelagic fishes at offshore wind farms: a synthesis of 4 years of research. *Hydrobiologia* 727: 121–136.

Reubens, J.T., De Rijcke, M., Degraer, S. & Vincx, M. (2014b) Diel variation in feeding and movement patterns of juvenile Atlantic cod at offshore wind farms. *Journal of Sea Research* 85: 214–221.

Righton, D., Quayle, V.A., Hetherington, S. & Burt, G. (2007) Movements and distribution of cod (*Gadus morhua*) in the southern North Sea and English Channel: results from conventional and electronic tagging experiments. *Journal of the*

Marine Biological Association of the United Kingdom 87: 599–613.

Russ, G.R., Stockwell, B. & Alcala, A.C. (2005) Inferring versus measuring rates of recovery in no-take marine reserves. *Marine Ecological. Progress Series* 292: 1–12.

Russell, D.J., Brasseur, S.M., Thompson, D., Hastie, G.D., Janik, V.M., Aarts, G., McClintock, B.T., Matthiopoulos, J., Moss, S.E. & McConnell, B. (2014) Marine mammals trace anthropogenic structures at sea. *Current Biology* 24: R638–R639.

Seaman, W. (2007) Artificial habitats and the restoration of degraded marine ecosystems and fisheries. *Hydrobiologia* 580: 143–155.

Slabbekoorn, H., Bouton, N., van Opzeeland, I., Coers, A., ten Cate, C. & Popper, A.N. (2010) A noisy spring: the impact of globally rising underwater sound levels on fish. *Trends in Ecology & Evolution* 25: 419–427.

Stanley, D.R. & Wilson, C.A. (1996) Abundance of fishes associated with a petroleum platform as measured with dual-beam hydroacoustics. *ICES Journal of Marine Science* 53: 473–475.

Stenberg, C., Støttrup, JG., van Deurs, M., Berg, C.W., Dinesen, G.E., Mosegaard, H., Grome, T.M. & Leonhard, S.B. (2015) Long-term effects of an offshore wind farm in the North Sea on fish communities. *Marine Ecology Progress Series* 528: 257–265.

Thomsen, F., Gill, A.B., Kosecka, M., Andersson, M.H., Andre, M., Degraer, S., Folegot, T., Gabriel, J., Judd, A., Neumann, T., Norro, A., Risch, D., Sigray, P., Wood, D. & Wilson, B. (2015) Study of the environmental impacts of noise, vibrations and electromagnetic emissions from marine renewables. Final Report to European Commission, Directorate-General for Research and Innovation. RTD-K3-2012-MRE. Retrieved 18 May 2018 from https://publications.europa.eu/en/publication-detail/-/publication/01443de6-effa-11e5-8529-01aa75ed71a1/language-en

Turner, K., Righton, D. & Metcalfe, J.D. (2002) The dispersal patterns and behaviour of North Sea cod (*Gadus morhua*) studied using electronic data storage tags. *Hydrobiologia* 483: 201–208.

van Ginneken, V.J.T. & Maes, G.E. (2005) The European eel (*Anguilla anguilla*, Linnaeus), its lifecycle, evolution and reproduction: a literature review. *Reviews in Fish Biology and Fisheries* 15: 367–398.

Vandendriessche, S., Derweduwen, J. & Hostens, K. (2015) Equivocal effects of offshore windfarms in Belgium on soft substrate epibenthos and fish assemblages. *Hydrobiologia* 756: 19–35.

Voegeli, F.A., Smale, M.J., Webber, D.M., Andrade,Y. & O'Dor, R.K. (2001) Ultrasonic telemetry, tracking and automated monitoring technology for sharks. *Environmental Biology of Fishes* 60: 267–281.

Westerberg, H. & Lagenfelt, I. (2008) Sub-sea power cables and the migration behaviour of the European eel. *Fisheries Management and Ecology* 15: 369–375.

Wilding, T., Gill, A.B., Boon, A., Sheehan, E., Dauvin, J.-C., Pezy, J.-P., O'Beirn, F., Janas, U., Rostin, L. & De Mesel, I. (2017) Turning off the DRIP ('data-rich, information-poor') – rationalising monitoring with a focus on marine renewable energy developments and the benthos. *Renewable and Sustainable Energy Reviews* 74: 848–859.

Wilhelmsson, D. (2009) Aspects of offshore renewable energy and the alterations of marine habitats. PhD thesis, Stockholm University.

Wilhelmsson, D. & Malm, T. (2008) Fouling assemblages on offshore wind power plants and adjacent substrata. *Estuarine Coastal and Shelf Science* 79: 459–466.

Wilhelmsson, D., Öhman, M.C., Ståhl, H. & Shlesinger, Y. (1998) Artificial reefs and dive tourism in Eilat, Israel. *Ambio* 27: 764–766.

Wilhelmsson, D., Malm, T. & Öhman, M. (2006) The influence of offshore wind power on demersal fish. *ICES Journal of Marine Science* 63: 775–784.

Wilhelmsson, D., Malm, T., Thompson, R., Tchou, J., Sarantakos, G., McCormick, N., Luitjens, S., Gullström, M., Patterson Edwards, J.K., Amir, O. & Dubi, A. (2010) *Greening Blue Energy: Identifying and managing the biodiversity risks and opportunities of offshore renewable energy*. Gland, Switzerland: IUCN.

Willsteed, E., Gill, A.B., Birchenough, S.N.R. & Jude, S. (2017) Assessing the cumulative environmental effects of marine renewable energy developments: establishing common ground. *Science of the Total Environment* 577: 19–32.

Winter, H., Aarts, G. & van Keeken, O.A. (2010) Residence time and behaviour of sole and cod in the offshore wind farm Egmond aan Zee (OWEZ). IMARES Report C038/10. Report No. OWEZ_R_265_T1_20100916. IMARES IJmuide.

Worzyk, T. (2009) *Submarine Power Cables: Design, installation, repair, environmental aspects*. Berlin: Springer.

CHAPTER 6

Marine mammals

GEORG NEHLS, ANDREW J.P. HARWOOD and
MARTIN R. PERROW

Summary

Marine mammals naturally occur in probably every offshore wind farm (OWF) worldwide. During construction, operation and decommissioning they are exposed to various pressures, of which the high levels of noise produced during pile driving of foundations into the seabed are the strongest, although other sources of noise including vessels may also have some effect. Noise may result in auditory effects with disturbance and resultant displacement, sound masking and even hearing loss, as well as non-auditory effects such as stress and physical injury. Other possible effects include the risk of collision with service vessels and changes to available habitats, including changes in hydrodynamics. Conversely, reef and refuge effects may actively benefit some species, owing to increases in prey associated with wind-farm underwater structures or reduction in fishing pressure and disturbance. Studies have tended to focus on a limited range of more abundant species in European shelf seas, including Harbour Porpoise and Grey and Harbour Seals. However, given increases in the geographic extent of developments and potential for developments in deeper water, a greater range of species may become the focus of future studies. A number of studies have demonstrated disturbance and partial displacement of Harbour Porpoises up to distances of about 20 km during pile-driving activities, which are reversible within 1–3 days. Operational wind farms are frequently used by Harbour Porpoise and both seal species and while some attraction has been shown for seals, overall impacts on marine mammals during operation seem to be low. Understanding of the impacts from OWFs and their interaction with other pressures and cumulative impacts, at both the individual and population levels, needs to be improved using appropriate studies and tools such as population modelling to link impacts to changes in vital rates.

Introduction

The polyphyletic marine mammal group is comprised of 130 species (Society for Marine Mammalogy, Committee on Taxonomy 2016). All marine mammals are today exposed in some way to anthropogenic pressures, leading to the International Union for Conservation of Nature and Nature Resources (IUCN 2017) classifying more than 25% of species as vulnerable (11%), endangered (13%) or critically endangered (2%), primarily as a result of a wide range of impacts that have also changed over time (Magera *et al.* 2013). While in former times dedicated hunting was the main cause for declining marine mammal populations, more recently, fisheries by-catch and a wide range of indirect human pressures in the marine environment, such as water pollution, ocean acidification, climate change and overfishing of prey, have caused adverse impacts (Read 2008; Davidson *et al.* 2012).

Following the evolution of motor-powered shipping in the late eighteenth century, levels of anthropogenic noise in the marine environment increased steadily and now include a variety of sources; for example, sonar, seismic exploration using airguns and offshore installations such as oil and gas rigs. Noise apparently continues to increase in many places, as documented by a rise of around 10 dB in parts of the Pacific since the 1960s (Andrew *et al.* 2002). Cetaceans are often assumed to be susceptible to noise pollution as a result of their use of sound for communication, navigation, orientation, predator avoidance and prey detection (Wartzok & Ketten 1999), all exploiting the fact that sound can travel rapidly over long distances in water where other senses such as vision may be limited by murky or dark environments. Some marine mammals, such as the odontocete cetaceans, use high-frequency sounds (150 Hz to 180 kHz) generated by a complex system of air sacs and the melon, a lipid-filled sac in the forehead (Au 1993; Richardson *et al.* 1995). Echolocation provides a means of navigating through the environment and facilitates prey capture. In contrast, the very large pelagic baleen whales (mysticetes) do not use echolocation and instead produce low-frequency sounds in the range of 10–200 Hz (7 Hz to 22 kHz) to communicate or 'sing' over huge distances (Southall *et al.* 2007; Cranford & Krysl 2015). Differently from cetaceans, pinnipeds vocalise and hear both above and below the water, and need to balance abilities in both media to support vital behaviours (Reichmuth *et al.* 2013) although, according to Southall *et al.* (2007), pinnipeds appear to be sensitive to a broader range of frequencies in water (75 Hz to 75 kHz) than in air (75 Hz to 30 kHz).

The noise emitted from offshore wind farms (OWFs) is principally related to construction and is particularly loud if steel foundations are driven into the seabed by hydraulic hammers (pile driving). This occupies a key part of site-specific Environmental Impact Assessments (EIAs) as well as other regulatory processes concerned with likely effects upon species protected under international conventions and national legislation (Box 6.1). The main species that may be affected by OWFs in Europe, which is the current centre of OWF activity (see Jameson *et al.*, Chapter 1), are the more abundant species of shallow shelf seas, especially Harbour Porpoise *Phocoena phocoena* and Harbour Seal *Phoca vitulina*, as well as Grey Seal *Halichoerus grypus* and Bottlenose Dolphin *Tursiops truncatus* (Figure 6.1). Many cetacean species favouring deeper waters are precluded from current development areas, although in other locations such as found in the USA, species listed under the Endangered Species Act 1973, including the Northern Right Whale *Eubalaena glacialis*, Humpback Whale *Megaptera novaeangliae* and various rorquals, may come into focus (Bailey *et al.* 2014).

Box 6.1   Key legislation protecting marine mammals in Europe and the USA

In Europe, the key legislation for nature conservations is the Habitats Directive, Council Directive 92/43/EEC of 21 May 1992 on the conservation of natural habitats and of wild fauna and flora. All cetaceans are listed in Annex IV(a) of the Habitats Directive (EU 1992) and as such all Member States are required to establish a system of strict protection under Article 12(1). Bottlenose Dolphin *Tursiops truncatus*, Harbour Porpoise *Phocoena phocoena*, Harbour (or Common) Seal *Phoca vitulina* and Grey Seal *Halichoerus grypus* are also included as Annex II species requiring the designation of Special Areas of Conservation (SACs). With respect to offshore wind farms, Article 12 of the Habitats Directive is of special importance as it prohibits (a) all forms of deliberate capture or killing of specimens of these species in the wild and (b) deliberate disturbance of these species, particularly during the period of breeding, rearing, hibernation and migration. In the context of the directive, the term killing includes injury which might impair survival and is thus relevant in relation to noise immissions and hearing impairment of cetaceans. The Habitats Directive further requires Member States to conduct appropriate assessments of any project, which may affect protected areas of the Natura 2000 network.

Cetaceans further receive international protection under Appendix II or III of the Convention on the Conservation of European Wildlife and Natural Habitats (Bern Convention 1979). Sixteen species of cetacean are also protected under the Convention on the Conservation of Migratory Species of Wild Animals (Bonn or CMS Convention 1979, which came into force in 1985).

The international Agreement on the Conservation of Small Cetaceans in the Baltic and North Seas 1992 (ASCOBANS) and the Agreement on the Conservation of Cetaceans of the Black Sea, Mediterranean Sea and Contiguous Atlantic area 2001 (ACCOBAMS) provide further protection under the auspices of the Bonn Convention. The Convention for the Protection of the Marine Environment of the North-East Atlantic (OSPAR) has also prescribed a series of programmes of Ecological Quality Objectives (EcoQOs), such as a project aimed at reducing by-catch of Harbour Porpoises to below 1.7% of the best population estimate.

In the USA, the Marine Mammal Protection Act (MMPA) of 1972 as Amended (NMFS 2015) protects all marine species and makes it illegal to 'take', that is harass, feed, hunt, capture, collect or kill any marine mammals (or part of a marine mammal) without a permit as managed by the federal government. The Endangered Species Act (ESA) of 1973 (US Fish & Wildlife Service 2003) provides for the conservation of species that are endangered or threatened throughout all or a significant portion of their range, and the conservation of supporting ecosystems.

Scope

This chapter provides an overview of current knowledge regarding the effects and impacts of OWFs on marine mammals. A review was conducted through internet searches, using key search terms such as 'marine mammals', 'offshore wind farms', 'impact assessments', 'disturbance', 'noise' and 'piling', and combinations of these words. This further expanded the literature already held by the authors as a result of the experience and activities of their

Figure 6.1 Marine mammals typically considered in studies investigating impacts associated with offshore wind farms in Europe. Clockwise from top right: Harbour Porpoise *Phocoena phocoena*, Bottlenose Dolphin *Tursiops truncatus*, Harbour Seal *Phoca vitulina* and Grey Seal *Halichoerus grypus*. Effects on large whales, such as Humpback Whale *Megaptera novaeangliae* (top left), are likely to require consideration in other areas of the world or if developments are planned in deeper waters. (Martin Perrow)

consultant/research organisations in relation to marine mammals and OWFs. The review is not exhaustive, but aims to provide an overview of the state of knowledge at the time of writing, particularly on noise exposure, and to identify critical gaps in our understanding.

Wind-farm related processes that may affect marine mammals include seismic surveys to determine geological conditions; installation of foundations, particularly through noisy pile driving; increased vessel activity during surveys and construction, leading to disturbance and even possible collision; suspension of sediments and disturbance of the seabed, including through cable laying; release of pollutants; and decommissioning (Figure 6.2). The presence of turbines will also be associated with other pressures, such as noise and vibration and changed local hydrodynamics, but with the benefit of providing artificial reefs and sanctuary from fishing activity, both leading to the accumulation of potential prey

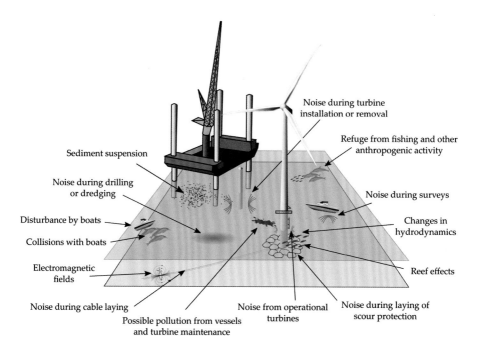

Figure 6.2 Range of effects that may influence marine mammals during the life of a wind farm.

(Figure 6.2). This suite of negative and positive effects form the basis of the *Themes* presented below, with the exclusion of disturbance of the seabed, which is generally assumed to be short term, and the release of pollutants, which has only rarely been documented (see Rees & Judd, Chapter 2) and with no demonstrable effect upon marine mammals. Nor does this review specifically consider other potential effects, such as electromagnetic fields from cables, scour protection, active and passive corrosion protection of the foundations, lighting of turbines and sonar transponders as navigational aids for submarines. All such factors are so far considered to cause no more than a negligible, if any, effect on marine mammals, but rather contribute to a changing environment in areas where OWFs are built.

The effects associated with wind-farm construction and decommissioning are thought to be numerous but short lived, while those associated with operation and maintenance are less variable but longer lived over the 25 years or more of the project. The *Themes* generally document effects sequentially throughout the lifetime of the project, although there is little information regarding the effects of decommissioning or repowering wind farms as only a few of the first wind farms, including Yttre Stengrund Wind Farm in Sweden (2015), Lely Wind Farm in the Netherlands (2016) and Vindeby in Denmark (2017), have been decommissioned (Topham & McMillan 2017).

The pathways responsible for any effects and impacts are detailed, with particular emphasis on noise-related effects and impacts due to the importance of hearing for many species (Box 6.2). The information presented has generally focused on a limited number of species that are of particular importance within Europe, particularly the Harbour Porpoise and Harbour Seal. As such, the chapter is naturally biased towards these two species but the experience gathered from studies on existing wind farms may provide valuable material to inform potentially critical projects in areas where such experience is not available (Tougaard & Mikaelsen 2017).

Finally, cumulative effects associated with the construction, operation or decommissioning of multiple OWFs within an effect range of each other are also considered.

Box 6.2 Potential effects and impacts of anthropogenic noise upon marine mammals

Potential effects and impacts of anthropogenic noise upon marine mammals range from negligible effects in relation to a single short disturbance event, to direct physical and even fatal injury. The latter may occur from exposure to very high levels of sound such as from shock waves from explosions damaging internal organs or air-filled body cavities (Young 1991). Noise thus may cause auditory and non-auditory effects.

A framework of potential impacts proposed by Richardson *et al.* (1995) has been widely adopted, including in wind-farm studies. This suggests four zones of auditory noise impact for marine mammals: (1) zone of audibility; (2) zone of responsiveness; (3) zone of masking; and (4) zone of hearing loss, discomfort and injury.

The zone of audibility is the threshold at which a noise is detectable above background noise and above the hearing threshold of an animal, while the zone of responsiveness refers to the threshold above which a behavioural reaction (positive or negative) to the noise is elicited. Above a certain noise level this may result in disturbance in form of cessation or modification of normal activities such as feeding, social interactions and vocalisations, and displacement including animals leaving the water in the case of seals (Richardson *et al.* 1995; Nowacek *et al.* 2007; Southall *et al.* 2007).

Masking occurs where noise inputs make vocalisations less likely to be detected by conspecifics or other species, thereby inhibiting the detection of predator, prey or environmental signals (Clark *et al.* 2009). Masking can occur at received levels below those required to generate a behavioural response. The extent of interference will be determined by spectral, temporal and spatial overlap between the masking noise and the sender/receiver.

A zone of hearing loss and injury may occur as a result of sound pressure high enough to cause trauma or loss of auditory senses. A temporary hearing threshold

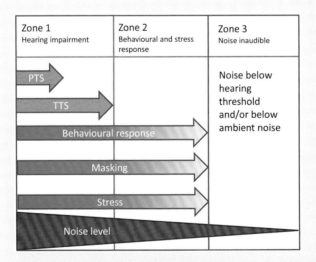

Figure 6.3 Effect zones of underwater noise in relation to distance to the source and thus received noise levels. PTS and TTS: permanent threshold and temporary threshold shifts in hearing, respectively. (Adapted from Gomez *et al.* 2016)

shift (TTS) is the result of a metabolic exhaustion of sensory cells in the cochlea caused by intense sounds (NMFS 2018). TTS may temporally reduce the animal's ability to perceive biologically significant sounds and therefore has a similar effect as masking, although TTS persists for a period after sound exposure during which both effects can occur (Richardson *et al.* 1995). Within limits, TTS is reversible (e.g. Nachtigall *et al.* 2003; 2004). Intense narrowband sounds affect hearing in similar frequency bands (up to an octave higher) whereas broadband sound has an impact mainly in a medium-frequency band of the sound spectrum (Schlundt *et al.* 2000; Knust *et al.* 2003; Kastak *et al.* 2005). In pinnipeds, it has been shown that an increase in noise duration has a greater effect on TTS than an increase in received level (Kastak *et al.* 2005).

Repeated, prolonged or chronic exposure to high sound levels as well as brief exposure to extremely loud noise or short signal rise times, such as in underwater explosions, can inflict structural damage to sensory cells, resulting in permanent hearing threshold shift (PTS) or acoustic trauma (Richardson *et al.* 1995).

It is important to acknowledge that the different zones overlap and not all zones may be applicable to a given noise source (Figure 6.3). Behavioural responses may occur over the whole range where noise is audible, although the response is usually related to the loudness. Within the range of audibility of a signal, two zones remain which are to be distinguished: a zone where hearing impairment and all other effects occur, and a zone where only behavioural and stress responses as well as masking occur.

Monitoring and mitigation of potential impacts fall outside the scope of this chapter, but are considered in further detail by Scheidat and Porter (Chapter 2) and by Thomsen and Verfuß (Chapter 7) in Volume 4 of this series.

Themes

Site preparations

Geophysical surveys to characterise the seabed before wind-farm development may use seismic devices, which are known to be disruptive to marine mammals (Gordon *et al.* 2003). However, as it is necessary to penetrate only a few metres into the substratum, low-energy systems such as pingers, parametric sub-bottom profilers and chirpers are used (Fugro Marine GeoServices Inc. 2017), the effects of which on marine mammals are generally seen to be trivial, in contrast with the high-intensity airguns required to search for oil and gas deposits (Gordon *et al.* 2003; Cerchio *et al.* 2014).

Before the start of construction of any OWF, the seabed is also checked for unexploded ordnance (UXO), which is either removed or exploded at the site. In Germany, the use of a bubble curtain is required to mitigate the noise impact of a detonated UXO, but if no mitigation measures are applied such explosions have a high potential to harm marine mammals. Given that UXO is frequently found in European waters, some authors (e.g. von Benda-Beckmann *et al.* 2015; Aarts *et al.* 2016) consider this to be an important potential impact, although no specific studies of actual effects appear to have been undertaken.

Before any piling activity begins, construction and guard vessels arrive at the site and acoustic harrassing devices (AHDs), such as pingers and seal scarers, may be used to deter marine mammals from the construction site in order to reduce the risk of hearing damage. In their study on seven German OWF projects, Brandt *et al.* (2018) observed significant decreases in detections of Harbour Porpoise at distances of up to 10 km before piling commenced. This was independent of piling or deterrence measures. The most likely explanation for this is the effect of increased shipping activity during preparation works (see *Effects of service vessels*, below), which was enhanced at low wind speeds by increasing sound propagation (see also Dragon *et al.* 2016). The effects of AHDs before and during piling were also exacerbated at lower wind speed, indicating that the effects of wind and sea state on sound propagation may be underestimated.

Seal scarers were originally designed to deter seals from fish farms, but are now the main deterrent used before offshore pile driving. They are rather loud and typically transmit sound in the frequency range 10–40 kHz, well within the range of best hearing in Harbour Porpoise, causing an aversion response at noise levels as low as 113 dB. The onset of a response is much lower compared to noise at lower frequencies and an aversion response has been recorded at distances of up to 7 km (Brandt *et al.* 2013; 2014). The effect of the seal scarer thus extends beyond the range needed to prevent hearing damage in Harbour Porpoise (see below), and in projects where noise mitigation during pile driving is applied, the seal scarer may become the dominant cause of porpoise displacement. The use of a specially developed porpoise-scarer may reduce such undesirable large effects in future (Kastelein *et al.* 2017a).

Potential for hearing damage from piling

Offshore turbines are placed on a variety of foundations and their potential impacts on marine mammals differ accordingly. Until now, monopiles, tripods or jacket foundations anchored in the seafloor by large steel piles, have been the most widely used (see Jameson *et al.*, Chapter 1) and it is these that generate significant noise (Box 6.3). The size of turbines has also increased considerably from the first commercial large-scale wind farm Horns Rev 1 in 2002, employing 2 MW turbines, to the usual standard of 5–6 MW today. The first 8 MW turbines have now been installed and 10 MW turbines are expected in the near future. Foundation size has increased accordingly from a monopile diameter of 1 m in the first projects to 8 m in recent projects. Gravity-base foundations and suction buckets have been installed in European wind farms and floating foundations are being developed for use in deeper water (see Jameson *et al.*, Chapter 1). These alternatives to monopiles

Box 6.3   Underwater noise from offshore pile driving

Most offshore turbines have been installed on steel foundations that are anchored by large steel piles driven into the seabed by hydraulic hammers. While for jacket and tripod foundations piles of a moderate diameter of 2–3 m are usually used, monopiles have reached diameters of 8 m and the use of 10 m monopiles has been envisaged. The piles are driven into the seabed to depths of 20–40 m with several thousand blows of up to 4,000 kJ at a frequency of 30–40 blows/minute.

Impact pile driving of large steel foundations radiates substantial levels of low-frequency impulsive noise into the water column, which can propagate over large distances. Pile-driving noise consists of short pulses of 0.1–0.3 seconds.

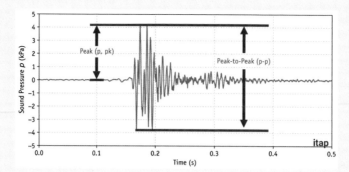

Figure 6.4 Underwater sound pressure impulse of a pile-driving blow, recorded at a distance of approximately 400 m. The peak level in this example is 20 log(2,400/10⁻⁶) dB=187.6 dB, whereas the peak-to-peak level is 20 log((2,400+2,290)/10⁻⁶) dB=193.4 dB. SEL: sound exposure level. (Nehls & Bellmann 2016)

A common quantity for describing pile-driving noise is sound exposure level (SEL), as defined in the following equation:

$$SEL = 10 \log \left(\frac{1}{T_0} \int_{T_1}^{T_2} \frac{p(t)^2}{p_0^2} dt \right)$$

The averaging start and stop times T_1 and T_2 are chosen arbitrarily, but in a way that the sound event lies in between T_1 and T_2 (Figure 6.4). T_0 is 1 second; that is, the SEL is the level of a continuous sound with a duration of 1 second and the same sound energy as the impulse. The SEL is independent of the blow rate.

Peak sound pressure (p_{peak}), often expressed as peak level (L_{peak})=20 log(p_{peak}/p_0), where p_{peak}=max |p(t)|; that is, the highest absolute sound pressure observed. Some authors, however, prefer a 'peak-to-peak level' (Matuschek & Betke 2009).

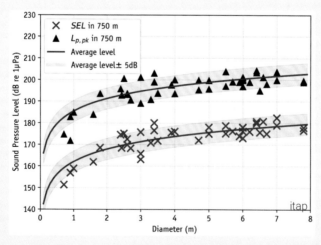

Figure 6.5 Relationship between pile diameter and noise levels expressed as sound exposure level (SEL) and peak level as L_{peak} at 750 m from unmitigated offshore pile driving. (Modified from Nehls & Bellmann 2016)

Pile-driving noise is made up of rather low frequencies, with the main energy below 100 Hz. At higher frequencies that are more audible to marine mammals, noise levels are low and fall below background levels at rather short distance.

Noise levels from pile driving depend on various factors, the most important being piling energy, pile diameter and water depth. Pile diameter and water depth define the surface emitting noise into the water column. As piling energy is related to pile diameter (larger piles need higher energy), pile diameter is a good predictor of actual noise immission. As piling energy may be kept low to reduce noise immission, there is some variability between projects, although a clear positive relationship between pile diameter and noise levels is apparent; but the increase in noise levels off for large piles (Figure 6.5).

Noise immission from offshore pile driving is measured at a standard distance of 750 m according to ISO 18406:2017 (Robinson & Theobald 2017). At 750 m, noise levels from large piles exceed values of 200 dB$_{peak}$ and 180 dB$_{SEL}$.

Underwater noise pressure travels at a speed of 1,480 m/s and propagates over large distances. Noise attenuates with increasing distance from the source owing to physical spreading and attenuation in the water column, seafloor and sea surface. The structure of the seafloor and water depth are important variables when modelling noise propagation around a construction site. Wind conditions are also important as, at higher wind speeds, air bubbles from waves in the upper water column attenuate noise (Dragon *et al.* 2016).

have the important benefit of needing no or much reduced pile-driving activity, thereby reducing the noise impact upon marine mammals.

Richardson *et al.* (1995) provided a conceptual framework on how anthropogenic underwater noise may affect marine mammals along the gradient of decreasing noise levels with increasing distance from the source, with the potential for displacement farther away from the source and hearing damage through a temporary threshold shift (TTS) or even permanent threshold shift (PTS) close to the source (Box 6.2). The onset of physical impairment differs between species groups. Criteria for assessing the onset of TTS and PTS have been collated in a lengthy process for six species groups in the USA (Southall *et al.* 2007; NMFS 2018). The proposed noise thresholds refer to cumulative exposures and are frequency weighted, in that they assume that noise energy adds up over exposure time, or equally over the number of blows during a piling operation, and that the frequency-dependent hearing ability of an animal defines its sensitivity to noise from different sources. It is further concluded that compared to continuous noise, impulsive noise leads to hearing damage at lower levels.

Thresholds for Harbour Porpoise

Harbour Porpoises are found to be more sensitive to hearing damage than other high-frequency cetaceans, with an onset of TTS recorded at a broadband level of 164 dB sound exposure level (SEL) by Lucke *et al.* (2009) for a captive animal exposed to impulsive noise from an airgun. Work from Kastelein *et al.* (2017b), also on captive Harbour Porpoises exposed to airgun noise, revealed considerably higher thresholds for the onset of TTS (SEL$_{cum}$ 188–191 dB), but the application of frequency weighing to the noise data revealed little difference between the studies.

Relating the noise exposure criteria to noise propagation modelling of OWFs reveals that the onset of hearing damage may be reached at distances of a few hundred metres for PTS and up to 5 km for TTS (Nehls *et al.* 2014; Mackenzie Maxon 2015) from a single strike when large monopiles are driven into the seabed. It needs, however, to be taken into account that a full piling operation consists of a few thousand blows, and thus cumulative exposures to the piling noise need to be considered. Under the assumption that Harbour Porpoises would remain stationary rather than moving away from a construction site, a cumulative noise dose sufficient to cause PTS may be reached at distances of 5–10 km (Nehls *et al.* 2014; Mackenzie Maxon 2015). Harbour Porpoises do, however, move away from loud noise sources and are usually deterred from the vicinity of the construction site before the start of piling. Taking uncertainties about swimming speed and direction into account, as well as uncertainties about noise propagation and possibly variable sensation levels (Nachtigall *et al.* 2016), it is not yet possible to predict accurately how many individuals will receive noise levels inducing either form of hearing damage. An expert group formed to give advice to several Danish OWFs concluded that the risk of hearing damage is considerably lower when taking into account soft-start of the piling operation, use of deterrents and porpoises fleeing from the noise source, compared with assessments assuming static exposures. However, the study could not rule out that pile driving will cause PTS to Harbour Porpoises without applying noise mitigation (Energinet 2015). As Harbour Porpoises are present in almost all European OWF sites in considerable densities, it has to be considered in impact assessments that there is a high risk of causing PTS and almost certainly TTS to Harbour Porpoises if projects are realised without appropriate mitigation measures. Studies on the behavioural response by tracking of tagged porpoises exposed to impulsive noise (van Beest *et al.* 2018) will provide data required to fully assess the noise dose experienced by Harbour Porpoise at offshore construction sites.

Thresholds for seals

The sensitivity of seals in relation to hearing impairment from underwater noise is considered to be much lower than that of cetaceans, and in phocids such as Harbour and Grey Seals the thresholds for the onset of TTS and PTS is estimated at 170 dB SEL and 185 dB SEL (M-weighted), respectively, which is 30 dB higher than that for high-frequency cetaceans. Such noise levels will only be recorded close to a piling operation (Box 6.2 and Box 6.3). However, tracking of Harbour Seals and auditory modelling by Hastie *et al.* (2015) in relation to the constriction of wind farms in the Greater Wash in the UK predicted SELs resulting in high risks of auditory damage, with all seals predicted potentially to suffer TTS and 50% to have PTS on a number of occasions. Such effects have the potential to influence individual fitness and the ability to function normally, with the prospect of consequences on seal populations.

Disturbance and displacement from piling

Harbour Porpoises and other marine mammals often respond aversively to anthropogenic noise, with the response becoming stronger with increasing noise levels. Responses of marine mammals to anthropogenic noise sources are variable and depend not only on noise strength, but also on the characteristics of a noise source and the context of the disturbance (Ellison *et al.* 2011). Although pile driving creates a relatively uniform noise, the different hearing abilities of marine mammals and likely different sensitivities mean that there is no uniform response by different species. As a result, cetaceans, specifically Harbour Porpoises, and seals are considered separately below.

The response of Harbour Porpoise

Most knowledge on the response to pile driving originates from studies on Harbour Porpoise in the North Sea. The first OWFs in Denmark and Germany caused effects on Harbour Porpoise over a large range. Reduced porpoise activity was recorded up to a distance of about 20 km during and shortly after piling (Tougaard *et al.* 2009a; Brandt *et al.* 2011; Haelters *et al.* 2012; Dähne *et al.* 2013; Diederichs *et al.* 2014). This corresponds with the suggestion of Bailey *et al.* (2010) that behavioural disturbance of Bottlenose Dolphins could similarly occur over large distances (up to 50 km) in relation to the pile-driving of two 5 MW turbines installed in the Moray Firth in north-east Scotland. However, the early projects involving Harbour Porpoise were constructed without noise mitigation and additional aspects of the construction work may have contributed to the strong response. For example, at Alpha Ventus OWF in Germany, piles were first vibrated up to 9 m into the substrate before being piled with a hydraulic hammer, using between 11,383 and 25,208 strokes lasting for 376–802 minutes to install some piles to a depth of 30 m. Pile-driving measurements employing a gradient design with acoustic monitoring equipment spaced along a transect away from the impact area were conducted during installation of one pile. A short-ramp procedure was employed with a duration of about 5 minutes, after which 449 blows over 30 minutes were utilised to penetrate to a depth of 21 m. Piling generated peak levels at 720 m from the pile of 196 dB re 1 µPa, while the SEL, as defined in Box 6.3, reached a maximum of 176 dB re 1 µPa 2s (Dähne *et al.* 2013). In keeping with the results from other studies, Brandt *et al.* (2011) found that porpoise acoustic activity fell by 100% in the hour after pile driving and did not return to normal for between 24 and 72 hours at a distance of 2.6 km from the site, with recovery time reducing with distance. An impact was detectable to 17.8 km from the site, but was not detectable at 22 km where activity increased. Up to around 5 km from the site, recovery times tended to exceed pauses in piling.

An important aspect of the response of Harbour Porpoise to anthropogenic noise is that the response becomes weaker with decreasing noise levels. In a detailed study using

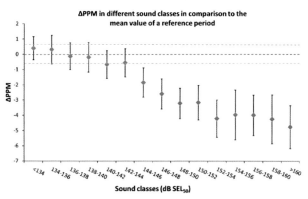

Figure 6.6 Response of Harbour Porpoise *Phocoena phocoena* to pile driving of 40 tripod foundations at the Trianel Borkum wind farm in the German North Sea (Diederichs *et al.* 2014). The values give the change in porpoise detections as porpoise positive minutes (PPM), which is minutes including at least one recording of a harbour porpoise click train. Noise levels are measured in sound classes of SEL$_{50}$, which is median sound exposure level in decibels of a given number of pulsed sounds such as hammer blows.

passive acoustic monitoring (PAM) during the construction of 40 tripod foundations at the Trianel Borkum wind farm in the German North Sea, Diederichs *et al.* (2014) related the strength and duration of porpoise response to pile-driving noise levels. Reduced porpoise activity was recorded up to a noise level of 144 dB SEL. The response at 144 dB SEL was, however, rather weak and a marked reduction in activity, probably representing strong to full displacement, was evident at noise levels above 152 dB SEL (Figure 6.6)

In a further study, Brandt *et al.* (2018) investigated the response of Harbour Porpoises to pile driving during the construction of seven OWFs in the German Bight of the North Sea. All projects applied noise-mitigation measures (see *Concluding remarks* below for details in Germany) but as these were still under development noise reduction was moderate. Non-parametric analyses revealed a clear gradient in the decline in porpoise detections at different noise-level classes. Compared to a baseline period of 25–48 hours before piling, porpoise detections declined by over 90% at noise levels above 170 dB, but only by about 25% at noise levels between 145 and 150 dB. Below 145 dB, this decline was smaller than 20% and thus was not clearly related to noise emitted by the piling process. The duration of the effect after piling was about 20–31 hours in the close vicinity of the construction site (up to 2 km) and decreased with increasing distance. Project-specific estimates ranged between 16 and 46 hours.

The generation of lower noise levels below a threshold may explain the weaker effect on Harbour Porpoise evident in other studies. For example, PAM during construction of two wind turbines off north-east Scotland by Thompson *et al.* (2010) and an investigation during impact and vibration piling at the Nigg Energy Park within the Moray Firth Special Area of Conservation in 2014 (Graham *et al.* 2017) did not detect a clear response. In the latter study, the results may have been confounded by low statistical power or the lack of ecological drivers, including availability of key habitat.

The response of seals

The response of seals to piling activity is complex in the sense that seals occur in and below the water surface as well as above it, with different sensitivities. In support of the theoretical considerations of Thomsen *et al.* (2006), early studies in Denmark showed a significant reduction of 31–60% in the numbers of seals using a haul-out some 10 km away during piling of Horns Rev (Edrén *et al.* 2004). At Scroby Sands in the UK, where the haul-out is less than 2 km from the OWF, aerial surveys showed a significant decline in the numbers of Harbour Seals without full recovery 2 years after piling (Box 6.4), suggesting displacement of animals to other areas outside the typical range from the haul-out. In contrast, Grey Seals showed a continued year-on-year increase in numbers after construction. The presence of this larger species competing for space and prey resources may have contributed to the failure of the smaller Harbour Seal to recover. Monitoring was not, however, linked to specific pile-driving events and therefore short-term disturbance and displacement could not be discounted in either species. Thus, where the two were linked, as at Horns Rev, boat-based surveys recorded a decline in the use of the wind-farm area during the construction phase, with no Harbour Seals present inside it on days with pile driving (Teilmann *et al.* 2006).

Where seals are individually tagged, this should increase the chance of detecting specific responses to short-lived events. Nevertheless, no clear changes in distribution of tagged Harbour or Grey Seals at sea were detected at Nysted by Teilmann *et al.* (2006), although this was thought to be partly because tagged seals rarely used the wind-farm site, with only 0.41% and 0.07% of location fixes for Harbour and Grey Seals respectively, were recorded within it. In contrast, at Egmond aan Zee OWF in the Netherlands, it was

Box 6.4   Monitoring the response of seals hauled out on Scroby Sands before, during and after construction of the Scroby Sands wind farm

Scroby Sands offshore wind farm, comprising 30 2 MW Vesta V80 turbines, was installed by E.ON UK Renewables Offshore Wind Ltd adjacent to Scroby Sands, a dynamic sand-bar system some 3 km offshore of Great Yarmouth, UK, in late 2003 and into 2004 (Figure 6.7). The site was the second to become operational in the UK. The main bank of Scroby Sands, some 2 km south of the wind farm, was emergent at low water and had been known to support one of the few colonies of Harbour Seals *Phoca vitulina* along the English east coast for over a century. In contrast, Grey Seals *Halichoerus grypus* only colonised the area in the late 1950s.

Figure 6.7 Seals hauled out on the main bank of Scroby Sands close to Scroby Sands offshore wind farm. (Martin Perrow)

According to Food and Environment Protection Act (FEPA) licence conditions, Skeate *et al.* (2012) used aerial surveys aboard a Cessna 150 aerobat aircraft flying at around 300 m to assess the response of Harbour and Grey Seals using haul-outs for a five-year period from 2002 to 2006 before, during and after the construction of Scroby Sands. In each year, between 10 and 15 surveys were undertaken at low water at regular intervals in the period from April to October inclusive, according to licence requirements, meaning that no surveys were undertaken during the period of pile driving from 21 October 2003 to 1 January 2004. During each survey, a series of images designed to cover the entire emergent bank was taken with a digital SLR camera. The resultant images were displayed in Abode Photoshop, where the seals could be identified to species level through a combination of size, body shape, muzzle shape, colour and haul-out pattern, with further identification of sex and age in relation to pups whenever possible (Figure 6.8).

A total of 7,409 seal sightings were made, comprising 4,364 Harbour Seals (59%), 2,735 Grey Seals (37%) and 310 unidentified seals (4%). The mean counts of Harbour Seals, including both adults and pups presumed to have been born on the bank, and

Figure 6.8 Aerial image of typical patterns of Harbour Seal (left of line) and Grey Seal (right of line) haul-outs with colour-coded identification of species and age for illustrative purposes in the former, as follows: yellow=Harbour Seal adult; orange=Harbour Seal pup; pink=Grey Seal bull; light blue=Grey Seal cow. (Air Images Ltd)

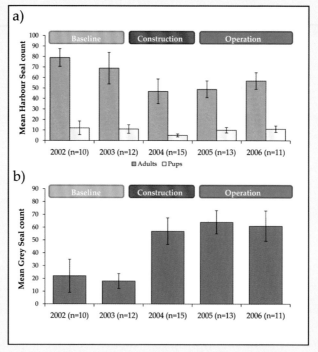

Figure 6.9 Variations in mean (±1 SE) annual haul-out counts of (a) Harbour Seal and (b) Grey Seal relative to the construction of Scroby Sands wind farm. Note the decline in Harbour Seals as construction piling was undertaken between October 2003 and March 2004. Numbers had not recovered 2 years after construction.

Grey Seals, in each monitoring year are shown in Figure 6.9. During modelling, Skeate *et al.* (2012) accounted for a number of environmental factors that could influence counts, including wind direction and speed and tidal height, as well as seasonal trends. Significant changes in seal numbers between years could not be explained by other factors. The decline in Harbour Seals during the year of construction without full recovery by the end of the monitoring period was reasoned to be induced by piling noise, but there was also a potential link to increased vessel traffic, to which hauled-out Harbour Seals may be sensitive (Andersen *et al.* 2012; Jansen *et al.* 2015). In contrast, Grey Seals showed a continued year-on-year increase in numbers after construction, suggesting no effect of construction noise, although short-term disturbance and displacement during the period of piling could not be discounted as monitoring was not specifically tied to discrete construction events. The increase in Grey Seals was linked with the wider success of the species in the area that also introduced the possibility of competition with Harbour Seals for resources, including space at the haul-out, which may have contributed to the failure of Harbour Seals to recover their numbers. Recovery could also have been retarded by the prospect of Harbour Seals being displaced to other haul-outs beyond their typical foraging range of up to 45 km, thereby reducing the prospect of an individual encountering and recolonising Scroby Sands (Thompson *et al.* 1996).

suggested that tagged Harbour Seals avoided the study area in the construction phase by at least 40 km (Lindeboom *et al.* 2011). Similarly, in their study of 24 tagged Harbour Seals in the Greater Wash, a key Round 2 development area for wind farms in the UK, Hastie *et al.* (2015) showed the closest that seals came to active pile driving of the Lincs OWF UK varied between 4.7 and 40.5 km. Here, 31 monopiles (5.2 m diameter) were installed between May 2011 and 2012 using intermittent piling periods. A ramp-up procedure was employed during the first hour of piling, followed by 4–5 hours of continuous piling during which the median strike interval of the hydraulic hammer was 2 seconds.

In an analysis of the response of tagged seals to construction of both Lincs and Sheringham Shoal OWFs, Russell *et al.* (2016) compared at-sea telemetry data from 19 Harbour Seals before any construction with data from 23 individuals during the construction of Lincs and after piling was complete at Sheringham Shoal in 2012. Two spatial analyses were used to compare the historical data with the 2012 data, and non-piling with piling data in 2012 alone. The results suggested a close-to-significant increase in the use of Sheringham Shoal in 2012 (up to May) compared to the baseline, although this was linked to a more general increase to the west of Sheringham Shoal rather than the wind farm driving the observed change. A significant increase in the use of the Lincs OWF site subject to piling was also attributed to a general increase in the use of the wider area. Lincs OWF is within 20 km of the main haul-outs and pupping grounds for Harbour Seals in the Wash, and individuals continued to move in and out of the estuary during construction. However, during piling, Russell *et al.* (2016) confirmed significant displacement of seals up to 25 km from the centre of the wind farm, but recovery time, defined as the time to return to an impacted area, was only 2 hours after piling. Thus, the gaps in piling of a few hours or days observed at Lincs (Hastie *et al.* 2015) seemed to allow unhindered travel and foraging, as reflected by the lack of an impact on local population growth.

Underwater noise immission during the operational phase

During operation, noise is generated by maintenance activities and by the turbines themselves. Machinery noise is the main contributor to underwater noise, with vibrations transmitted from the nacelle to the foundations, where they are radiated to the seabed and water. In contrast, airborne noise is almost completely reflected from the water surface. Turbine-related noise depends on the foundation type as well as on the size and type of turbine, but little information has been published on this so far.

Tougaard *et al.* (2009b) made recordings of underwater wind turbine noise (100 Hz to 150 kHz) at between 14 and 40 m from foundations at three wind farms (Middelgrunden and Vindeby in Denmark and Bockstigen-Valar in Sweden) with turbines ranging between 450 kW and 2 MW. They found that absolute noise levels were low [109–127 dB re 1 μPa (rms)] for the range of sounds produced up to a frequency of 20 kHz. The noise produced was of low intensity and frequency and was thought to be audible to Harbour Seals at distances up to a few kilometres, but only to 14–40 m from the bases for Harbour Porpoises. There was very limited capability for such noise to mask signals, let alone injure the animals (Box 6.2). Similarly, in a study using simulated wind-turbine noise, Lucke *et al.* (2007) suggested that the potential masking effects for Harbour Porpoise would be limited to short range in the open sea.

The operational noise immission from modern larger turbines also does not exceed ambient noise by much (20 dB re 1 μPa at 12 m/s windspeed) (Norro & Degraer 2016). A modelling exercise indicated that operational noise may be audible to marine mammals, especially those with good hearing abilities at lower frequencies, such as Common Minke Whale *Balaenoptera acutorostrata*, over considerable distances of up to 20 km (Marmo 2013). However, there is no indication so far that this would lead to disturbance.

Effects of service vessels

During operation, commercial boat traffic will generally be excluded from the area, although maintenance vessels will service the wind farm throughout its life. While larger heavy-lift vessels may be required to perform more complex maintenance tasks, such as swapping gearboxes, crew transfer vessels such as high-speed catamarans typically around 20–24 m in length will typically visit each turbine around six times per year during routine minor service activities (Gellatly 2013).

The noise level generated by boats depends on their design and the speeds at which they travel, and further noise is emitted by their sonars. Vessel noise may be audible to many species of marine mammals at considerable distance, with the potential to lead to a range of chronic effects including changes in behaviour, sound masking and displacement from important areas (Richardson *et al.* 1995; Morton & Symonds 2002; Jansen *et al.* 2015). Specific changes in behaviour and communication to increased vessel activity and noise have also been shown for Harbour Porpoise and especially Bottlenose Dolphin (Nowacek *et al.* 2001; Jensen *et al.* 2009; La Manna *et al.* 2013).

For pinnipeds, Jones *et al.* (2017) evaluated the co-occurrence of Grey and Harbour Seals and shipping traffic around the British Isles and modelled acoustic exposure to individual Harbour Seals that was validated with acoustic recorders. Co-occurrence rates were highest within 50 km of the coast, close to seal haul-outs. Areas with high risk of exposure included 11 out of 25 Special Areas of Conservation (SACs) (Box 6.1). Predicted cumulative M-weighted SELs for 70% of the Harbour Seals had upper bounds that exceeded levels that may induce TTS. Seals may also be disturbed from haul-outs by shipping, which appears to relate to visual as well as noise stimuli (Andersen *et al.* 2012; Jansen *et al.* 2015), with Harbour Seals seemingly more sensitive than Grey Seals where they occur sympatrically (Box 6.4).

The increase in vessel traffic during the life of a wind farm also increases the potential for marine mammals to be struck by vessels. Surprisingly perhaps, the review by van Waerebeek *et al.* (2007) of nearly 250 reported vessel collisions showed that 19 species of potentially small agile cetaceans including Bottlenose Dolphin and Harbour Porpoise had been involved in at least one incident. However, it is the larger cetaceans (Dolman *et al.* 2006; Panigada *et al.* 2006), sirenians (Beck *et al.* 1982) and some pinnipeds (Goldstein *et al.* 1999) that may be vulnerable.

Wind-farm service vessels are designed to travel relatively quickly and thus they may be expected to pose a higher risk than slower vessels. However, there appear to be no reports of instances or strandings which might hint at collisions with service vessels. In UK wind farms, the restriction of the use of vessels with ducted propellers, over concern that these were contributing to Harbour Seal mortalities where animals had corkscrew wounds, proved to be unfounded as predation by Grey Seal was shown to be responsible (Onoufriou *et al.* 2016).

With respect to both disturbance through noise and collision risk, servicing a wind farm invariably leads to additional ship traffic in an area between the wind farm and a nearby port. This ship traffic may pass areas of high value for marine mammals and marine protected areas. This factor has not usually been addressed in EIAs, but should be considered in future.

Reefs and refuges

The physical presence of turbine bases and associated structure such as scour protection, coupled with other OWF infrastructure such as substations (see Jameson *et al.*, Chapter 1), has a range of effects upon coastal processes including changes in water flow and sediment transport (see Rees & Judd, Chapter 2) as a result of the presence of the structures themselves and any altered seabed topography. Moreover, changes in atmospheric conditions and air pressure above the sea surface are predicted to lead to significant changes in ocean dynamics, including upwelling of nutrients affecting biological productivity (see Broström, Chapter 3). However, little is yet known of these potential effects compared to the change in plant and animal communities resulting from the rapid colonisation of those structures present producing the so-called 'reef effect' (Leonhard *et al.* 2006; Petersen & Malm 2006; Linley *et al.* 2007). The initial colonisation of species within lower trophic levels is quickly followed by larger invertebrates such as crabs and lobsters and small fish, thereby attracting larger predatory fish (see Dannheim *et al.*, Chapter 4; Gill & Wilhelmsson, Chapter 5) and potentially top predators such as marine mammals or seabirds (see Vanermen & Stienen, Chapter 8). In some cases, however, wind-farm structures have been shown to have a limited effect on fish aggregations relative to seasonal changes or weather conditions (van Hal *et al.* 2017).

In some countries, such as Germany and Denmark, commercial fishing is banned from wind farms, although this is not the case in others such as the UK and now the Netherlands, where it was formerly not allowed (Lindeboom *et al.* 2011). Nevertheless, some fisheries may be limited owing to safety restrictions or simply because of difficulties using specific gear within the confines of a wind farm. Where fishing is excluded or significantly reduced, the abundance of fish within a farm may be expected to increase as a result of reduced mortality rates of target species and by-catch (Stenberg *et al.* 2011; Lindeboom *et al.* 2011; Wilhelmsson & Langhamer 2014). However, outside the wind farm, the benefits of such refuges, including spillover effects to adjacent areas, are more uncertain (Gell & Roberts 2003).

As with other factors, the ability of wind farms to act as reefs or refuges will depend on a wide variety of site-specific circumstances and it is most likely that such effects will be strongly influenced by the size of the wind-farm areas. The potential for wind farms to operate as reefs and refuges for different groups of marine mammals, specifically Harbour Porpoise and seals, are reviewed below. This recognises that any effect may be greater in larger wind-farm areas and may thus become more prominent in the future.

Harbour Porpoises in operational wind farms

Rather few published studies have analysed whether porpoises occur in higher or lower numbers within operational OWFs. In 2005 and 2006, Diederichs *et al.* (2008) investigated the presence of Harbour Porpoises in the Danish OWFs of Horns Rev 1 and Nysted 1 by passive acoustic measurements using T-PODs. No differences inside and outside either wind farm could be detected. Moreover, in Horns Rev, no difference between porpoise detections at different distances from single turbines could be found and this wind farm did not seem to influence the presence of Harbour Porpoises at all. In comparison, at Nysted, only a weak effect was found between different distances of the T-PODs from single turbines, with more recordings farther than 700 m compared to less than 150 m. This effect was only apparent when no additional variables that could also affect Harbour Porpoise activity were included in the analysis. In 2006, a shutdown of the Nysted wind farm for a week for maintenance did not lead to changes in the utilisation of the wind farm by Harbour Porpoises (Figure 6.10).

Nevertheless, there did appear to be an effect on the 24-hour cycle of Harbour Porpoise recordings. Especially in 2005, a pronounced diurnal rhythm was noted, with most recordings during the night close to single turbines in both wind farms. A converse pattern was observed at more than 900 m away from single turbines during the day at Horns Rev, when a maximum of porpoise recordings was recorded. No clear pattern between day and night could be found more than 700 m away from single turbines in Nysted, however. Moreover, in 2006, the diurnal pattern changed in both areas and the

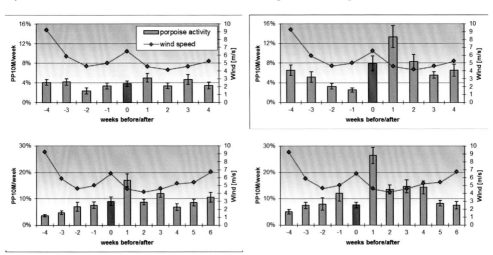

Figure 6.10 Presence of Harbour Porpoises along four transects of T-PODS inside (left) and outside (right) of the offshore wind farm Nysted, Denmark, in weeks before and after a shutdown of turbines (week 0, red bar).

differences between the distance groups was no longer very pronounced. Differences in the diurnal cycle of Harbour Porpoise activity were thought to reflect differences in the fish community close to single turbines. Overall, it was concluded that operational OWFs are regularly incorporated into Harbour Porpoise habitat.

Furthermore, a significant increase in Harbour Porpoise activity inside the operational site, relative to the baseline conditions at the Dutch OWF Egmond aan Zee (Scheidat *et al.* 2011; Lindeboom *et al.* 2011), may have been due to a reef effect resulting from prey aggregations around turbine bases. However, the pattern observed could equally suggest potential for a refuge effect from nearby shipping lanes, consistent with the general impression that Harbour Porpoise tends to avoid or at least not be attracted to vessels (see *Effects of service vessels*, above).

Seals in operational wind farms

In relation to seals, Russell *et al.* (2014) present telemetry data suggesting that both Grey and Harbour Seals trace underwater wind-farm infrastructure as well as other anthropogenic structures, such as pipelines. An example of the tracks of one Harbour Seal that visited Sheringham Shoal wind farm in each of its of 13 foraging trips from haul-outs over 30 km away in the Greater Wash is shown in Figure 6.11. This clearly shows movements consistent with foraging activity directly between turbine bases, many of which (77 of the 90 present) have extensive rocky scour protection extending up to 11 m from the bases. Harbour Seals in the Wash are known to take a range of prey, from small demersal fish to crabs (Hall *et al.* 1998). The latter are abundant in the site, judging by the successful pot fishery for Edible Crab *Cancer pagurus* and European Lobster *Homarus gammarus* that operates around and within the wind farm. The telemetry study reinforced the regular observation of both Harbour and Grey Seals (Figure 6.12) within the operational wind farm during detailed ornithological studies (see Harwood *et al.* 2017).

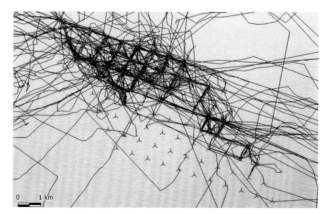

Figure 6.11 Tracks of a satellite-tagged Harbour Seal *Phoca vitulina* around Sheringham Shoal in the Greater Wash, UK, with the turbines and substations (circles) shown in red. While tagged, the seal visited the wind farm on each of its 13 trips to sea. (Deborah Russell)

Cumulative and population effects

The number of wind farms has increased rapidly since the turn of the millennium, albeit often within limited areas, and further expansions are expected in the coming years. As a consequence, there is a rising likelihood of marine mammals becoming exposed to

Figure 6.12 Grey Seal *Halichoerus grypus* with a captured ray, probably a Thornback Ray *Raja clavata*, in the vicinity of the Sheringham Shoal wind farm (Greater Wash, UK). The operational site appears to have become increasingly attractive to foraging seals, consistent with a reef effect. (Martin Perrow)

pile-driving noise and other pressures related to OWFs. Understanding how pressures from OWFs interact with pressures from other projects and existing pressures from other sources such as fisheries or pollution is crucial to assess any potential population consequences. It is, however, not straightforward to assess cumulative effects because the nature of the various pressures varies greatly. While pile-driving noise certainly contributes to overall ocean noise budgets and the effects may well accumulate with other noise sources, such as seismic surveys or shipping, the function of OWFs as refuges from fisheries is very different from the impacts of fisheries on the food resources of marine mammals.

Indeed, there is a clear need to understand better the in-combination interaction of wind farms with other different pressures and stressors, such as incidental by-catch, prey depletion, other sources of anthropogenic noise such as sonar, increasing boat traffic, pollution, and the installation of underwater turbines or tidal energy devices (Wilson *et al.* 2006; Davidson *et al.* 2012; Lusseau *et al.* 2012). Especially where populations are small, compared to large populations of widely dispersed species, the expansion of OWFs in particular areas is more likely to cause population effects (Dolman *et al.* 2006; Tougaard & Mikaelsen 2017).

There have been several attempts to determine possible population impacts of OWFs, all focusing on the responses of marine mammals to underwater noise from pile driving. For example, Thompson *et al.* (2013) described a framework to assess population-level impacts of disturbance from piling noise on a protected Harbour Seal population near to proposed wind-farm developments in north-east Scotland. They used models of seal distribution combined with models of noise propagation to estimate received sound levels integrated with data on potential impacts to infer displacement and auditory injury. These impacts were linked to possible changes in vital rates using expert judgement and applied to population models under different construction scenarios. Although the authors acknowledged the inherent uncertainty in the adopted approach, the results suggested that, despite short-term reduction in abundance during and immediately after construction, the following recovery resulted in no clear difference between impact and baseline scenarios after 25 years.

Moreover, two different approaches to possible population consequences of pile driving have been developed, namely the Interim Population Consequences of Disturbance (iPCoD) (King *et al.* 2015; Harwood *et al.* 2016) and Disturbance Effects of Noise on the Harbour Porpoise Population in the North Sea (DEPONS) (van Beest *et al.* 2015). Although both models simulate population dynamics based on the birth and survival rates of individual animals, they model survival in a different ways. Whereas iPCoD uses average survival rates derived from data from North Sea animals, in the DEPONS model, survival emerges from the individuals' ability continuously to find food. The models also differ in the way they model the consequences of exposure to noise and the kinds of output they can provide (Nabe-Nielsen & Harwood 2016).

While the DEPONS model focuses on Harbour Porpoise, iPCOD (King *et al.* 2015; Harwood *et al.* 2016) incorporates stage-structured stochastic population models for five species, comprising Harbour Seal, Grey Seal, Harbour Porpoise, Bottlenose Dolphin and Common Minke Whale, in UK waters. Attempts were also made to capture major sources of uncertainty in model parameters. Population models were run to forecast the effects of planned developments on the dynamics of populations over a specified number of years. The models can be used to investigate the effects of multiple developments under different scenarios, including cumulative effects of several developments, and in-combination effects, including mortality associated with incidental by-catch.

Brandt *et al.* (2016) used the iPCoD model to estimate disturbance consequences of wind-farm construction on the population level based on the results from aerial survey data and C-POD data obtained during the construction of seven OWFs in the German Bight. Using conservative input parameters for construction effects arising from the study, thereby increasing the chances for the model to predict a population decline, the risk of a decline of 1% of the population in the German Bight was estimated to be below 30%. The predicted median decline is below the 1% generally considered as critical for all chosen time periods, and varies between 0.9% for the piling period and 0.2% for 12 years after piling had finished.

As a result, there were no indications for such a population decline of Harbour Porpoises over the five-year study period arising from analyses of daily C-POD data and aerial survey data at a larger scale. Despite extensive construction activities over the study period and an increase in these over time, there was no negative trend in acoustic porpoise detections or densities within any of the subareas studied. In some areas, C-POD data even detected a positive trend from 2010 to 2013. Even though clear negative short-term effects (1–2 days in duration) of OWF construction were found on acoustic porpoise detections and densities, there is no indication that Harbour Porpoises within the German Bight are negatively affected by wind-farm construction at the population level. This is despite sound-mitigation techniques still being under development during the study period and further improved thereafter.

A major uncertainty about predicting marine mammal responses to pile driving is the fact that responses have only been measured through animal densities or acoustic activities in relation to piling events. However, new tracking technologies allow researchers to take a much closer look at individual behaviour and an animal's responses to loud sound sources (van Beest *et al.* 2018), and will improve the input data for models analysing population impacts.

Concluding remarks

Marine mammals are subject to a large range of anthropogenic effects and impacts that are growing in scope and magnitude, with those associated with wind farms among the most novel. With respect to marine mammals, underwater noise generated during construction, specifically pile driving, is considered to pose the greatest risk of a negative impact on these animals. As noted by Richardson *et al.* (1995), it is difficult to measure and quantify the effects of sound on marine mammals owing to the technical complexities associated with measurements and study design, although various types of modelling studies show promise (Nabe-Nielsen *et al.* 2014; Hastie *et al.* 2015; Harwood *et al.* 2016).

Many species of marine mammals are endangered and are protected by law. In addition, there is considerable public awareness about marine mammals and possible negative impacts are thus high on the agenda when discussing further expansions of offshore wind energy. From the available knowledge it appears, however, that the most relevant impacts of OWFs on marine mammals are associated with the use of steel piles as foundations, while other types, such as suction buckets, gravity foundations and floating foundations, will produce less noise and thus fewer impacts. Regulations on OWF development specific for marine mammals have thus been primarily focused on reducing pile-driving noise. Although technologies for noise mitigation are well developed and have been shown to reduce underwater noise immissions and also the disturbance to marine mammals (Nehls *et al.* 2016; Thomsen & Verfuß, Volume 4, Chapter 7), they have only been applied in a few countries so far, primarily Germany and more recently Denmark, as few countries have implemented regulations on underwater noise. Regulating underwater noise immission from pile driving may be justified by the risk of hearing impairments it poses to marine mammals. Recent assessments concluded that a permanent hearing impairment could result from cumulative noise exposures in Harbour Porpoise when large monopiles are driven into the seabed (Nehls *et al.* 2014; Mackenzie Maxon 2015). Under EU regulations this would be a violation of the obligations concerning strictly protected species.

In Germany, a precautionary SEL criterion of 160 dB re 1 μPa^2 s outside a radius of 750 m during piling has been included in the licensing process based on noise levels required for the onset of TTS in Harbour Porpoises (Lucke *et al.* 2009). This SEL is based on a single impulse criterion and does not account for multiple exposures. The rationale behind this German noise threshold is that Harbour Porpoise will be displaced by deterrents and soft-start of the piling operation and reach a safe distance from the construction site before underwater noise immission reaches a level that may harm the animals. In other countries, the use of deterrents is regarded as sufficient to avoid hearing impairment. Furthermore, there is no uniform approach on whether a TTS or a PTS has to be avoided.

Noise mitigation not only reduces the risk of exposing marine mammals to noise that may impair their hearing abilities, but also reduces disturbance and displacement. However, there is no uniformity in the approach to regulate disturbance, as the extent and duration of disturbance need to be assessed in relation to the importance of an area for marine mammals.

Whether operational wind farms are preferentially used by marine mammals depends on a number of factors, with the relative value of the wind farm as a foraging area compared to the surrounding habitat perhaps being a key consideration. Moreover, as indicated for Harbour Porpoise at Egmond aan Zee (Scheidat *et al.* 2011), reef and refuge effects may be indistinguishable from each other. The benefits of reef and refuge effects will also only apply to species able to cope with the operational noise and boat traffic in the wind farm. Furthermore, even where effects of reef and refuges from fisheries operate, it remains uncertain whether this simply represents a redistribution of available prey or an increase in prey availability over the larger area.

To date, wind farms have been focused on shallow coastal waters containing relatively few, and generally small species of odontocetes and pinnipeds. As development expands into different regions and moves farther offshore into deeper waters occupied by different species, especially the great whales, the pressure will be on to undertake more elegant and robust studies to understand better the potential impacts of wind farms.

Acknowledgements

We would like to thank E.ON for funding the work carried out at Scroby Sands OWF and SCIRA Offshore Energy Limited for access to data relating to monitoring at the Sheringham Shoal OWF. We would also like to thank Dr Debbie Russell at the Sea Mammal Research Unit (SMRU) at the University of St Andrews for supplying Figure 6.8, Eleanor Skeate of ECON Ecological Consultancy Ltd for supplying information for Box 6.4 and Michael Bellmann of ITAP for supplying Figure 6.5. Matthias Schultze, Miriam Brandt and Ansgar Diederichs of BioConsult SH also made various useful inputs to the manuscript.

References

Aarts, G., von Benda-Beckmann, A., Lucke, K., Sertlek, H., van Bemmelen, R., Geelhoed, S., Brasseur, S., Scheidat, M., Lam, F., Slabbekoorn, H. & Kirkwood, R. (2016) Harbour porpoise movement strategy affects cumulative number of animals acoustically exposed to underwater explosions. *Marine Ecology Progress Series* 557: 261–275.

Andersen, S.M., Teilmann, J., Dietz, R., Schmidt, N.M. & Miller, L.A. (2012) Behavioural responses of harbour seals to human induced disturbances. *Aquatic Conservation: Marine and Freshwater Ecosystems* 22: 113–121.

Andrew, R.K., Howe, B.M., Mercer, J.A. & Dzieciuch, M.A. (2002) Ocean ambient sound: comparing the 1960s with the 1990s for a receiver off the California coast. *Acoustics Research Letters Online* 3: 65–70.

Au, W.W.L. (1993) *The Sonar of Dolphins*. New York: Springer.

Bailey, H., Senior, B., Simmons, D., Rusin, J., Picken, G. & Thompson, P.M. (2010) Assessing underwater noise levels during pile-driving at an offshore windfarm and its potential effects on marine mammals. *Marine Pollution Bulletin* 60: 888–897.

Bailey, H., Brookes, K.L. & Thompson, P.M. (2014) Assessing environmental impacts of offshore wind farms: lessons learned and recommendations for the future. *Aquatic Biosystems* 10: 8. doi: 10.1186/2046-9063-10-8.

Beck, C., Bonde, R. & Rathbun, G. (1982) Analyses of propeller wounds on manatees in Florida. *Journal of Wildlife Management* 46: 531–535.

Brandt, M.J., Diederichs, A., Betke, K. & Nehls, G. (2011) Responses of harbour porpoises to pile driving at the Horns Rev II offshore wind farm in the Danish North Sea. *Marine Ecology Progress Series* 421: 205–216.

Brandt, M.J., Höschle, C., Diederichs, A., Betke, K., Matuschek, R. & Nehls, G. (2013) Seal scarers as a tool to deter harbour porpoises from offshore construction sites. *Marine Ecology Progress Series* 475: 291–302.

Brandt, M.J., Hansen, S., Diederichs, A. & Nehls, G. (2014) Do man-made structures and water depth affect the diel rhythms in click recordings of harbor porpoises (*Phocoena phocoena*)? *Marine Mammal Science* 30: 1109–1121.

Brandt, M.J., Dragon, A.-C., Diederichs, A., Schubert, A., Kosarev, V., Nehls, G., Wahl, V., Michalik, A., Braasch, A., Hinz, C., Ketzer, C., Todeskino, D., Gauger, M., Laczny, M. & Piper, W. (2016) Effects of offshore pile driving on harbour porpoise abundance in the German Bight 2009–2013. Oldenburg, Neu Broderstorf & Husum: IBL Umweltplanung,

Institut für Angewandte Ökosystemforschung & BioConsult SH. Retrieved 25 June 2018 from https://www.offshore-stiftung.de/sites/offshore-link.de/files/documents/Study_Effects%20of%20offshore%20pile%20driving%20on%20harbour%20porpoise%20abundance%20in%20the%20German%20Bight_0.pdf

Brandt, M.J., Dragon, A.C. , Diederichs, A., Bellmann, M.A., Wahl, V., Piper, W., Nabe-Nielsen, J. & Nehls, G. (2018) Disturbance of harbour porpoises during construction of the first seven offshore wind farms in Germany. *Marine Ecology Progress Series* 596: 213–232.

Cerchio, S., Strindberg, S., Collins, T., Bennett, C. & Rosenbaum, H. (2014) Seismic surveys negatively affect humpback whale singing activity of Northern Angola. *PLoS ONE* 9: e86464. doi: 10.1371/journal.pone.0086464.

Clark, C.W., Ellison, W.T., Southall, B.L., Hatch, L., van Parijs, S.M., Frankel, A. & Ponirakis, D. (2009) Acoustic masking in marine ecosystems: intuitions, analysis, and implication. *Marine Ecology Progress Series* 395: 201–222.

Cranford, T.W. & Krysl, P. (2015) Fin whale sound reception mechanisms: skull vibration enables low-frequency hearing. *PLoS ONE* 10(1): e0116222. doi: 10.1371/journal.pone.0116222.

Dähne, M., Gilles, A., Lucke, K., Peschko, V., Adler, S., Krügel, K., Sundermeyer, J. & Siebert, U. (2013) Effects of pile-driving on harbour porpoises (*Phocoena phocoena*) at the first offshore wind farm in Germany. *Environmental Research Letters* 8: 025002. doi: 10.1088/1748-9326/8/2/025002

Davidson, A.D., Boyer, A.G., Kim, H., Pompa-Mansilla, S., Hamilton, M.J., Costa, D.P., Ceballos, G. & Brown, J.H. (2012) Drivers and hotspots of extinction risk in marine mammals. *Proceedings of the National Academy of Sciences of the United States of America* 109: 3395–3400.

Diederichs, A., Hennig, V. & Nehls, G. (2008) Investigations of the bird collision risk and the responses of harbour porpoises in the offshore wind farms Horns Rev, North Sea, and Nysted, Baltic Sea, in Denmark. Part II: Harbour porpoises. Final Report 2008. Hamburg & Husum: Universität Hamburg & BioConsult SH. Retrieved 5 June 2018 from http://bioconsult-sh.de/site/assets/files/1397/1397.pdf

Diederichs, A., Pehlke, H., Nehls, G., Bellmann, M., Gerke, P., Oldeland, J., Grunau, C., Witte, S. & Rose, A. (2014) Entwicklung und Erprobung des Großen Blasenschleiers zur Minderung der Hydroschallemissionen bei Offshore-Rammarbeiten. Schlussbericht. OWP Borkum West II: Baumonitoring und Forschungsprojekt HYDROSCHALL-OFF BW II. Husum, Lübeck & Oldenburg: BioConsult SH, Hydrotechnik Lübeck & itap. Retrieved 5 June 2018 from https://www.researchgate.net/profile/Hendrik_Pehlke/publication/272172115_Entwicklung_und_Erprobung_des_Grossen_Blasenschleiers_zur_Minderung_der_Hydroschallemissionen_bei_Offshore-Rammarbeiten_Schlussbericht/links/54ddb5fd0cf28a3d93fa245b/Entwicklung-und-Erprobung-des-Grossen-Blasenschleiers-zur-Minderung-der-Hydroschallemissionen-bei-Offshore-Rammarbeiten-Schlussbericht.pdf

Dolman, S.J., Williams-Grey, V., Asmutis-Silvia, R. & Isaac, S. (2006) Vessel collisions and cetaceans: what happens when they don't miss the boat. A WDCS Science Report. Chippenham: Whale and Dolphin Conservation Society. Retrieved 26 June 2018 from http://www.wdcs.co.uk/media/submissions_bin/shipstrikes.pdf

Dragon, A.-C., Brandt, M.J., Diederichs, A. & Nehls, G. (2016) Wind creates a natural bubble curtain mitigating porpoise avoidance during offshore pile driving. *Proceedings of Meetings on Acoustics* 27: 070022. doi.org/10.1121/2.0000421

Edrén, S.M.C., Teilmann, J., Dietz, R. & Carstensen, J. (2004) Effect from the construction of Nysted offshore wind farm on seals in Rødsand seal sanctuary based on remote video monitoring. Technical Report. Copenhagen: Ministry of the Environment, Denmark. Retrieved 26 June 2018 from https://www.researchgate.net/profile/Rune_Dietz/publication/268411508_Effects_from_the_Construction_of_Nysted_Offshore_Wind_Farm_on_Seals_in_Rodsand_Seal_Sanctuary_Based_on_Remote_Video_Monitoring/links/54be60360cf218da9391e822/Effects-from-the-Construction-of-Nysted-Offshore-Wind-Farm-on-Seals-in-Rodsand-Seal-Sanctuary-Based-on-Remote-Video-Monitoring.pdf

Ellison, W. T., Southall, B. L., Clark, C. W. & Frankel, A. S. (2011) A new context-based approach to assess marine mammal behavioral responses to anthropogenic sounds. *Conservation Biology* 26: 21–28.

Energinet (2015) Underwater noise and marine mammals. Rev.4, 17 April 2015. Retrieved from http://naturstyrelsen.dk/media/162610/underwater-noise-and-marine-mammals_2392023_rev4.pdf

European Union (1992) Council Directive 92/43/EEC of the Council of the European Communities of 21 May 1992 on the conservation of natural habitats and of wild fauna and flora. *Official Journal of the European Communities*, L206/7 (22.7.92). Retrieved 7 September 2016 http://ec.europa.eu/environment/nature/legislation/habitatsdirective/index_en.htm

Fugro Marine GeoServices Inc. (2017) Geophysical and geotechnical investigation methodology assessment for siting renewable energy facilities on the Atlantic OCS. OCS Study No. BOEM 2017-049. Herndon, VA: US Department of the Interior, Bureau of Ocean Energy Management, Office of Renewable Energy Programs. Retrieved 26 June 2018 from https://www.boem.gov/G-and-G-Methodology-Renewable-Energy-Facilities-on-the-Atlantic-OCS/

Gell, F.R. & Roberts, C.M. (2003) Benefits beyond boundaries: the fishery effects of marine reserves. *Trends in Ecology & Evolution* 18: 448–455.

Gellatly, B. (2013) *Operations & Maintenance Special Report.* London: WindPower Offshore.

Goldstein, T., Johnson, S.P., Philips, A.V., Hanni, K.D., Fauquier, D.A. & Golland, F.M.D. (1999) Human-related injuries observed in live stranded pinnipeds along the central California coast 1986–1998. *Aquatic Mammals* 25: 43–51.

Gomez, C., Lawson, J.W., Wright, A.J., Buren, A.D. Tollit, D. & Lesage, V. (2016) A systematic review on the behavioural responses of wild marine mammals to noise: the disparity between science and policy. *Canadian Journal of Zoology* 94: 801–819.

Gordon, J.C.D., Gillespie, D., Potter, J., Frantzis, A., Simmonds, M.P., Swift, R. & Thompson, D. (2003) A review of the effects of seismic surveys on marine mammals. *Marine Technology Society Journal* 37: 16–34.

Graham, I.M., Pirotta, E., Merchant, N.D., Farcas, A., Barton, T.R., Cheney, B., Hastie, G.D. & Thompson, P.M. (2017) Responses of bottlenose dolphins and harbor porpoises to impact and vibration piling noise during harbor construction. *Ecosphere* 8(5): e01793. doi: 10.1002/ecs2.1793

Haelters, J., Van Roy, W., Vigin, L. & Degraer, S. (2012) The effect of pile driving on harbour porpoises in Belgian waters. In Degraer, S., Brabant, R. & Rumes, B. (eds) *Offshore Windfarms in the Belgian Part of the North Sea: Heading for an understanding of environmental impacts.* Brussels: Royal Belgian Institute of Natural Sciences, Management Unit of the North Sea Mathematical Models, Marine Ecosystem Management Unit. pp. 127–143.

Hall, A.J., Watkins, J. & Hammond, P.S. (1998) Seasonal variation in the diet of harbour seals in the south-western North Sea. *Marine Ecology Progress Series* 170: 269–281.

Harwood, J., King, S., Booth, C., Donovan, C., Schick, R.S., Thomas, L. & New, L. (2016) Understanding the population consequences of acoustic disturbance for marine mammals. In Popper, A.N. & Hawkins, A. (eds) *The Effects of Noise on Aquatic Life II.* New York: Springer. pp. 417–423.

Harwood, A.J.P., Perrow, M.R., Berridge, R., Tomlinson, M.L. & Skeate, E.R. (2017) Unforeseen responses of a breeding seabird to the construction of an offshore wind farm. In Köppel, J. (ed.) *Conference on Wind Energy and Wildlife Interactions Presentations from the CWW2015 conference.* Cham: Springer International Publishing. pp. 19–41.

Hastie, G.D., Russell, D.J.F., McConnell, B., Moss, S., Thompson, D. & Janik, V.M. (2015) Sound exposure in harbour seals during the installation of an offshore wind farm: predictions of auditory damage. *Journal of Applied Ecology* 52: 631–640.

International Union for Conservation of Nature and Nature Resources (2017) IUCN Red List of Threatened Species (Version 2017.1). Retrieved 24 August 2017 from http://www.iucnredlist.org

Jansen, J.K., Brady, G.M., ver Hoef, J.M. & Boveng, P.L. (2015) Spatially estimating disturbance of harbour seals (*Phoca vitulina*). *PLoS ONE* 10: e0129798. doi: 10.1371/journal.pone.0129798

Jensen, F.H., Bejder, L., Wahlberg, M., Aguilar De Soto, N., Johnson, M.P. & Madsen, P.T. (2009) Vessel noise effects on delphinid communication. *Marine Ecology Progress Series* 395: 161–175.

Jones, E.L., Hastie, G.D., Smout, S., Onoufriou, J., Merchant, N.D., Brookes, K.L. & Thompson, D. (2017) Seals and shipping: quantifying population risk and individual exposure to vessel noise. *Journal of Applied Ecology* 54: 1930–1940.

Kastak, D., Southall, B.L., Schusterman, R.J. & Reich-muth-Kastak, C. (2005) Underwater temporary threshold shift in pinnipeds: Effects of noise level and duration. *Journal of the Acoustical Society of America* 118: 3154–3163.

Kastelein, R.A., Helder-Hoek, L., van de Voorde, S., von Benda-Beckmann, A.M., Lam, F.-P.A., Jansen, E., de Jong, C.A.F. & Ainslie, M.A. (2017a) Temporary hearing threshold shift in a harbor porpoise (*Phocoena phocoena*) after exposure to multiple airgun sounds. *Journal of the Acoustical Society of America* 142: 2430–2442.

Kastelein, R.A., Huybrechts, J., Covi, J. & Helder-Hoek, L. (2017b) Behavioral responses of a harbor porpoise (*Phocoena phocoena*) to sounds from an acoustic porpoise deterrent. *Aquatic Mammals* 43: 233–244.

King, S., Schick, R.S., Donovan, C., Booth, C.G., Burgman, M., Thomas, L. & Harwood, J. (2015) An interim framework for assessing the population consequences of disturbance. *Methods in Ecology and Evolution* 6: 1150–1158.

Knust, R., Dahlhoff, P., Gabriel, J., Heuers, J., Hüppop, O. & Wendeln, H. (2003) Untersuchungen zur Vermeidung und Verminderung von Belastungen der Meeresumwelt durch Offshore-Windenergieanlagen im küstenfernen Bereich der Nord-und Ostsee: Offshore-WEA. Final Report No. 20097106, UBA-FB 000478. Berlin: Umweltbundesamt, Umweltforschungsplan des Bundesministeriums für Umwelt, Naturschutz und Reaktorsicherheit.

La Manna, G., Manghi, M., Pavan, G., Lo Mascolo, F. & Sarà, G. (2013) Behavioural strategy of common bottlenose dolphins (*Tursiops truncatus*) in response to different kinds of boats in the waters of Lampedusa Island (Italy). *Aquatic Conservation: Marine and Freshwater Ecosystems* 23: 745–757.

Leonhard, S.B., Pedersen, J., Klaustrup, M. & Hvidt, C.B. (2006) Benthic communities at Horns Rev before, during and after construction of Horns Rev offshore wind farm. Final Report. Annual Report 2005. Vattenfall. p. 78. Retrieved 26 June 2018 from https://corporate.vattenfall.dk/globalassets/danmark/om_os/horns_rev/benthic-communities-at-horns.pdf

Lindeboom, H.J., Kouwenhoven, H.J., Bergman, M.J.N., Bouma, S., Brasseur, S., Daan, R., Fijn, R.C., de Haan, D., Dirksen, S., van Hal, R., Lambers, R.H.R., ter Hofstede, R., Krijgsveld, K.L., Leopold, M. & Scheidat, M. (2011) Short-term ecological effects of an offshore wind farm in the Dutch coastal zone; a compilation. *Environmental Research Letters* 6(3): 1–13.

Linley, E.A.S., Wilding, T.A., Black, K., Hawkins, A.J.S. & Mangi, S. (2007) Review of the reef effects of offshore wind farm structures and their potential for enhancement and mitigation. Report to the Department Business, Enterprise and Regulatory Reform (BERR). PML Applications Ltd in association with the Scottish Association for Marine Sciences (SAMS). Contract No. RFCA/005/0029P. Retrieved 26 June 2018 from http://webarchive.nationalarchives.gov.uk/+/http:/www.berr.gov.uk/files/file43528.pdf

Lucke, K., Lepper, P.A., Hoeve, B., Everaarts, E., van Elk, N. & Siebert, U. (2007) Perception of low-frequency acoustic signals by a harbour porpoise (*Phocoena phocoena*) in the presence of simulated offshore wind turbine noise. *Aquatic Mammals* 33: 55–68.

Lucke, K., Siebert, U., Lepper, P.A. & Blanchet, M.-A. (2009) Temporary shift in masked hearing thresholds in a harbor porpoise (*Phocoena phocoena*) after exposure to seismic airgun stimuli. *Journal of the Acoustical Society of America* 125: 4060–4070.

Lusseau, D., Christiansen, F., Harwood, J., Mendes, S., Thompson, P.M., Smith, K. & Hastie, G.D. (2012) *Assessing the* risks to marine mammal populations from renewable energy devices – an interim approach. Countryside Council for Wales and NERC Knowledge Exchange Programme. Peterborough: Joint Nature Conservation Committee. Retrieved 26 June 2018 from https://www.abdn.ac.uk/lighthouse/documents/CCW_JNCC_NERC_workshop_final_report.pdf

Mackenzie-Maxon, C. (2015) Smalands Farvandet Offshore Wind Farm underwater noise. Ramboll report to Energinet. Retrieved from http://naturstyrelsen.dk/media/162580/underwater-noise_smaalandsfarvandet.pdf

Magera, A.M., Mills Flemming, J.E., Kaschner, K., Christensen, L.B. & Lotze, H.K. (2013) Recovery trends in marine mammal populations. *PLoS ONE* 8 (10): e77908. doi: 10.1371/journal.pone.0077908.

Marmo, B. (2013) Modelling of noise effects of operational offshore wind turbines including noise transmission through various foundation types. Marine Scotland. Edinburgh: Scottish Government. Retrieved 26 June 2018 from https://tethys.pnnl.gov/sites/default/files/publications/Marmo_et_al_2013.pdf

Matuschek, R. & Betke, K. (2009) Measurements of construction noise during pile driving of offshore research platforms and wind farms. In Boone, M. (ed.) *Proceedings of the NAG/DAGA International Conference on Acoustics, Rotterdam, March 2009*. pp. 262–265. Retrieved 26 June 2018 from https://tethys.pnnl.gov/sites/default/files/publications/Matuschek_and_Betke_2009.pdf

Morton, A. & Symonds, H. (2002) Displacement of *Orcinus orca* (L.) by high amplitude sound in British Columbia, Canada. *ICES Journal of Marine Science* 59: 71–80.

Nabe-Nielsen, J. & Harwood, J. (2016) Comparison of the iPCoD and DEPONS models for modelling population consequences of noise on harbour porpoises. Scientific Report from DCE – Danish Centre for Environment and Energy, Aarhus University. No. 186. Retrieved 26 June 2018 from http://dce2.au.dk/pub/SR186.pdf

Nabe-Nielsen, J., Sibly, R.M., Tougaard, J., Teilmann, J. & Sveegaard, S. (2014) Effects of noise and by-catch on a Danish harbour porpoise population. *Ecological Modelling* 272: 242–251.

Nachtigall, P.E., Pawloski, J.L. & Au, W.W.L. (2003) Temporary threshold shifts and recovery following noise exposure in the Atlantic bottlenose dolphin (*Tursiops truncatus*). *Journal of the Acoustical Society of America* 113: 3425–3429.

Nachtigall, P.E., Supin, A.Y., Pawlowski, J.L. & Au, W.W.L. (2004) Temporary threshold shifts after noise exposure in the bottlenose dolphin (*Tursiops truncatus*) measured using evoked auditory potentials. *Marine Mammal Science* 20: 673–687.

Nachtigall, P.E., Supin, A.Y., Pacini, A.F. & Kastelein, R.A. (2016) Conditioned hearing sensitivity change in the harbor porpoise (*Phocoena phocoena*). *Journal of the Acoustical Society of America* 140: 960–967.

National Marine Fisheries Service (2018) 2018 Revisions to: Technical guidance for assessing the effects of anthropogenic sound on marine mammal hearing (version 2.0): underwater thresholds for onset of permanent and temporary threshold shifts. US Department of Commerce, NOAA. NOAA Technical Memorandum NMFS-OPR-59. Retrieved 26 June 2018 from https://www.fisheries.noaa.gov/national/marine-mammal-protection/marine-mammal-acoustic-technical-guidance

Nehls, G., Mueller-Blenkle, C., Dorsch, M., Girardello, M., Gauger, M., Laczny, M., Meyer-Loebbecke, A. & Wengst, N. (2014) Horns Rev 3 Offshore Wind Farm. Marine mammals. Orbicon report to Energinet. Retrieved 26 June 2018 from https://ens.dk/sites/ens.dk/files/Vindenergi/marine_mammals_v3.pdf

Nehls, G. & Bellmann, M. (2016) Weiterentwicklung und Erprobung des 'Großen Blasenschleiers' zur Minderung der Hydroschallemissionen bei Offshore-Rammarbeiten. Final Report. Project No. 0325645A/B/C/D. Husum & Oldenburg: BioConsult & itap. Retrieved 26 June 2018 from http://hydroschall.de/wp-content/uploads/2016/03/BMWi-FKZ0325645ABCD_Weiterentwicklung-und-Erprobung-des-gro%C3%9Fen-Blasenschleiers.pdf

Nehls, G., Rose, A., Diederichs, A., Bellmann, M. & Pehlke, H. (2016) Noise mitigation during pile driving efficiently reduces disturbance of marine mammals. *Advances in Experimental Medicine and Biology* 875: 755–762.

Norro, A., Degraer, S. (2016) Quantification and characterisation of Belgian offshore wind farm operational sound emission at low wind speeds. In: Degraer, S., Brabant, R., Rumes, B. & Vigin, L. (eds) *Environmental Impacts of Offshore Wind Farms in the Belgian Part of the North Sea: Environmental Impact Monitoring Reloaded*. Royal Belgian Institute of Natural Sciences, OD Natural Environment, Marine Ecology and Management Section. Retrieved 26 June 2018 from http://odnature.naturalsciences.be/downloads/mumm/windfarms/winmon_report_2016.pdf

Nowacek, S.M., Wells, R.S. & Solow, A.R. (2001) Short-term effects of boat traffic on bottlenose dolphins, *Tursiop truncations*, in Sarasota Bay, Florida. *Marine Mammal Science* 17: 673–688.

Nowacek, D.P., Thorne, L.H., Johnston, D.W. & Tyack, P.L. (2007) Responses of cetaceans to anthropogenic noise. *Mammal Review* 37: 81–115.

Onoufriou, J., Jones, E., Hastie, G. & Thompson, D. (2016) Investigations into the interactions between harbour seals (*Phoca vitulina*) and vessels in the inner Moray Firth. *Scottish Marine and Freshwater Science* 7(24). doi: 10.7489/1805-1.

Panigada, S., Pesante, G., Zanardelli, M., Capoulade, F., Gannier, A. & Weinrich, M.T. (2006) Mediterranean fin whales at risk from fatal ship strikes. *Marine Pollution Bulletin* 52: 1287–1298.

Petersen, J.K. & Malm, T. (2006) Offshore windmill farms: threats to or possibilities for the marine environment. *AMBIO: A Journal of the Human Environment* 35: 75–80.

Read, A.J. (2008) The looming crisis: interactions between marine mammals and fisheries. *Journal of Mammalogy* 89: 541–548.

Reichmuth, C., Holt, M.M., Mulsow, J., Sills, J.M. & Southall, B.L. (2013) Comparative assessment of amphibious hearing in pinnipeds. *Journal of Comparative Physiology A* 199: 491–507.

Richardson, W.J., Greene, C.R., Jr, Malme, C.I. & Thomson, D.H. (1995) *Marine Mammals and Noise*. San Diego, CA: Academic Press.

Robinson, S.P. & Theobald, P. (2017) An international standard for the measurement of underwater sound radiated from marine pile-driving. *Journal of the Acoustical Society of America* 141: 3847.

Russell, D.J., Brasseur, S.M., Thompson, D., Hastie, G.D., Janik, V.M., Aarts, G., McClintock, B.T., Matthiopoulos, J., Moss, S.E. & McConnell, B. (2014) Marine mammals trace anthropogenic structures at sea. *Current Biology* 24: R638–R639.

Russell, D.J.F., Hastie, G.D., Thompson, D., Janik, V.M., Hammond, P.S., Scott-Hayward, L.A.S., Matthiopoulos, J., Jones, E.L. & McConnell, B.J. (2016) Avoidance of wind farms by harbour seals is limited to pile driving activities. *Journal of Applied Ecology* 53: 1642–1652.

Scheidat, M., Tougaard, J., Brasseur, S., Carstensen, J., van Polanen Petel, T., Teilmann, J. & Reijnders, P. (2011) Harbour porpoises (*Phocoena phocoena*) and wind farms: a case study in the Dutch North Sea. *Environmental Research Letters* 6/2, p: 025102.

Schlundt, C.E., Finneran, J.J., Carder, D.A. & Ridgway, S.H. (2000) Temporary shift in masked hearing thresholds of bottlenose dolphins, *Tursiops truncatus*, and white whales, *Delphinapterus leucas*, after exposure to intense tones. *Journal of the Acoustical Society of America* 107: 3496–3508.

Skeate, E.R., Perrow, M.R. & Gilroy, J.J. (2012) Likely effects of construction of Scroby Sands offshore wind farm on a mixed population of harbour *Phoca vitulina* and grey *Halichoerus grypus* seals. *Marine Pollution Bulletin* 64: 872–881.

Society for Marine Mammalogy, Committee on Taxonomy (2017) List of marine mammal species and subspecies. Anacortes, WA: Society for Marine Mammalogy. Retrieved 26 June 2018 from https://www.marine-mammalscience.org/species-information/list-marine-mammal-species-subspecies/

Southall, B.L., Bowles, A.E., Ellison, W.T., Finneran, J.J., Gentry, R.L., Greene, C.R., Kastak, D., Ketten, D.R., Miller, J.H., Nachtigall, P.E., Richardson, W.J., Thomas, J.A. & Tyack, P.L. (2007) Marine mammal noise exposure criteria: initial scientific recommendations. *Aquatic Mammals* 33: 411–521.

Stenberg, C., van Deurs, M., Støttrup, J.G., Mosegaard, H., Grome, T.M., Dinesen, G.E., Christensen, A., Jensen, H., Kaspersen, M., Berg, C.W., Leonhard, S.B., Skov, H., Pedersen, J., Hvidt, C.B. & Klaustrup, M. (2011) Effect of the Horns Rev 1 offshore wind farm on fish communities. Follow-up seven years after construction. Leonhard, S.B., Stenberg C. & Støttrup, J.G. (eds) DTU Aqua Report No. 246-2011. Charlottenlund: DTU Aqua. Institut for Akvatiske Ressourcer. Retrieved 26 June 2018 from http://orbit.dtu.dk/files/7615058/246_2011_effect_of_the_horns_rev_1_offshore_wind_farm_on_fish_communities.pdf

Teilmann, J., Tougaard, J., Carstensen, J., Dietz, R. & Tougaard, S. (2006) Summary on seal monitoring 1999–2005 around Nysted and Horns Rev Offshore Wind Farms. Report to Energi E2 A/S and Vattenfall A/S. Report No. 2389313244. National Environmental Research Institute. Retrieved 26 June 2018 from https://corporate.vattenfall.dk/globalassets/danmark/om_os/horns_rev/summary-on-harbour-porpoise-m.pdf

Thompson, P.M., Mcconnell, B.J., Tollit, D.J., Mackay, A., Hunter, C. & Racey, P.A. (1996) Comparative distribution, movements and diet of harbour and grey seals from Moray Firth, N. E. Scotland. *Journal of Applied Ecology* 33: 1572–1584.

Thompson, P.M., Lusseau, D., Barton, T., Simmons, D., Rusin, J. & Bailey, H. (2010) Assessing the responses of coastal cetaceans to the construction of offshore wind turbines. *Marine Pollution Bulletin* 60: 1200–1208.

Thompson, P.M., Hastie, G.D., Nedwell, J., Barham, R., Brookes, K.L., Cordes, L.S., Bailey, H. & McLean, N. (2013) Framework for assessing impacts of pile-driving noise from offshore wind farm construction on a harbour seal population. *Environmental Impact Assessment Review* 43: 73–85.

Thomsen, F., Lüdemann, K., Kafemann, R. & Piper, W. (2006) Effects of offshore wind farm noise on marine mammals and fish. Hamburg: Biola, on behalf of COWRIE Ltd. Retrieved from https://www.thecrownestate.co.uk/media/5935/km-ex-pc-noise-062006-effects-of-offshore-windfarm-noise-on-marine-mammals-and-fish.pdf

Topham, E. & McMillan, D. (2017) Sustainable decommissioning of an offshore wind farm. *Renewable Energy* 102: 470–480.

Tougaard, J. & Mikaelsen, M.A. (2017) Taiwanese white dolphins and offshore wind farms. Scientific Report No. 245. Aarhus University, DCE – Danish Centre for Environment and Energy. Retrieved from http://dce2.au.dk/pub/SR245.pdf

Tougaard, J., Carstensen, J., Teilmann, J., Skov, H. & Rasmussen, P. (2009a) Pile driving zone of responsiveness extends beyond 20 km for harbor porpoises (*Phocoena phocoena* (L.)). *Journal of the Acoustical Society of America* 126: 11–14.

Tougaard, J., Henriksen, O.D. & Miller, L.A. (2009b) Underwater noise from three types of offshore wind turbines: estimation of impact zones for harbor porpoises and harbor seals. *Journal of the Acoustical Society of America* 125: 3766–3773.

van Beest, F.M., Nabe-Nielsen, J., Carstensen, J., Teilmann, J. & Tougaard, J. (2015) Disturbance effects on the harbour porpoise population in the North Sea (DEPONS): status report on model development. Scientific Report from DCE – Danish Centre for Environment and Energy. No. 140. Retrieved 25 June 2018 from http://dce2.au.dk/pub/SR140.pdf

van Beest, F.M., Teilmann, J., Hermannsen, L., Galatius, A., Mikkelsen, L., Sveegaard, S., Balle, J.D., Diet,z R. & Nabe-Nielsen, J. (2018) Fine-scale movement responses of free-ranging harbour porpoises to capture, tagging and short-term noise pulses from a single airgun. *Royal Society Open Science* 5: 170110. van Hal, R., Griffioen, A.B. & van Keeken, O.A. (2017) Changes in fish communities on a small spatial scale, an effect of increased habitat complexity by an offshore wind farm. *Marine Environmental Research* 126: 26–36.

van Waerebeek, K., Baker, A.N., Félix, F., Gedamke, J., Iñiguez, M., Sanino, G.P., Secchi, E., Sutaria, D., van Helden, A. & Wang, Y. (2007) Vessel collisions with small cetaceans worldwide and with large whales in the Southern Hemisphere, an initial assessment. *Latin American Journal of Aquatic Mammals* 6(1). doi: 10.5597/lajam00109.

von Benda-Beckmann, A.M., Aarts, G., Sertlek, H.Ö., Lucke, K., Verboom, W.C., Kastelein, R.A., Ketten, D.R., van Bemmelen, R., Lam, F.-P.A., Kirkwood, R.J. & Ainslie, M.A. (2015) Assessing the impact of underwater clearance of unexploded ordnance on harbour porpoises (*Phocoena phocoena*) in the southern North Sea. *Aquatic Mammals* 41: 503–523.

Wartzok, D. & Ketten, D.R. (1999) Marine mammal sensory systems. In Reynolds, J.E. III & Rommel, S.A. (eds) *Biology of Marine Mammals*. Washington, DC: Smithsonian Institution Press. pp. 117–175.

Wilhelmsson, D. & Langhamer, O. (2014) The influence of fisheries exclusion and addition of hard substrata on fish and crustaceans. In Shields, M.A. & Payne, I.L. (eds) *Marine Renewable Energy Technology and Environmental Interactions. Humanity and the Sea Series*. Dordrecht: Springer. pp. 49–60.

Wilson, B., Batty, R., Daunt, F. & Carter, C. (2006) Collision risks between marine renewable energy devices and mammals, fish and diving birds. Report to the Scottish Executive. No. PA37 1QA. Oban: Scottish Association for Marine Science. Retrieved 25 June 2018 from https://depts.washington.edu/nnmrec/workshop/docs/Wilson_Collisions_report_final_12_03_07.pdf

Young, G.A. (1991) Concise methods for predicting the effects of underwater explosions on marine life. Silver Spring, MD: Naval Surface Warfare Center. Retrieved 25 June 2018 from http://www.dtic.mil/dtic/tr/fulltext/u2/a241310.pdf

CHAPTER 7

Migratory birds and bats

OMMO HÜPPOP, BIANCA MICHALIK, LOTHAR BACH,
REINHOLD HILL and STEVEN K. PELLETIER

Summary

The proportion of terrestrial bird species and bats migrating over the sea is much higher than one might assume. Especially during autumn migration, remarkable numbers of birds and bats can be recorded offshore. Consequently, the rapid, worldwide advance of offshore wind power facilities may pose a significant threat to migrating birds and bats in terms of increased energy expenditure and higher collision risk. The extent and the behaviour of birds and bats at offshore wind farms (OWFs) are, however, dependent on a variety of intrinsic, environmental, site- and species-specific conditions. During daytime and under good visibility as well as during clear nights, many bird species avoid entering OWFs, fly well above rotor height or fly between the turbine rows. Under these conditions, effects on flight energetics are mainly negligible and the risk of collision is generally low. In deteriorating weather situations, flight altitudes of birds at sea normally decrease and artificially lit structures may yield particular attraction. As a consequence, energy expenditure as well as collision risk may increase. Bats are observed to fly low over the sea but may also migrate at higher altitudes. However, some bats change their altitude rapidly when they approach tall vertical obstacles and, onshore, bats have been observed to feed near the turbine blades or to roost at the nacelles. Technical and logistical constraints associated with assessing actual numbers of bird and bat collisions with offshore wind turbines remain challenging. Consequently, there are only very few studies in this context highlighting which species or species groups may be particularly vulnerable, especially under adverse circumstances.

Introduction

Migration is a widespread phenomenon in all major branches of the animal kingdom. It can take a variety of temporal, spatial and population-related (such as differences between sex and age classes) forms and has been described by biologists in many ways.

Migration can be obligate, facultative, and partial, and sometimes varies between different populations and even individuals within a species. It allows animals to take advantage of temporally predictable foraging and breeding conditions in different areas that cannot be used simultaneously. A complex combination of internal physiological rhythms and external cues such as photoperiod, weather, food, energy, and other resources and social stimuli influence decisions to migrate at a given time and site (Alerstam 1990; Dingle & Drake 2007; Dingle 2014; Berdahl *et al.* 2017; Gnanadesikan *et al.* 2017).

Every day and night worldwide, there are birds and bats on the move. For example, 2.1 billion songbirds and near-passerine birds migrate between the Palearctic and Africa (Hahn *et al.* 2009; BirdLife International 2010) to reach their breeding grounds and then go back to their non-breeding areas.

Globally, there are about nine major bird flyways, or general flight paths of migrating birds, connecting the northern and southern hemispheres (BirdLife International 2010). A flyway encompasses the whole life cycle of a migratory bird (Boere & Stroud 2006). Although originally developed for waterbirds, the flyway concept is generally useful for the study and description of bird migration (BirdLife International 2010). All flyways include large bodies of marine water (Figure 7.1). In the Western Palaearctic, the most prominent flyway is the East Atlantic flyway (red in Figure 7.1), which links a discontinuous band of breeding grounds, stretching in the Arctic from eastern Canada to Siberia, as well as over central, northern, and western Europe, with the non-breeding areas in Europe on to southern Africa (BirdLife International 2010). The Wadden Sea forms a particularly important key stopover site for waterbirds migrating along this flyway (Boere & Piersma 2012).

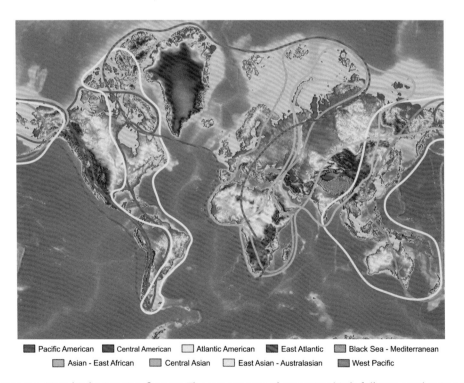

Pacific American   Central American   Atlantic American   East Atlantic   Black Sea - Mediterranean
Asian - East African   Central Asian   East Asian - Australasian   West Pacific

Figure 7.1 Main bird migration flyways. The movements of migrating birds follow a predominantly north-south axis linking geographically distinct breeding and non-breeding areas. (After Boere & Stroud 2006; BirdLife International 2010; Map: BigMap by OpenStreetMaps)

A few birds from the high American Arctic are also part of the huge East Atlantic flyway but most North American bird species utilise a system of three different American flyways, again in huge numbers. Some will avoid following the Central American land bridge and instead directly cross the open Gulf of Mexico (Atlantic American flyway, light blue in Figure 7.1). The East Asian-Australasian flyway encompasses large parts of East Asia and extends from Arctic Russia and North America to the southern limits of Australia and New Zealand (yellow in Figure 7.1), with over 50 million migratory waterbirds, including 8 million waders. Huge numbers of passerine and near-passerine migrants use this flyway too, many travelling to south-east Asia from breeding grounds in eastern and central Siberia (BirdLife International 2010).

Many migrating birds and bats almost inevitably meet anthropogenic offshore structures such as wind farms or oil and gas-producing platforms at least twice a year, especially in the North and Baltic Seas or in the Gulf of Mexico. The dimension of worldwide offshore wind energy development raises several concerns over its effects upon wildlife. Nearly 88% (12,631 MW or 12.6 GW) of the worldwide offshore wind capacity is installed in waters off the European coasts (GWEC 2017). Overall, in 2016, there were 81 European offshore wind farms (OWFs) consisting of 3,589 turbines in ten north-west European countries. Installations in the North Sea account for 72% of all offshore wind capacity in Europe (GWEC 2017). The remaining 12% of the global offshore wind installations are located mainly off the coasts of China, Japan, and South Korea (GWEC 2017).

Governments of other countries are also setting ambitious targets for offshore wind power development. For example, beyond the first Rhode Island OWF which consists of five turbines and became operational in 2016, there are currently 17 projects off the Atlantic, Pacific and Great Lakes coasts in the USA under various stages of development (AWEA 2017; GWEC 2017). Worldwide, the trend is moving towards larger areas and increasingly taller and more powerful wind turbines, resulting in an increase of OWFs in both the horizontal and vertical scales. For example, the average rotor diameter of offshore wind turbines is expected to rise from currently slightly over 105 m in 2017 to around 125 m in the early 2020s with a simultaneous increase in power from an average of 6 to 12 MW (Deign 2018). Individual turbine types will soon reach a height of 260 m and a rotor diameter of 220 m, with a rotor swept area equivalent to 38,000 m^2 or seven American football fields (GE 2018). There are presently no foreseeable limits on development.

Despite this rapid evolution of offshore technology, relatively little is known about the vulnerability of birds and bats around offshore structures. Depending on the kind of structure, species, location, seasons, weather and further factors, birds and bats may be unaffected, displaced or potentially even attracted to turbines, and accordingly more or less subjected to collision, or barotrauma in the case of bats (Barclay et al. 2017a), with fatal consequences.

Scope

This chapter reviews our understanding of the effects of anthropogenic offshore structures on migratory birds and bats and weighs their potential impact at both the taxonomic and geographic level. Main bird groups considered here include waterfowl, waders and other terrestrial birds, especially passerines. Effects on seabirds, most of which are also migratory in some way, are explicitly detailed by Vanermen & Stienen (Chapter 8) and King (Chapter 9).

We start with a brief overview of the characteristics of the migration of the named bird groups (Figure 7.2), as well as of bats; and of the factors affecting offshore migration.

Figure 7.2 (a) Waterfowl (e.g. Dark-bellied Brent Goose *Branta bernicla*), (b) waders (e.g. Grey Plover *Pluvialis squatarola*, Sanderling *Calidris alba*, Purple Sandpiper *Calidris maritima*, Dunlin *Calidris alpina*, Ruff *Philomachus pugnax*, Bar-tailed Godwit *Limosa lapponica*, Ruddy Turnstone *Arenaria interpres*), and (c) passerines (e.g. Snow Bunting *Plectrophenax nivalis*) are the major groups of birds (other than seabirds) detected during visual offshore observations in the North Sea. [(a, b) Jochen Dierschke; (c) Martin Perrow]

The chapter then has two major foci: (1) to compile the knowledge on the behaviour of birds and bats at anthropogenic offshore structures; and (2) to summarise the results on collisions at anthropogenic offshore structures. Most available evidence comes from studies at offshore or nearshore study sites in north-western Europe, mainly in the North and Baltic Seas. Since these sites also concentrate the vast majority of global OWFs, this review sets a geographic focus on north-western Europe, but also prospects the potential effects of planned offshore installations elsewhere. It is mainly based on peer-reviewed publications, supplemented by publicly accessible environmental impact studies, and can be seen as an extension of the studies by Hüppop *et al.* (2004; 2006; 2016) and Hill *et al.*

(2014a) for birds, and Hüppop and Hill (2016) and Peterson *et al.* (2016) for bats. Even though relatively little is known about the effects of anthropogenic offshore structures on migrating birds, even less is known for bats. As a consequence, this review is somewhat bird focused, but infers conclusions on bats from the little available evidence as well as from comparison with existing knowledge on night-migrating passerines. Compared to the relatively broad knowledge of the effects and impacts of wind turbines onshore (Perrow 2017), many aspects offshore still remain only speculative.

Themes

Characteristics of offshore migration of birds

Since the early voyages of Columbus, large numbers of passerines and shorebirds migrating from eastern North America to the Caribbean and South America have been reported from ships crossing the western North Atlantic (Williams *et al.* 1977). In Europe, the first written evidence of offshore landbird migration goes back to the sixteenth century, when 'incredible swarms of birds' were described for the tiny remote island of Helgoland in the North Sea (Stresemann 1967). For its inhabitants, the huge number of birds was a welcome supplement to their poor diets. Later, bird collectors and scientists became aware of the surprising dimension of offshore bird movements (Gätke 1895; Dierschke *et al.* 2011; Hüppop & Hüppop 2011). Offshore migration was confirmed by innumerable observations on other, sometimes far more remote islands, in both the old and the new world (Gosse 1847; Gundlach 1861; Drury & Keith 1962; Williams *et al.* 1977; DeSante & Ainley 1980; Williams & Williams 1990; Yong *et al.* 2015), on lighthouses or lightvessels (e.g. Blasius 1893–1894; Clarke 1912; Hansen 1954; Gauthreaux & Belser 2006; Vaughan 2009).

The intensity of offshore bird migration is generally higher in autumn than in spring and it peaks during the night and the early morning hours (e.g. Hüppop *et al.* 2006; Fijn *et al.* 2015). Migration intensities offshore vary substantially from day to day and are concentrated on a few days and nights, namely in areas with rapidly changing weather conditions such as the North Sea (Gruber & Nehls 2003; Hüppop *et al.* 2006), whereas the flux over, for example, the Caribbean Islands is more equalised (Richardson 1976). Accurate projections of the numbers of passing migrants are difficult to make and depend considerably on the method of observation (Molis *et al.*, Volume 4, Chapter 5). For example, long-term daytime migration counts at the offshore research platform FINO 1, situated in the North Sea about 45 km off the nearest coastal island (Hill *et al.* 2014a), provided estimates of approximately 140,000 birds, of which 15,000 were passerines, crossing the platform at an imaginary line of 6–20 km orthogonal to the north-east–south-west main migration direction each year (Hill *et al.* 2014a). Likewise, Dierschke (2003) estimated from visual daytime observations that about 1 million migrating seabirds, waterfowl and waders passed the remote island of Helgoland each year in a radius of 5–10 km. Of course, visual observations cannot cover nocturnal migration, and Fijn *et al.* (2015) calculated total numbers from daytime and night-time radar observations within an OWF about 10–18 km off the Dutch coast. Their data suggest more than 3 million birds (echoes) passing the whole wind-farm area of about 27 km² annually.

The extent of bird migration over the sea is largely dependent on the ecological needs and, as for example in raptors (Kerlinger 1985; Malmiga *et al.* 2014; Agostini *et al.* 2015; Nourani & Yamaguchi 2017), on the morphology of the flight apparatus of the respective species or species group. Whereas seabirds commonly avoid flying over land as they depend on open water for foraging and/or resting, songbirds, shorebirds and other terrestrial species mainly migrate along the coast as they are not able to rest at sea. At the German Bight of the North Sea, for example, littoral migration seems to be generally more pronounced than offshore migration (Hüppop et al. 2006; 2010). In contrast, over the Gulf of Mexico, steady wind currents may have shaped a relatively narrow band of intensive migration across the sea to arrive at a preferred stopover location at the northern Gulf of Mexico coast in spring (Gauthreaux et al. 2006; Cohen et al. 2017). At the Great Lakes in east-central North America, a great proportion of migratory birds crosses directly, but at some locations birds seem to avoid lake crossing (Diehl et al. 2003).

During the day, migrating landbirds can often be observed to follow the coastline and to refrain from crossing open water (Drury & Keith 1962; Hüppop et al. 2010). Nocturnal migrants seem generally less sensitive to a land/sea transition and rather cross water bodies in broad front migration (Myres 1964; Bruderer & Liechti 1998; Diehl et al. 2003; Archibald et al. 2017). However, several exceptions were found where coasts guide the routes of nocturnal migrants, such as in the Netherlands, Germany, and Nova Scotia (Richardson 1978a; Jellmann 1988; Buurma 1995). Over the course of a night, the proportion of migrants redirecting along or towards the coast or even reversing their flight direction may increase (Bruderer & Liechti 1998; Fortin et al. 1999; Hüppop et al. 2006; Horton et al. 2016; Nilsson & Sjöberg 2016; Archibald et al. 2017).

Radar observations revealed that the density of migrating landbirds is highest near continental margins, but significant migrations have also been detected more than 3,000 km from land (Williams & Williams 1990). Migrating terrestrial bird species repeatedly cross large bodies of water such as the Mediterranean Sea (Fortin et al. 1999; Meyer et al. 2000; Agostini et al. 2005; 2015) or the Gulf of Mexico (Robinson et al. 1996); the latter including the tiny, 3–4 g Ruby-throated Hummingbird Archilochus colubris. Birds are also known to 'island hop' before making the 800 km Caribbean Sea crossing to South America (BirdLife International 2010), and within the Atlantic American flyway, a few small songbird species even fly large distances over the western Atlantic Ocean to avoid detours along the coast (DeLuca et al. 2015).

Soaring birds, such as storks, pelicans, and many (broad-winged) raptors, are generally dependent for lift on thermals created by the differential warming of the land's surface or updrafts created by topography (Kerlinger 1985; Baisner et al. 2010; Malmiga et al. 2014; Miller et al. 2016; Nourani & Yamaguchi 2017). Most long-distance migratory birds of prey appear especially reluctant to fly even short distances over water, and instead migrate mainly, if not entirely, over land and circumvent larger bodies of water. They tend to concentrate in large numbers at isthmuses, islands and other geographic bottlenecks to take the shortest sea crossings (Williams et al. 1977; Looft & Busche 1981; Kerlinger 1984; 1985; Meyer et al. 2000; Corso 2001; Agostini et al. 2005; 2015; Berndt et al. 2005; Germi 2005; Bildstein 2006; Goodrich & Smith 2008; Hilgerloh 2009; Baisner et al. 2010; Farmer et al. 2010; Miller et al. 2016; Nourani & Yamaguchi 2017). For the largest species, such as Griffon Vulture Gyps fulvus, even crossing the short distance of 14 km over the Strait of Gibraltar is an enormous challenge, with many birds attempting passage for weeks before crossing (Bildstein et al. 2009). Such barriers can be associated with a high mortality in large soaring birds, mainly in inexperienced young (Oppel et al. 2015). Hence, it is not surprising that these species seek refuge at offshore structures when confronted with adverse weather (Mote 1969; Moore 2000). The 'island effect', whereby birds are attracted by small islands offering refuge, was offered by Skov et al. (2016) as a plausible mechanism for the apparent attraction of raptors to wind farms when crossing part of the Baltic Sea, particularly under adverse wind conditions. Such attraction enhances the threat to a group of species known to be at high risk of collision with wind turbines (e.g. Drewitt & Langston 2008; Thaxter *et al.* 2017; de Lucas & Perrow 2017). Smaller raptor species with longer, narrower wings, such as falcons and accipiters, are more adapted to flapping flight than larger species and are more likely to cross large water bodies, with some regularly making long-distance movements across open water (Kerlinger 1985; Meyer *et al.* 2000; Bildstein 2006; Goodrich & Smith 2008; Concepcion *et al.* 2017).

There is less awareness of the offshore migration of night-migrating passerines. During migration, many diurnal bird species become generally nocturnal. Roughly two-thirds of the European bird species migrate mainly or exclusively at night (Martin 1990). Nocturnal migration saves time for refuelling at stopover sites during daytime (Newton 2008) and reduces predation risk by raptors (Newton 2008; but see Alerstam 1990) and, during the night, winds are less turbulent (Kerlinger & Moore 1989; Alerstam 2011; Shamoun-Baranes *et al.* 2017). Nocturnally migrating passerine birds can be frequently detected at offshore

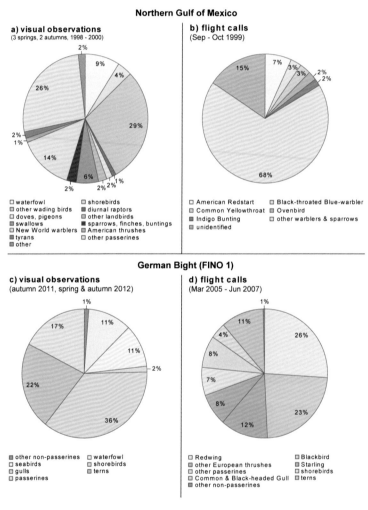

Figure 7.3 Species composition of birds recorded at offshore platforms in the Northern Gulf of Mexico (a, b) and in the German North Sea (c, d), as by means of visual observations (a, c) and by automated flight call recordings (b, d). 'Other wading birds' in (a) comprise herons, egrets, ibises and the like, unidentified calls in (b) include *Catharus* thrushes, orioles, tanagers and non-passerines such as herons. [Data in (a) from Russell (2005), (b) from Farnsworth & Russell (2007), (c) from Hill *et al.* (2014a) with supplemental information from the authors, and (d) from Hüppop *et al.* (2012)]

platforms (Figure 7.3), for example in the North Sea (e.g. Müller 1981; Hüppop *et al.* 2006; 2012; 2016; Hüppop & Hilgerloh 2012), the Baltic Sea (e.g. Sjöberg *et al.* 2015) and the Gulf of Mexico (Russell 2005; Farnsworth & Russell 2007), or at remote islands such as Helgoland (Dierschke *et al.* 2011), the Caribbean Islands (Williams *et al.* 1977), and islands in the Yellow Sea, Japan and south-east Asia (e.g. Yong *et al.* 2015).

The majority of nocturnally migrating birds typically departs within the first 1–3 hours after sunset (e.g. Drost 1931; Alerstam 1990; Åkesson *et al.* 1996; Schmaljohann *et al.* 2009; 2011; Müller *et al.* 2016) when navigation-relevant twilight cues become available (e.g. Cochran *et al.* 2004; Muheim *et al.* 2006; Sjöberg & Muheim 2016). There is however, considerable variation in observed nocturnal departure times, mainly due to an individual combination of intrinsic and current extrinsic factors. As such, individual body condition and ambient wind conditions define the main drivers of departure decisions in general (Müller *et al.* 2016) (see Figure 7.4) and at ecological barriers in particular (Schmaljohann *et al.* 2009; Müller *et al.* 2016).

Depending on the observation method (see Molis *et al.*, Volume 4, Chapter 5), time of day and season, distance from the coast, the geographic location in general and the behaviour of the species, the recorded composition of bird species offshore can vary substantially. It is thus rather difficult to state which bird species or groups of species dominate the range of species observed at sea. Radar echoes of birds flying over the open oceans were identified as seabirds, waterfowl, shorebirds and even passerines (Williams & Williams 1990). Sea-watching studies generally revealed that seabirds, gulls, terns, ducks, and geese, as well as waders at more coastal sites, dominate the species composition at sea during daytime (e.g. Edelstam 1972; Camphuysen & van Dijk 1983; Dierschke 2003; Hüppop *et al.* 2004; 2006; 2010; Jakobsen 2008; Ellermaa & Lindén 2015).

Figure 7.3 shows two examples of bird species composition recorded with different techniques from offshore platforms in the northern Gulf of Mexico, with varying distances to the nearest shore of 8–204 km and water depths of approximately 11–535 m, and in the German North Sea at about 45 km from the coast and a water depth of almost 30 m (FINO 1). Daytime observations revealed that waterfowl and wading birds comprise about half to four-fifths of the birds recorded offshore (Figure 7.3a,c). The higher proportion of songbirds observed at the northern Gulf of Mexico platforms compared to their proportion at FINO 1 is partly due to the additional recording of birds using the platform for stopover (Russell 2005). With acoustic registrations at night, the pattern was seemingly different, with about 76–85% of the recorded birds being passerines (Figure 7.3b,d). This apparent discrepancy can, to some extent, be explained by the recording techniques. A noticeable difference between the flight-call recordings at the two locations may be the lack of gulls and terns at the platform in the northern Gulf of Mexico (Figure 7.3b). The two examples in Figure 7.3, however, illustrate, on the one hand, what one would expect to encounter offshore, namely seabirds and waterbirds. On the other hand, passerines and other mainly terrestrial birds are more frequently encountered offshore during their nocturnal migration than one might assume.

Wind speed and direction are important factors in bird flight speeds and energetics. To save time and energy during migration, birds are expected to select time periods and flight altitudes with favourable wind conditions, which means tailwinds or at least winds of low speed. Indeed, migrants adjust their flight altitude accordingly (Liechti 2006; Schmaljohann *et al.* 2009; Kahlert *et al.* 2012; Kemp *et al.* 2013). Likewise, diurnal migrants at the Strait of Messina in the Mediterranean selected the first optimal wind layer they encountered when climbing, even when better winds occurred at higher altitudes (Mateos-Rodriguez & Liechti 2012).

Flight altitudes of birds often exceed 4,000 m (Bruderer 1997; Alerstam & Gudmundsson 1999; Liechti 2006). Long-ranging radar detected bird migration off the coasts of North and South America at considerable altitudes, up to 4,000 m (Williams & Williams 1990). In contrast, more than one-third of the birds recorded offshore in the North Sea were flying in the lowest 200 m above sea level as determined by modified marine radar (Hüppop et al. 2004; 2006; Fijn et al. 2015). Since the radar signal from the lowest flying birds was not distinguishable from reflections of the water surface, the number of low-flying birds was most likely to be underestimated. In contrast, very high-flying individuals were likely underdetected because they were too far away from the radar (Hüppop et al. 2006). Visual observations revealed that considerable numbers of diurnally migrating bird species or species groups, such as waterfowl, waders, or songbirds, often fly at very low altitudes over the sea (Krüger & Garthe 2001; Dierschke & Daniels 2003; Hüppop et al. 2004; 2006; Kahlert et al. 2012) (Figure 7.2). The low flight altitudes of swans and geese over the sea observed visually have been confirmed by Global Positioning System (GPS) altitude data (Griffin et al. 2011). Radar observations along the Dutch coast revealed that waders starting to fly close to the shoreline at relatively low altitudes of 5–135 m during the day continue to do so during the night and at greater distances from the coast (Dirksen et al. 1996).

In sunny conditions, soaring raptors often gain height by uplift before crossing the sea in gliding flight, often well above rotor height. But when forced to cross open water in less favourable conditions with no uplift and adverse winds, they generally fly at a lower altitude (Berndt et al. 2005; Corso 2001; Germi 2005; Baisner et al. 2010; Malmiga et al. 2014). During poor weather such as fog, drizzle, rain, low air pressure, headwinds, and strong winds in general, flight altitudes as well as the overall numbers of birds at sea generally decrease (Gruber & Nehls 2003; Hüppop et al. 2004; 2006; Aumüller et al. 2011; Kahlert et al. 2012). Flight heights also change with time of day, which may reflect temporal changes in intraspecific behaviour, such as in thrushes (Myres 1964) or the spectrum of species involved. Over the North and Baltic Seas, general median flight altitude as measured by radar is lowest during the day and highest at night, mostly in the first hours after sunset, which means soon after taking off (Myres 1964; Bourne 1980; Hüppop et al. 2004; Kahlert et al. 2012). Flight altitude then decreases continuously throughout the night and briefly increases again around sunrise. Diehl et al. (2003) and Archibald et al. (2017) reported similar 'dawn ascents' from the Great Lakes mirroring the experiences of Richardson

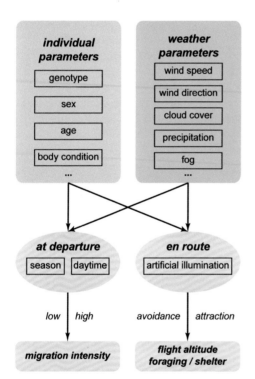

Figure 7.4 Presumed main factors influencing bird and bat migration over the sea. The intensity of migration is generally dependent on season and time of day, with the individual decision to depart being shaped by intrinsic factors as well as, for example, by the current weather or food situation. En route weather as well as individual body condition and the length of the intended flight determine the flight altitude and the readiness to respond to artificially illuminated structures at sea in order to forage or seek shelter.

(1978a) off eastern North America. At a site in the North Sea close to the Weser estuary where a near-shore wind farm was planned, Walter and Todeskino (2005) could not find any general differences in flight altitudes between daytime and night-time by means of radar. Instead, they observed considerable variation in flight altitudes between days and seasons and discussed the possible influence of weather parameters. Overall, in their study, most birds flew below 500 m (up to 89%) and even below 200 m (up to 61%). In conclusion, many species or species groups generally fly in large numbers within or just below the rotor-swept area of offshore wind turbines.

The main factors determining offshore bird migration are summarised schematically in Figure 7.4. All the factors shown can be influenced by background factors such as food availability at breeding, stopover or wintering sites, other weather parameters and the availability of adequate stopover habitats (Richardson 1978b; 1990; Alerstam 1990; Cresswell 2014; Cooper *et al.* 2015; and many others).

Characteristics of offshore migration of bats

Although the first experiments to mark individual bats started in the early 19th century in both North America and Europe (Allen 1921; Eisentraut 1934; Hutterer *et al.* 2015), many facts about bat migration remain largely unknown and systematic offshore studies (Cryan & Brown 2007; Ahlén *et al.* 2009; Bach *et al.* 2009; Frey *et al.* 2012; Lagerveld *et al.* 2014; Rydell *et al.* 2014; Fritzén 2015; Hüppop & Hill 2016; Peterson *et al.* 2016) are rare. Whereas birds need to feed regularly at their stopover and wintering sites, some bat species hibernate and thus can spend the winter in areas without access to food (Fleming & Eby 2003; Hutterer *et al.* 2005). Accordingly, migration distances rarely exceed 100 km (Hutterer *et al.* 2005). Nevertheless, at least six European (Fleming & Eby 2003; Hutterer *et al.* 2005) and three North American (Cryan 2003; Hatch *et al.* 2013) bat species are known to migrate distances up to 4,000 km. Numbers of migrating bats are less well known than numbers of migrating birds but, for example, more than 20 million Mexican Free-tailed Bats *Tadarida brasiliensis mexicana* change from their maternity roosts in the south-western USA and northern Mexico to their wintering sites farther south in Mexico (Wiederholt *et al.* 2013). In the Netherlands, probably 50,000–100,000 Nathusius' Pipistrelles *Pipistrellus nathusii* migrate in autumn (Lina & Reinhold 1997).

The main migration periods of bats in the northern hemisphere are from April to June and from late July to October (Cryan 2003; Hutterer *et al.* 2005; Rydell *et al.* 2014; Peterson *et al.* 2016). Within Europe, the main migratory direction is from north-east to south-west or south from summer to winter, respectively (Hutterer *et al.* 2005), whereas in North America, movements broadly occur along the north-south axis and tend to be variable on a species-specific and/or population-specific level (Cryan 2003; Cryan *et al.* 2014b; Peterson *et al.* 2016).

Our general awareness of regular movements of bats over sea has long been hindered by their nocturnality. The first evidence of bat migration off the coasts of Europe goes back to von Dalla Torre (1889), who mentioned the occurrence of numerous bats on the island of Helgoland during migration times. Early published records of North American bats at sea in the Atlantic include reports not only of single bats, but also of flocks of up to 200 individuals passing by (e.g. Nichols 1920; Norton 1930; Thomas 1921; Allen 1923; Griffin 1940; Carter 1950). These, and numerous later records of bats from platforms, ships and remote islands (e.g. Cryan & Brown 2007; Skiba 2007; Ahlén *et al.* 2009; Pelletier *et al.* 2013; Lagerveld *et al.* 2014; Petersen *et al.* 2014; Fritzén 2015; Hüppop & Hill 2016; Peterson *et al.* 2016), render it highly unlikely that all movements can be explained by vagrancy or wind drift alone, as is often assumed.

Bats crossing large bodies of open water have been recorded over the North Sea (Skiba 2007; Lagerveld *et al.* 2014; Petersen *et al.* 2014; Hüppop & Hill 2016), Baltic Sea (Ahlén *et al.* 2009; Rydell *et al.* 2014; Fritzén 2015), the Mediterranean Sea (Amengual *et al.* 2007; Pereira *et al.* 2009) and the Atlantic Ocean (Hatch *et al.* 2013; Pelletier *et al.* 2013; Petersen *et al.* 2014; Peterson *et al.* 2016), as well as off the Pacific coast (Cryan & Brown 2007; Pelletier *et. al.* 2013). A satellite-tracked Black Flying-fox *Pteropus alecto* flew from Papua New Guinea to west Australia and back, crossing 150 km of open sea (Breed *et al.* 2010). Also, several Large Flying-foxes *Pteropus vampyrus* were recorded crossing the Straits of Malacca (around 50 km of open sea) between Malaysia and Sumatra (Epstein *et al.* 2009). Nonetheless, bat migration presumably concentrates along the coastlines (Šuba *et al.* 2012; Rydell *et al.* 2014). Migrating bats sometimes accumulate in large numbers at stopovers on peninsulas or islands before they embark on open sea crossings (Ahlén *et al.* 2009; Rydell *et al.* 2014), where one can expect large numbers and hence a higher risk of collision with anthropogenic structures. Radio telemetry revealed that during autumn migration, half of the tagged Silver-haired Bats *Lasionycteris noctivagans* departed from Long Point across Lake Erie (minimum crossing distance 38 km) while half departed along the shoreline (McGuire *et al.* 2012). Conversely, Hatch (2015) observed that Silver-haired Bats and Eastern Red Bats *Lasiurus borealis* travel east to west along the Lake Erie shoreline as opposed to flying longer distances over open water.

Acoustic studies on coastlines, coastal islands, fixed offshore structures such as buoys and platforms, and US National Oceanic and Atmospheric Administration (NOAA) research vessels found that bat activity decreased with distance from the North American mainland. However, ship-based surveys documented bat activity up to 130 km off the southern New England coast, with a mean distance of all shipboard calls at 60 km (Peterson *et al.* 2016). The seasonality of the appearance of certain bat species in Bermuda highlights their extraordinary sense of navigation and their ability to travel regularly distances of more than 1,000 km over open water (Hatch *et al.* 2013). However, only a few systematic studies enable quantitative conclusions to be reached on the offshore species composition (see examples in Figure 7.5) or the proclivity of individual species to travel over open water. Acoustic surveys conducted off the North American Atlantic coast and Great Lakes found that species differed in their seasonal activity patterns, with Eastern Red Bats detected over an extended period from July to October, Hoary Bats *Lasiurus cinereus* (Figure 7.6) in mid-August and Silver-haired Bats in early September (Peterson *et al.* 2016). Eastern Red Bat was the most widespread species (Figure 7.5), occurring at 97% of all locations monitored and accounting for 40% of all identified bat passes. Silver-haired and Hoary Bats were detected in small numbers, but at a high percentage of sites (89% and 95%, respectively). *Myotis* species were detected at some of the most remote sites and were the most commonly identified species group at coastal sites (Peterson *et al.* 2016). At most offshore sites studied in Europe, Nathusius' Pipistrelle was the most often recorded species farther offshore (Figure 7.5).

Despite the distance from the mainland, the intensity of bat migration over the sea seems to be highly dependent on low or moderate wind speeds (Cryan & Brown 2007; Ahlén *et al.* 2009; Lagerveld *et al.* 2014; Hüppop & Hill 2016; Peterson *et al.* 2016). Dense cloud cover seems to be an important predictor of bat occurrence offshore, such as at a remote island (Cryan & Brown 2007) or an isolated offshore platform (Hüppop & Hill 2016). The influence of precipitation, fog, and air pressure on bat occurrence at sea is somewhat unclear or at least site specific. Departures of radio-tagged migratory tree bats close to Lake Erie were more likely to occur on nights with higher barometric pressure, lower wind differential (difference between wind gust and sustained wind speed) and a higher dew point (Hatch 2015). Whereas lack of precipitation and high ambient pressure

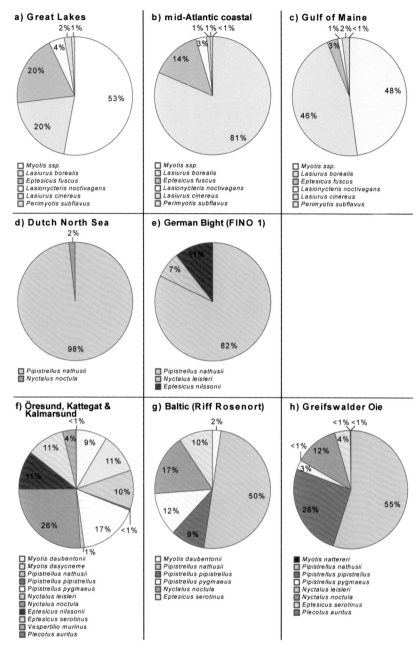

Figure 7.5 Examples of species composition of bats recorded off North American (a–c) and European coasts (d–h). Data in (a)-(c) are from a long-term automated recording survey at 31 study sites in North America 4–41.6 km from the nearest coast and at seven coastal sites in 2009–2014 (Peterson *et al.* 2016). Data in (d) stem from autonomous recordings at two wind farms 15–23 km from the Dutch North Sea coast in 2012 (Lagerveld *et al.* 2014). Data in (e) stem from long-term autonomous acoustic recordings at the research platform FINO 1 in the German Bight, 45 km from the nearest coast, in 2004–2015 (Hüppop & Hill 2016). Data in (f) stem from systematic observations of bats over the sea in Öresund, Kattegat and the southern Baltic Sea, at distances of 11.8–19.1 km from the nearest coast, in 2005, 2006 and 2008 (Ahlén *et al.* 2009). Data in (g) and (h) stem from autonomous recordings in the German Baltic Sea in 2012 at the working platform Riff Rosenort, 2.2 km from coast (g), and at the island Greifswalder Oie, about 10 km from the coast (Seebens *et al.* 2013).

Figure 7.6 A surprisingly broad spectrum of bats has been detected offshore, including (a) Hoary Bat *Lasiurus cinereus* in the coastal North Atlantic and (b) Particoloured Bat *Vespertilio murinus* in the Baltic Sea. [(a) Jens Rydell; (b) Stefan Weiel]

reliably predicted the occurrence of bats at two wind farms approximately 15–23 km off the Dutch coast (Lagerveld *et al.* 2014), low stratus and/or low barometric pressure determined bat occurrence farther offshore (Cryan & Brown 2007; Hüppop & Hill 2016). Figure 7.4 summarises environmental factors of influence on offshore bat as well as bird migration.

Because of methodological constraints (see Molis *et al.*, Chapter 5, Volume 4 of this series), it is difficult to estimate the number of bats crossing the sea or entering an OWF (Lagerveld *et al.* 2014; Hüppop & Hill 2016). Based on literature data on anecdotal observations and direct ultrasonic detections at the islands of Borkum and Helgoland, Skiba (2007) suggested that approximately 5,300 bats, mainly Nathusius' Pipistrelle *Pipistrellus nathusii*, could cross the German Bight each autumn, but also conceded that this number could be vastly underestimated.

Compared to birds, knowledge on flight altitudes of migrating bats is more limited, and even more so over open water. This is again due to technical and logistical constraints, since the equipment used for recording bats depends mainly on the species-specific sound pressure of echolocation calls and environmental conditions, and only covers distances up to 100 m under the best conditions, and more typically 30 m or less (Ahlén *et al.* 2009; Hüppop & Hill 2016; Peterson *et al.* 2016; Molis *et al.*, Volume 4, Chapter 5). Ahlén *et al.* (2009) studied migrating and foraging bats over the Baltic Sea by direct observation, radar, thermal imaging, and acoustic recording. All migrating bats observed from the vessel flew at relatively low altitudes. Even some of the normally higher flying Common Noctules *Nyctalus noctula* flew lower than 10 m above the sea surface, although most of them flew higher.

Offshore, bats presumably use supporting tailwinds (Hatch *et al.* 2013; Smith & McWilliams 2016) on a larger scale and correspondingly many may fly much higher, and thus beyond the detection range of all recording equipment, as suggested by Hüppop and Hill (2016) for species in the southern North Sea and Hatch *et al.* (2013) for Eastern Red Bats off the North American east coast, where at least five out of 11 individuals flew at a height of more than 200 m. Onshore, Brazilian Free-tailed Bats *Tadarida brasiliensis* were found feeding in large numbers up to 1,118 m above ground level, with peak activity at 400–500 m (McCracken *et al.* 2008). Radar echoes of dense groups of this species rose to altitudes of more than 3,000 m (Williams *et al.* 1973). Recordings by bat detectors sent aloft on helium-filled kite balloons confirm that bats were actively pursuing insects at altitudes of 300–800 m (Fenton & Griffin 1997, and several unpublished observations therein), and a thermal camera in southern Sweden revealed that one of the species known to migrate,

Common Noctule, forages at altitudes up to 1,200 m (Ahlén *et al.* 2009). In conclusion, it can be assumed that offshore registrations of bat calls are only partly representative of the overall offshore migration of bats, missing possibly stronger migration at altitudes that are beyond the coverage by the available equipment, but well within the area of the rotor blades (Hüppop & Hill 2016; Molis *et al.*, Volume 4, Chapter 5).

No technology is available to investigate offshore bat migration at higher altitudes, principally because the use of tethered balloon or kite studies is not feasible at sea and discrimination of single migrating bats from birds by radar is not yet possible (Drake & Reynolds 2012; Aschwanden *et al.* 2015; Molis *et al.*, Volume 4, Chapter 5). Bat detectors can be installed on top of tall offshore structures, but since at least some bat species are attracted by light during migration (see discussion in Hüppop & Hill 2016; Voigt *et al.* 2017) or change their altitude rapidly when they are near tall vertical obstacles (Ahlén *et al.* 2009; Molis *et al.*, Volume 4, Chapter 5), this would provide only little information on undisturbed migratory flight altitudes. It does, however, emphasise that as for onshore turbines (e.g. Barclay *et al.* 2017a; 2017b), bats may be attracted to offshore wind turbines because of their isolated prominence and/or ability to provide stopover resting and perhaps even foraging sites (Pelletier *et al.* 2013; Peterson *et al.* 2016). The continued miniaturisation of very-high-frequency nanotags along with the corresponding deployment of automated radio-telemetry stations has proven ability to track individual bats, as well as small shorebirds and passerines, across portions of their entire migration routes (Smolinsky *et al.* 2013; Peterson *et al.* 2014; Hatch 2015; Weller *et al.* 2016; Taylor *et al.* 2017). A further expansion of this technology via an increased number of strategically positioned stations offers new opportunities to assess the temporal and regional movement patterns of individual species (Lagerveld *et al.* 2017; Molis *et al.*, Volume 4, Chapter 5).

Do offshore wind farms act as barriers?

Many migrating birds avoid entering or crossing wind farms (Desholm & Kahlert 2005; Masden *et al.* 2009; Plonczkier & Simms 2012; Aumüller *et al.* 2013; Dierschke *et al.* 2016). For example, Environmental Impact Assessment studies in the context of wind-turbine construction in the German Bight revealed a significant reduction of species numbers in the direction of a wind farm (Aumüller *et al.* 2013; Hill *et al.* 2014b). In divers *Gavia* spp. and Northern Gannets *Morus bassanus*, common migrating seabirds in the German Bight, the observed migration event rate reduced to one-third in the direction of a wind farm (Aumüller *et al.* 2013). On the other hand, Great Cormorant *Phalacrocorax carbo* and European Shag *Phalacrocorax aristotelis* often show strong attraction to OWFs (Dierschke *et al.* 2016; Vanermen & Stienen, Chapter 8), whereas that of several gull species (Hill *et al.* 2014b; Schulz *et al.* 2014) and Red-breasted Merganser *Mergus serrator* is weaker, although still classed as attraction. From this, one might conclude that wind farms have only small barrier effects on migrants of these species.

In general, avoidance rates strongly depend on external conditions (Chamberlain *et al.* 2006). During the day and under good visibility, flocks of migrating seaducks and geese actively change their flight path at a distance of at least 1–2 km to avoid entering a wind farm (Desholm & Kahlert 2005; Pettersson 2005; Petersen *et al.* 2006; Krijgsveld *et al.* 2011). At Kalmarsund (Baltic Sea), this avoidance behaviour led to displacement in the overall migration pattern of waterfowl by about 2 km after construction of the wind farm (Pettersson 2005). Deviations from the migratory trajectories, however, result in an increase in flight distance and thus in energy expenditure, although avoiding one wind farm will not seriously increase energy expenditure. Masden *et al.* (2009) measured the detour taken by Common Eiders *Somateria mollissima* when facing a single wind farm in the Baltic Sea. The estimated length of the additional flight path was about 500 m, a distance which seems to be negligible considering a total migration route of 1,400 km (Masden *et al.* 2009). However, if birds encounter multiple wind farms, the cumulative energy expenditure sums up (Fox *et al.* 2006; Masden *et al.* 2009). In their example, Masden *et al.* (2009) calculated that for 100 detours, the energetic costs could sum up to 1% of a bird's body mass. Bearing in mind that, at dawn, nocturnally migrating birds reorient under some circumstances towards land (Bruderer & Liechti 1998; Smolinsky *et al.* 2013; Archibald *et al.* 2017), the additional energetic costs when re-encountering the same offshore or nearshore wind farm several times within one season may not be negligible.

The extent of a detour that passing birds have to make also depends on the distance of the wind farms to each other and on the configuration of the individual wind farms, which influences their permeability (Hüppop *et al.* 2006; Masden *et al.* 2012). According to movement models, diamond-shaped wind farms properly aligned to the birds' flight routes had a low permeability, which reduced the risk of collision, and modelled lengths of flown detours were relatively small (Masden *et al.* 2012).

During the night, birds migrating over the sea are more prone to enter an OWF (Desholm & Kahlert 2005; Krijgsveld *et al.* 2011; Schulz *et al.* 2014), although the proportion of birds actually entering a wind farm is generally low, at least for geese and seaduck species (Desholm & Kahlert 2005; Pettersson 2005; Krijgsveld *et al.* 2011; Plonczkier & Simms 2012; Vanermen & Stienen, Chapter 8), but seemingly higher for herons, ducks (other than seaduck), waders and songbirds (Krijgsveld *et al.* 2011; Krijgsveld *et al.* personal communication). When entering a wind farm, seaduck and geese, and presumably other species too, seem to reduce their risk of collision by adjusting their flight paths and flying in the corridors between the turbines (Desholm & Kahlert 2005; Petersen *et al.* 2006), or by increasing their flight altitude (Pettersson 2005; Plonczkier & Simms 2012).

Extrapolating from their radar observations at a Dutch OWF, Fijn *et al.* (2015) calculated that annually about 1.6 million birds passed through the rotor-swept zone (25–115 m above sea level) of the entire wind farm area of about 27 km². During the night, generally fewer birds were recorded in the rotor-swept zone, mainly owing to the greater proportion of normally high-flying night-migrating songbirds (Blew *et al.* 2008; Krijgsveld *et al.* 2011; Fijn *et al.* 2015).

Depending on the prevailing conditions, this general pattern can, however, vary considerably (Blew *et al.* 2008; Hill *et al.* 2014b; Schulz *et al.* 2014). With increasing wind speeds, for example, the proportion of medium-sized passerine birds at rotor height can increase as well (Krijgsveld *et al.* 2011). During daylight conditions, Blew *et al.* (2008) found about 10% to more than 30% of individually tracked songbirds, mainly pipits, wagtails, finches and Common Starlings *Sturnus vulgaris* reacting to the turbines. Actively rotating turbines also seem to provoke strong avoidance reactions in passing birds during the night (Schulz *et al.* 2014). At the Alpha Ventus test site in the German North Sea, the proportion of birds flying close to the turbines at rotor height was 13.6–30 times higher when the turbine blades were standing still than when they were turning (Schulz *et al.* 2014). Furthermore, during two consecutive nights of heavy songbird mass migration through the wind farm at altitudes below 200 m above sea level, significantly more birds were recorded near turbines and at rotor height during the night with non-operating turbines (Schulz *et al.* 2014). Likewise, in a Dutch wind farm, avoidance of individual turbines by local and migrating seabirds and non-marine birds, e.g., thrushes and geese, was higher at night when turbine blades were rotating (Krijgsveld *et al.* 2011).

The overall avoidance behaviour of night-migrating passerines to OWFs is presently unknown. We assume that in general barrier effects and thus effects on flight energetics may be neglected for passerines. The situation may, however, become quite different in poor weather, i.e., fog, drizzle, or rain, when birds could be attracted and disoriented by the lights of (static) offshore structures (Hill *et al.* 2014b; Schulz *et al.* 2014). Birds can get 'trapped' for hours, showing continuous disoriented flight around the lighted structure (Ballasus *et al.* 2009; Hope Jones 1980; Larkin & Frase 1988; Gauthreaux & Belser 2006), and may die from exhaustion or collision (Hüppop *et al.* 2016).

The behaviour of migrating bats at OWFs has been poorly studied. During their investigations in the southern Baltic Sea, Ahlén *et al.* (2009) frequently recorded bats foraging at sea close to the water surface but then rapidly increasing in altitude at tall vertical obstacles such as ships and wind turbines. Likewise, bats have been observed roosting on the nacelles of wind turbines about 6 km off the coast (Ahlén *et al.* 2009). Hüppop and Hill (2016) also recorded calls at the FINO 1 platform that indicate extensive exploratory behaviour and foraging, while Lagerveld *et al.* (2014) detected no feeding buzzes at two wind farms off the Dutch coast. From this sparse evidence, it could be concluded that OWFs do not pose energetic barriers to migrating bats, but instead yield some attraction. Migratory bat species have good visual acuity that may be sufficient for the detection of objects such as stars or illuminated offshore structures over longer distances (Eklöf *et al.* 2013). Species that have been observed catching insects near streetlights, including *Lasiurus* spp., *Eptesicus* spp., *Nyctalus* spp. and *Pipistrellus* spp. (Mathews *et al.* 2015), may therefore be attracted by safety and other lights offshore. Cryan and Brown (2007) observed *Lasiurus cinereus* in the spotlight of a lighthouse on a remote island off the American west coast and Voigt *et al.* (2017), using a light on-off treatment, observed that the activity of migrating *Pipistrellus nathusii* and *P. pygmaeus* on the Baltic coast increased by more than 50% in the light-on compared to the light-off treatment. Hence, it cannot be excluded that migrating bats also become 'trapped' by illuminated offshore structures in a similar manner to birds (Hüppop & Hill 2016).

The prospects of collision

Turbine collision fatalities are the key factor when assessing the ecological impact of wind farms on birds and bats (for the extremely diverse situation onshore, see de Lucas & Perrow 2017 and Barclay *et al.* 2017a). Offshore, collisions with artificial structures are extremely difficult to quantify. On the one hand, there are serious logistical and methodological constraints (Molis *et al.*, Volume 4, Chapter 5) and on the other hand, collision risk depends on a broad set of different ecological variables, which influence overall migration intensity, such as seasonal timing, weather and wind conditions, individual body conditions, intensity of migration, flight height, group of species and location (Figure 7.4). These variables may also mutually counteract or even cumulatively add up their possibly negative effects.

There may also be a difference in collision risk depending on the type of structure, for example static platforms with lattice towers and guy wires or power lines versus smooth, columnar wind turbines with rotating blades (Hill *et al.* 2014b; Brenninkmeijer & Klop 2017; de Lucas & Perrow 2017). As previously noted, lights also play an important role as a potential attractant to both birds and bats (e.g. Ballasus *et al.* 2009; Gauthreaux & Belser 2006; Hüppop & Hilgerloh 2012; Longcore *et al.* 2008; 2012; Voigt *et al.* 2017) (Box 7.1). What is striking, however, about the collision mortality data from onshore wind farms is the relative absence of large-scale fatality events, compared with those recorded at tall communication towers supported by guy wires, where collisions of hundreds of birds occur at times in a single night (Kerlinger *et al.* 2010).

It can be speculated that for birds the risk of collision with offshore wind turbines is greater at night. During the day, migrating birds generally appear to avoid crossing

Box 7.1 Collisions of birds with artificially lit offshore and nearshore structures: a brief chronology

Offshore and nearshore bird collisions have been reported from a variety of anthropogenic structures, mainly lighthouses and lightvessels, but also from platforms and brightly illuminated ships (Clarke 1912; Wiese *et al.* 2001; Ballasus *et al.* 2009; Vaughan 2009; Bocetti 2011; Ronconi *et al.* 2015). The vast majority of recorded fatalities were nocturnal migrants, mainly passerines, attracted by light under adverse conditions such as haze or drizzle. In principle, mass collisions are rare episodic events, but hundreds to thousands of birds may be affected in a single night. This box presents a brief chronology of evidence stretching back nearly 140 years.

Back in 1880, Joel Asaph Allen collected data on fatalities from 24 lighthouses on the coasts of the USA and concluded that the total number of fatalities, mainly songbirds, but also waders and other waterbirds, at all US lighthouses at that time must amount to many thousands annually (Allen 1880).

Barrington (1900) analysed data collected at Irish lighthouses and lightships from 1891 to 1897. From 1894 onwards, lightkeepers were to cut off and forward a leg and wing of any bird that was killed striking the lanterns. This resulted in almost 1,600 samples, with Common Blackbird *Turdus merula* (*n*=148), Sedge Warbler *Acrocephalus schoenobaenus* (*n*=142), Eurasian Skylark *Alauda arvensis* (*n*=130), Song Thrush *Turdus philomelos* (*n*=118), Common Whitethroat *Sylvia communis* (*n*=109), Northern Wheatear *Oenanthe oenanthe* (*n*=89), Willow Warbler *Phylloscopus trochilus* (*n*=82) and Common Starling *Sturnus vulgaris* (*n*=80) being the most numerous species.

Almost simultaneously, Rudolf Blasius started a systematic long-term survey on collisions at German lighthouses and lightvessels along the coasts of the North Sea and Baltic Sea. During the first 10 years of these studies (Blasius 1893–1894) at least 12,737 birds were killed, more regularly in autumn than in spring. Collision victims were predominantly Eurasian Skylarks (*n*=3,208), Common Starlings (*n*=2,728), thrushes *Turdus* spp. (*n*=1,961), European Robins *Erithacus rubecula* (*n*=1,726) and other songbirds.

Hansen (1954) compiled lots of data on birds killed at Danish lights from 1886 to 1939. In total, 40,700 individuals from 170 species were recorded as collision victims. Again, songbirds were by far the most numerous group (*n*=22,718), with Eurasian Skylarks equalling 55.8% of all fatalities.

Eagle Clarke spent a whole month on board of a lightship in the English Channel in 1903 and reported that under rainy and foggy conditions songbirds often flew round the vessel, and great numbers struck the glass of the lantern and were lost in the sea (Clarke 1912).

In a 41-year study at Long Point lighthouse, Lake Erie, Ontario, the total number of birds reported killed was 6,259 in spring and 11,899 in autumn (Jones & Francis 2003). Most numerous of the 121 species found were Swainson's Thrush *Catharus ustulatus* (9.2%), Common Yellowthroat *Geothlypis trichas* (8.7%), Ovenbird *Seiurus aurocapillus* (7.9%), Red-eyed Vireo *Vireo olivaceus* (7.0%) and Blackpoll Warbler *Setophaga striata* (5.5%). The change in beam characteristics in 1989 to a narrower and less powerful beam resulted in a drastic reduction in avian mortality at the lighthouse.

Illuminated oil, gas and research platforms may also attract birds in large numbers despite their much less intensive lights compared to lighthouses and lightvessels. Attracted birds may die after collision with infrastructure, and overexposure to heat

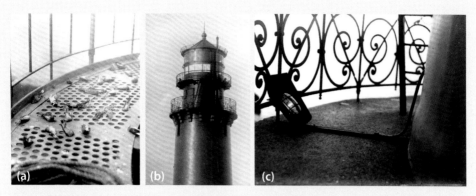

Figure 7.7 Measures to mitigate fatalities (a) at the Helgoland lighthouse (b) in the 1920s included the introduction of bird protection lamps (c), which made the head of the lighthouse visible to birds in darkness. (Archive, Institute of Avian Research "Vogelwarte Helgoland")

from or incineration in flares (Ronconi *et al.* 2015). For example, Bourne (1979) had assumed annual mortality in flares to be a few hundred birds per platform.

Müller (1981) estimated that within only 6 hours during a night in October, some 120,000 birds of at least 37 species passed a research platform in the German part of the North Sea. Undoubtedly attracted by the platforms lights, many birds crashed into the lamps along the catwalks surrounding the platform. A total of 248 fatalities were found, 52.5% of which were Song Thrushes *Turdus philomelos*, alongside Meadow Pipits *Anthus pratensis* (14.5%), and Eurasian Skylarks (12.1%) and other songbirds. Right after switching off the deck lights, the number of birds at the platform dropped dramatically, as did the number of fatalities.

On an illuminated research platform with an 81 m lattice tower in the southern North Sea, a total of 767 birds of 34 species, mainly thrushes, Common Starling and other passerines, were found on 45 of 160 visits from October 2003 to December 2007. Collisions were estimated to be at least 150 birds per year (Hüppop *et al.* 2016).

The episodic nature of bird fallout related to attraction to lights at offshore structures has made it difficult to quantify mortality. Moreover, many victims fall into the water and are washed away. In the Gulf of Mexico, it is estimated that the nearly 4,000 platforms may cause roughly 200,000 collision fatalities per year and some estimates of total annual bird mortality at oil and gas platforms in the North Sea are as high as 6 million birds (Ronconi *et al.* 2015). Drawing conclusions on the potential for collision mortality at offshore wind farms from these data demands extreme caution as the attraction of a single illuminated static structure in the open sea is not directly comparable to that of rotating turbines within an offshore wind farm (Hüppop *et al.* 2016).

Finally, it should be noted that there is considerable scope to reduce bird collision at illuminated structures. For example, at Helgoland, Hugo Weigold, the first head of the Vogelwarte Helgoland (Helgoland Bird Observatory) in the 1920s, introduced bird protection lamps that made the head of the lighthouse visible to birds at night, thereby reducing rates of nocturnal collision (Figure 7.7). Similarly, removing non-flashing lights from towers is one of the most effective and economically feasible means of achieving a significant reduction in avian fatalities at onshore communication towers (Gehring *et al.* 2009). However, this may not be possible for reasons of safety for shipping at offshore wind turbines.

through OWFs (see above), thereby limiting collision risk. Waterfowl, such as ducks and geese, appear to avoid collisions at night by increasing their flight distance from the turbines and flying in the corridors between turbines (Desholm & Kahlert 2005). However, very little is known about the behaviour of passerines, or their subsequent risk of collision, at offshore wind turbines (see *Do offshore wind farms act as barriers?* above). Since most passerine species mainly migrate at night, they are theoretically more vulnerable to collision than diurnally active birds (Hüppop *et al.* 2006).

Illuminated structures exhibit a particular attraction to night-migrating songbirds on land (Ballasus *et al.* 2009; Haupt 2009; Longcore *et al.* 2008; 2012; de Lucas & Perrow 2017) as well as at sea (Hüppop *et al.* 2006; Poot *et al.* 2008; Hüppop & Hilgerloh 2012). In combination with adverse weather conditions, illuminated structures could account for a major cause of mortality for birds migrating over the sea (Hill *et al.* 2014b; Hüppop *et al.* 2016). Fatality numbers are much higher at turbines in brightly illuminated regions or in migration bottleneck areas such as the Eemshaven area at the utmost Dutch North Sea coast (Brenninkmeijer & Klop 2017) although, in contrast, Kerlinger *et al.* (2010) detected no significant differences between fatality rates at onshore turbines with flight safety lights and at turbines without lighting. Whereas onshore turbines generally only have flashing red lights, for aviation safety, offshore turbines in some countries must also have bright, steady lights for shipping safety. Since these steady lights attract nocturnal migrants more than flashing ones, one might expect the collision risk for passerines to be considerably higher at offshore turbines than at onshore ones, especially under deteriorating weather conditions when birds are seeking refuge (Hüppop *et al.* 2016). Since terrestrial birds sometimes follow even tiny structures before flying out to sea (Hüppop *et al.* 2004), nearshore wind farms may act as 'guidelines', and as such attract birds and increase the risk of collision.

Besides direct collision, another potentially dangerous factor for birds and bats flying close to the turbines may be wind wakes of spinning blades (Fox *et al.* 2006; Kikuchi 2008; Baerwald *et al.* 2008; Barclay *et al.* 2017a). These effects are, however, difficult to estimate since they cannot be easily observed directly at night, and offshore, carcasses are difficult to sample for the analysis of proximate mortality causes. In summary, the effects of adverse weather conditions, such as rain or fog, low flight height and illumination may cumulatively result in a higher risk of collision compared with good weather conditions and an absence of illumination (Hüppop *et al.* 2006; 2016; Aumüller *et al.* 2011).

Despite a continued evolution of technologies, our current ability to quantify collisions at offshore wind turbines remains a challenge (Molis *et al.*, Volume 4, Chapter 5). The species composition of collision victims found at offshore structures (Box 7.1) frequently differs from that of visual or acoustic observations and from the species composition of fatalities at onshore wind turbines (de Lucas & Perrow 2017), suggesting that some species or groups of species may be more vulnerable to collisions than others. Carcasses found at the research platform FINO 1 were mainly thrushes, Common Starling and other passerines (Hüppop *et al.* 2006; 2016), although these species were most prominent within the nocturnal flight call recordings (Figure 7.3).

For nearshore sites, there has been only one study, in Northumberland (north-east England), quantifying the number of fatalities from a beached bird survey (Newton & Little 2009). Over an 11-year period, 3,748 beached bird carcasses were found close to the wind farm, but only 3% (114 individuals) of deaths were attributed to wind turbines. The most frequently recorded victims were seabirds: European Herring Gulls *Larus argentatus* (39 individuals), Great Black-backed Gulls *Larus marinus* (36), Black-headed Gulls *Chroicocephalus ridibundus* (7), and Common Eiders (15); while small passerines were probably under-represented (Newton & Little 2009).

Virtually nothing is presently known about collisions of bats with artificial offshore structures. Onshore, mortality rates of bats related to wind turbines vary considerably between sites and range from zero to 70 killed bats per turbine per year (Arnett *et al.* 2008; Rydell *et al.* 2010a; Brinkmann *et al.* 2011; Barclay *et al.* 2017a). Regarding the slow life history of bats with relatively long lives, low natural mortality, and low reproduction rates (Barclay & Harder 2003), increased mortality related to wind power poses a serious problem in bat conservation (Kunz *et al.* 2007; O'Shea *et al.* 2016). Mortality of bats at onshore wind turbines is highest during the migratory seasons (Rydell *et al.* 2010a) and annual mortality rates vary with distance of the wind parks to the bats' suspected migration routes (Rydell *et al.* 2010b and references therein). Bats rarely collide with tall, static anthropogenic structures (Crawford & Baker 1981; Barclay *et al.* 2017a) and the main causes of death in onshore wind parks are collisions with moving blades or barotrauma caused by a sudden change in air pressure (Rydell *et al.* 2010a; Barclay *et al.* 2017a). Onshore, collisions of bats with turbines blades occur predominantly while foraging for insects (Rydell *et al.* 2010b), searching for roosts at the nacelles (Cryan & Barclay 2009; Cryan *et al.* 2014a) or just passing the rotor-swept area, mainly during low wind and high temperatures (Arnett *et al.* 2011; Amorim *et al.* 2012).

Factors determining bat migration intensity and behaviour may also help to explain the variation in bat mortality at onshore wind farms. According to Cryan and Brown (2007), it is likely that, with the growing trend towards installation of wind farms in offshore waters, turbines will have an impact on bats in marine habitats as well as on land. For some bat species, such as Hoary Bat, this expanded development is considered likely to contribute to potentially serious population declines (e.g. Zimmerling & Francis 2016; Frick *et al.* 2017).

Migration of bats over the sea seems to be relatively predictable when viewed in context with specific weather parameters (Figure 7.4). Low wind speeds, low illumination by the moon, and dense cloud cover were the main predictors of bat occurrence at a remote North American island (Cryan & Brown 2007). In addition, most registrations of bats at an illuminated offshore research platform in the North Sea coincided with rain, fog, or low stratus (Hüppop & Hill 2016). Since records of bats in this long-term study were generally lower than at comparable sites in the North Sea, that is, on the offshore island of Helgoland (Skiba 2007; O. Hüppop, unpublished data), the authors concluded that a significant proportion of passing bats may not have been recorded because they flew at higher altitudes than detectable by the recording system (Hüppop & Hill 2016). This assumption, in combination with the behaviour of the bats at the platform, as inferred from their recorded echolocating signal, implies that at least some bat species seem to be attracted by illuminated offshore installations for foraging or temporary refuge (Hüppop & Hill 2016). At a nearshore wind farm in the Baltic Sea, bats have been observed to change their flight altitude to actively feed close to the turbine blades (Ahlén *et al.* 2009). This considerably increases the risk of collisions.

Current technical and logistical limitations (Molis *et al.*, Volume 4, Chapter 5) restrict our ability to directly study collisions of birds and bats with offshore wind turbines. During four seasons, Petersen *et al.* (2006) recorded 17 thermal video sequences automatically triggered by passing animals at a single turbine in the offshore wind farm at Nysted, Denmark. Of these 17 sequences, two were bats and overall, only one bird or bat was observed to fall following collision with the rotating turbine blades (Petersen *et al.* 2006).

Concluding remarks

The occurrence of migratory birds and bats and their behaviour at anthropogenic offshore structures depends on a variety of environmental conditions and is strongly species or species-group specific. Mostly under good visibility, some waterfowl show a strong aversion to OWFs (Desholm & Kahlert 2005; Plonczkier & Simms 2012). Their behaviour during poor visibility is, however, largely unknown. Although Plonczkier & Simms (2012) carried out radar observations 24 hours a day, the assignment to species was possible only by visual observation during the daytime. In contrast, Skov *et al.* (2016) suggest that raptors, which generally avoid sea crossings, may be attracted to OWFs in adverse wind conditions. Moreover, many nocturnally migrating songbird species and presumably also (some) bat species are attracted by illuminated offshore structures, at least under some circumstances. One might speculate that this also holds true for offshore wind turbines. It is of note that seabirds show a similar range of response from strict avoidance by divers, Northern Gannet, and some auks, to attraction by some gulls and cormorants (Vanermen & Stienen, Chapter 8).

Under good weather conditions, the clear majority of birds that enter an OWF can be observed to fly well above rotor height or to fly through the corridors between turbines. In contrast, bats usually fly close to the water surface but sometimes even rapidly increase their altitude to forage near the turbine blades. Energetic costs when avoiding a single wind farm are possibly negligible, but they may accumulate to some extent when facing multiple wind farms along the migration route. Disoriented birds can become 'trapped' by lighted structures and thus exposed to mortality from exhaustion.

The likelihood of migrating birds and bats entering an OWF is highest at night. On clear nights with favourable winds, most migrating birds will fly well above the rotor-swept area, resulting in a very low collision risk. Deteriorating weather conditions, however, often lead to decreased flight altitudes, which in combination with poor visibility and attraction to artificial lighting can result in dramatically increased collision risk with anthropogenic structures. Environmental factors that influence the overall offshore migration intensity also affect the overall risk of collision.

Although collisions may be recorded using a range of increasingly sophisticated equipment (Molis *et al.*, Volume 4, Chapter 5), quantification of the actual number of offshore collision victims is not yet possible. However, four main conclusions may be tentatively drawn from the very few mortality studies to date: (1) Newton and Little (2009) concluded from carcasses found along shorelines adjacent to a nearshore wind farm that at least 16.5–21.5 birds collided per year and turbine, although the great majority of these were seabirds, especially large gulls; (2) the type of anthropogenic offshore structure; that is, an isolated and illuminated, static platform compared to an array of wind turbines with rotating blades, can affect the total number as well as the species composition of collision victims (Hüppop *et al.* 2016); (3) the number of collision victims probably decreases with increasing distance from the coast; and (4) high numbers of collision victims can be expected at sites where migration concentrates, such as marine straits or close to peninsulas, not only of nocturnal migrants but also of (large) raptors (Skov *et al.* 2016). In Europe, night-migrating passerines, especially thrushes, Common Starlings, European Robins *Erithacus rubecula* and Eurasian Skylarks *Alauda arvensis*, and vireos, kinglets, and wood warblers in North America, suffer the highest risk of collision with offshore structures (Box 7.1). This impression can change with geographic location of the offshore structure, as well as with season and distance to the coast.

The collision risk to bats at offshore wind turbines remains extremely difficult to assess. While the overall risk to individual populations may ultimately be low, there is evidence that the presence of tall vertical structures in an otherwise open environment may serve as an attractant and may pose distinct seasonal migration risks (Ahlén et al. 2009; Cryan & Barclay 2009; Cryan et al. 2014a). Furthermore, the risks to birds and bats may rise in the future with the expected increase in the overall height and volume of the rotor swept zone (Deign 2018).

Wind turbines off the coasts of Europe currently provide the main proportion of global offshore wind capacity. It is thus not surprising that most knowledge on the effects of anthropogenic offshore structures on migrating birds and bats has come from past and ongoing investigations in Europe. However, the overall pattern of responses to OWFs as well as of the main influencing factors on offshore migration and collision risk may hold true on a global scale despite obvious variations in migration routes and offshore species group compositions. Our ever-evolving toolbox of remote monitoring and assessment technologies, particularly those involving advanced radars, nanotag telemetry, acoustics and infrared (both thermal and light supported), and high-definition imageries, as well as advanced analytical software, all hold promise for better understanding the potential risk of offshore wind energy facilities to bird and bat populations (Molis et al., Volume 4, Chapter 5). Proper implementation of these tools and techniques will help us to recognise seasonal periods and environmental conditions when the risk to individual species or species groups may be most elevated. Those findings, in conjunction with the adoption of a true adaptive operational management strategy, could thus lead directly to a significant reduction in mortality risk to birds and bats in the offshore environment.

Acknowledgements

Financial support was received from the Federal Agency for Nature Conservation through the project BIRDMOVE (grant no. 3515822100). We thank Jochen Dierschke and Stefan Weiel for providing photographs, and Franz Bairlein for useful comments on an earlier draft of the chapter.

References

Agostini, N., Panuccio, M. & Massa, B. (2005) Flight behaviour of honey buzzards (Pernis apivorus) during spring migration over the sea. Buteo 14: 3–9.

Agostini, N., Panuccio, M. & Pasquaretta, C. (2015) Morphology, flight performance, and water crossing tendencies of Afro-Palearctic raptors during migration. Current Zoology 61: 951–958.

Ahlén, I., Baagøe, H.J. & Bach, L. (2009) Behavior of Scandinavian bats during migration and foraging at sea. Journal of Mammalogy 90: 1318–1323.

Åkesson, S., Alerstam, T. & Hedenström, A. (1996) Flight initiation of nocturnal passerine migrants in relation to celestial orientation conditions at twilight. Journal of Avian Biology 27: 95–102.

Alerstam, T. (1990) Bird Migration. Cambridge: Cambridge University Press.

Alerstam, T. (2011) Optimal bird migration revisited. Journal of Ornithology 152: S5–S23.

Alerstam, T. & Gudmundsson, G.A. (1999) Migration patterns of tundra birds: tracking radar observations along the Northeast Passage. Arctic 52: 346–371.

Allen, A.A. (1921) Banding bats. Journal of Mammalogy 2: 53-57.

Allen, G.M. (1923) The red bat in Bermuda. *Journal of Mammalogy* 4: 61.

Allen, J.A. (1880) Destruction of birds by lighthouses. *Bulletin of the Nuttall Ornithological Club* 5: 131–138.

Amengual, B., López-Roig, M. & Serra-Cobo, J. (2007) First record of seasonal over sea migration of *Minopterus schreibersii* and *Myotis capaccinii* between Balearic Islands (Spain). *Acta Chiropterologica* 9: 319–322.

American Wind Energy Association (AWEA) (2017) Offshore wind: major milestones & achievements. American Wind Energy Association. Retrieved 13 November 2017 from http://www.awea.org/offshore-wind-milestones

Amorim, F., Rebelo, H. & Rodrigues, L. (2012) Factors influencing bat activity and mortality at a wind farm in the Mediterranean region. *Acta Chiropterologica* 14: 439–457.

Archibald, K.M., Buler, J.J., Smolinsky, J.A. & Smith, R.J. (2017) Migrating birds reorient toward land at dawn over the Great Lakes, USA. *The Auk* 134: 193–201.

Arnett, E.B., Brown, W.K., Erickson, W.P., Fiedler, J.K., Hamilton, B.L., Henry, T.H., Jain, A., Johnson, G.D., Kerns, J., Koford, R.R., Nicholson, C.P., O'Connell, T.J., Piorkowski, M.D. & Takersley, R.D., Jr (2008) Patterns of bat fatalities at wind energy facilities in North America. *Journal of Wildlife Management* 72: 61–78.

Arnett, E.B., Huso, M.M.P., Schirmacher, M. & Hayes, J.P. (2011) Altering turbine speed reduces bat mortality at wind-energy facilities. *Frontiers in Ecology and the Environment* 9: 209–214.

Aschwanden, J., Wanner, S. & Liechti, F. (2015) Investigation on the effectivity of bat and bird detection at a wind turbine. Final Report. Bird Detection. Sempach: Schweizerische Vogelwarte. Retrieved 3 December 2016 from http://www.batsandwind.com/pdf/Hanagasioglu%20et%20al.%202015_Investigation%20of%20effectiveness%20of%20DTBat%20and%20DTBird.pdf

Aumüller, R., Boos, K., Freienstein, S., Hill, K. & Hill, R. (2011) Beschreibung eines Vogelschlagereignisses und seiner Ursachen an einer Forschungsplattform in der Deutschen Bucht. *Vogelwarte* 49: 9–16.

Aumüller, R., Boos, K., Freienstein, S., Hill, K. & Hill, R. (2013) Do migrating birds avoid offshore wind turbines? A method to investigate and analyze reactions of diurnally migrating birds to offshore wind farms. *Vogelwarte* 51: 3–13.

Bach, L., Bach, P., Helge, A., Maatz, K., Schwarz, V., Teuscher, M. & Zöller, J. (2009) Fledermauszug auf Wangerooge – erste Ergebnisse aus dem Jahr 2008. *Natur- und Umweltschutz (Zeitschrift Mellumrat)* 8: 10–12.

Baerwald, E.F., D'Amours, G.H., Klug, B.J. & Barclay, R.M.R. (2008) Barotrauma is a significant cause of bat fatalities at wind turbines. *Current Biology* 18: R695–R696.

Baisner, A.J., Andersen, J.L., Findsen, A., Yde Granath, S.W., Madsen, K.Ø. & Desholm, M. (2010) Minimizing collision risk between migrating raptors and marine wind farms: development of a spatial planning tool. *Environmental Management* 46: 801–808.

Ballasus, H., Hill, K. & Hüppop, O. (2009) Gefahren künstlicher Beleuchtung für ziehende Vögel und Fledermäuse. *Berichte zum Vogelschutz* 46: 127–157.

Barclay, R.M.R. & Harder, L.D. (2003) Life histories of bats: life in the slow lane. In Kunz, T.H. & Fenton, M.B. (eds) *Bat Ecology*. Chicago, IL: University of Chicago Press. pp. 209–253.

Barclay, R.M.R., Baerwald, E.F. & Rydell, J. (2017a) Bats. In Perrow, M.R. (ed.) *Wildlife and Wind Farms, Conflicts and Solutions. Volume 1. Onshore: Potential Effects.* Exeter: Pelagic Publishing. pp. 191–221.

Barclay, R.M.R, Jacobs, D.S., Harding, C.T., McKechnie, A.E., McCulloch, S.D., Markotter, W., Paweska, J. & Brigham, R.M. (2017b) Thermoregulation by captive and free-ranging Egyptian rousette bats (*Rousettus aegyptiacus*) in South Africa. *Journal of Mammalogy* 98: 572–578.

Barrington, R.M. (1900) *The Migration of Birds as Observed at Irish Lighthouses and Lightships.* Dublin: Edward Posonby.

Berdahl, A., Westley, P.A.H. & Quinn, T.P. (2017) Social interactions shape the timing of spawning migrations in an anadromous fish. *Animal Behaviour* 126: 221–229.

Berndt, R.K., Hein, K., Koop, B. & Lunk, S. (2005) *Die Vögel der Insel Fehmarn.* Husum: Husum Druck- und Verlagsgesellschaft.

Bildstein, K.L. (2006) *Migrating Raptors of the World: Their ecology and conservation.* Ithaca, NY: Cornell University Press.

Bildstein, K.L., Bechard, M.J., Farmer, C. & Newcomb, L. (2009) Narrow sea crossings present major obstacles to migrating griffon vultures *Gyps fulvus. Ibis* 151: 382–391.

BirdLife International (2010) Spotlight on flyways. Presented as part of the BirdLife State of the world's birds website. Retrieved 13 March 2017 from http://datazone.birdlife.org/sowb/spotFlyway

Blasius, R. (1893–1894) Schlussfolgerungen aus den ornithologischen Beobachtungen an deutschen Leuchtthürmen in dem zehnjährigen Zeitraume von 1885–1894. *Ornis* 8: 593–620.

Blew, J., Hoffmann, M., Nehls, G. & Henning, V. (2008) Investigations of the bird collision risk and the responses of harbour porpoises in the offshore wind farms Horns Rev, North Sea and Nysted, Baltic Sea, in Denmark. Part I: Birds. Funded by the the German Federal Ministry for the Environment, Nature Conservation and Nuclear Safety. Final Report 2008. Retrieved 26 June 2017 from https://tethys.pnnl.gov/sites/default/files/publications/Bird_Collision_Risk_and_the_Responses_of_Harbour_Porpoises.pdf

Bocetti, C.I. (2011) Cruise ships as a source of avian mortality during fall migration. *The Wilson Journal of Ornithology* 123: 176–178.

Boere, G.C. & Piersma, T. (2012) Flyway protection and the predicament of our migrant birds: a critical look at international conservation policies and the Dutch Wadden Sea. *Ocean & Coastal Management* 68: 157–168.

Boere, G.C. & Stroud, D.A. (2006) The flyway concept: what it is and what it isn't. In Boere, G.C., Galbraith, C.A. & Stroud, D.A. (eds) *Waterbirds Around the World*. Edinburgh: The Stationery Office. pp. 40–47.

Bourne, W.R.P. (1979) Birds and gas flares. *Marine Pollution Bulletin* 10: 124–125.

Bourne, W.R.P. (1980) The midnight descent, dawn ascent and re-orientation of land birds migrating across the North Sea in autumn. *Ibis* 122: 536–540.

Breed, A.C., Field, H.E., Smith, C.S., Edmonston, J. & Meers, J. (2010) Bats without borders: long-distance movements and implications for disease risk management. *EcoHealth* 7: 204–212.

Brenninkmeijer, A. & Klop, E. (2017) Bird mortality in two Dutch wind farms: effects of location, spatial design and interactions with powerlines. In Köppel, J. (ed.) *Wind Energy and Wildlife Interactions*. Cham: Springer. pp. 99–116.

Brinkmann, R., Behr, O., Niermann, I. & Reich, M. (2011) *Entwicklung von Methoden zur Untersuchung und Reduktion des Kollisionsrisikos von Fledermäusen an Onshore-Windenergieanlagen. Ergebnisse eines Forschungsvorhabens.* Göttingen: Cuvillier.

Bruderer, B. (1997) The study of bird migration by radar. Part 2: Major achievements. *Naturwissenschaften* 84: 45–54.

Bruderer, B. & Liechti, F. (1998) Flight behaviour of nocturnally migrating birds in coastal areas: crossing or coasting. *Journal of Avian Biology* 29: 499–507.

Buurma, L.S. (1995) Long-range surveillance radars as indicators of bird numbers aloft. *Israel Journal of Zoology* 41: 221–236.

Camphuysen, C.J. & van Dijk, J. (1983) Zee- en kustvogels langs de Nederlandse kust, 1974–1979. *Limosa* 56: 81–230.

Carter, T.D. (1950) On the migration of the red bat (*Lasiurus borealis borealis*). *Journal of Mammalogy* 31: 349–350.

Chamberlain, D.E., Rehfisch, M.R., Fox, A.D., Desholm, M. & Anthony, S.J. (2006) The effect of avoidance rates on bird mortality predictions made by wind turbine collision risk models. *Ibis* 148: 198–202.

Clarke, M.E. (1912) *Studies in Bird Migration*. London: Gurney and Jackson.

Cochran, W.W., Mouritsen, H. & Wikelski, M. (2004) Migrating songbirds recalibrate their magnetic compass daily from twilight cues. *Science* 304: 405-408.

Cohen, E.B., Barrow, W.C., Buler, J.J., Deppe, J.L., Farnsworth, A., Marra, P.P., McWilliams, D.W., Wilson, R.R., Woodrey, M.S. & Moore, F.R. (2017) How do en route events around the Gulf of Mexico influence migratory landbird populations? *The Condor* 119: 327–343.

Concepcion, C.B., Dumandan, P.T., Silvosa, M.R., Bildstein, K.L. & Katzner, T.E. (2017) Species composition, timing, and weather correlates of autumn open-water crossings by raptors migrating along the East-Asian Oceanic Flyway. *Journal of Raptor Research* 51: 25–37.

Cooper, N.W., Sherry, T.W. & Marra, P.P. (2015) Experimental reduction of winter food decreases body condition and delays migration in a long-distance migratory bird. *Ecology* 96: 1933–1942.

Corso, A. (2001) Raptor migration across the Strait of Messina, southern Italy. *British Birds* 94: 196–202.

Crawford, R.L. & Baker, W.W. (1981) Bats killed at a North Florida television tower: a 25-year record. *Journal of Mammalogy* 62: 651–652.

Cresswell, W. (2014) Migratory connectivity of Palaearctic–African migratory birds and their responses to environmental change: the serial residency hypothesis. *Ibis* 156: 493–510.

Cryan, P.M. (2003) Seasonal distribution of migratory tree bats (*Lasiurus* and *Lasionycteris*) in North America. *Journal of Mammalogy* 84: 579–593.

Cryan, P.M. & Barclay, R.M.R. (2009) Causes of bat fatalities at wind turbines: hypotheses and predictions. *Journal of Mammalogy* 90: 1330–1340.

Cryan, P.M. & Brown, A.C. (2007) Migration of bats past a remote island offers clues toward the problem of bat fatalities at wind turbines. *Biological Conservation* 139: 1–11.

Cryan, P.M., Gorresen, P.M., Hein, C.D., Schirmacher, M.R., Diehl, R.H., Huso, M.M., Hayman, D.T.S., Fricker, P.D., Bonaccorso, F.J., Johnson, D.H., Heist, K. & Dalton, D.C. (2014a) Behaviour of bats at wind turbines. *Proceedings of the National Academy of Sciences of the United States of America* 111: 15126–15131.

Cryan, P.M., Stricker, C.A. & Wunder, M.B. (2014b) Continental-scale, seasonal movements of a heterothermic migratory tree bat. *Ecological Applications* 24: 602–616.

Deign, J. (2018) Wind Tech Trends: Offshore Turbine Capacity Could Double in Europe by 2024. There's no end in sight for wind turbine growth. Retrieved 16 May 2018 from https://www.greentechmedia.com/articles/read/offshore-wind-turbine-capacity-could-double-in-europe#gs.17tvL_Y

de Lucas, M. & Perrow, M.R. (2017) Birds: collision. In Perrow, M.R. (ed.) *Wildlife and Wind Farms, Conflicts and Solutions. Volume 1. Onshore: Potential Effects*. Exeter: Pelagic Publishing. pp. 155–190.

DeLuca, W.V., Woodworth, B.K., Rimmer, C.C., Marra, P.P., Taylor, P.D., McFarland, K.P., Mackenzie, S.A. & Norris, D.R. (2015) Transoceanic migration by a 12 g songbird. *Biology Letters* 11 (4): 20141045. doi: 10.1098/rsbl.2014.1045.

DeSante, D.F. & Ainley, D.G. (1980) The avifauna of the South Farallon Islands, California. *Studies in Avian Biology* 4: 1–104.

Desholm, M. & Kahlert, J. (2005) Avian collision risk at an offshore wind farm. *Biology Letters* 1: 296–298.

Diehl, R.H., Larkin, R.P. & Black, J.E. (2003) Radar observations of bird migration over the Great Lakes. *The Auk* 102: 278–290.

Dierschke, J., Dierschke, V., Hüppop, K., Hüppop, O. & Jachmann, K.F. (2011) *Die Vogelwelt der Insel Helgoland*. Helgoland: OAG Helgoland.

Dierschke, V. (2003) Quantitative Erfassung des Vogelzugs während der Hellphase bei Helgoland. *Corax* 19 (2): 27–34.

Dierschke, V. & Daniels, J.-P. (2003) Zur Flughöhe ziehender See-, Küsten- und Greifvögel im Seegebiet um Helgoland. *Corax* 19: 31–37.

Dierschke, V., Furness, R.W. & Garthe, S. (2016) Seabirds and offshore wind farms in European waters: avoidance and attraction. *Biological Conservation* 202: 59–68.

Dingle, H. (2014) *Migration. The biology of life on the move*, 2nd edn. New York: Oxford University Press.

Dingle, H. & Drake, V.A. (2007) What is migration? *BioScience* 57: 113–121.

Dirksen, S., Spaans, A.L. & van der Winden, J. (1996) Nachtelijke trek en vlieghoogtes van Steltlopers in het voorjaar over de noordelijke Havendam van Ijmuiden. *Sula* 10, 129–142.

Drake, V.A. & Reynolds, D.R. (2012) *Radar Entomology. Observing insect flight and migration.* Wallingford: CAB International.

Drewitt, A.L. & Langston, R.W.H. (2008) Collision effects of windpower generators and other obstacles on birds. *Annals of the New York Academy of Sciences* 1134: 233–266.

Drost, R. (1931) Über den Einfluß des Lichtes auf den Vogelzug, insbesondere auf die Tagesaufbruchszeit. *Proceedings of the VII. International Ornithological Congress*. Amsterdam. pp. 340–356.

Drury, W.H. & Keith, J.A. (1962) Radar studies of songbird migration in coastal New England. *Ibis* 104: 449–489.

Edelstam, C. (1972) Visible migration of birds at Ottenby, Sweden. *Vår Fågelvärld, Supplement 7*: 1–360.

Eisentraut, M. (1934) Markierungsversuche bei Fledermäusen. *Zeitschrift für Morphologie und Ökologie der Tiere* 28: 553–560.

Eklöf, J., Šuba, J., Petersons, G. & Rydell, J. (2013) Visual acuity and eye size in five European bat species in relation to foraging and migration strategies. *Environmental and Experimental Biology* 12: 1–6.

Ellermaa, M. & Lindén, A. (2015) Sügisränne põõsaspeal 2014. aastal. *Hirundo* 28: 20–49. Extended summary in English. Retrieved 22 November 2017 from www.eoy.ee/poosaspea/file.../19/article-2014-Hirundo.pdf

Epstein, J.H., Olival, K.J., Pulliam, J.R.C., Smith, C., Westrum, J., Hughes, T., Dobson, A.P., Zubaid, A., Rahman, S.A., Basir, M.M., Field, H.E. & Daszak, P. (2009) *Pteropus vampyrus*, a hunted migratory species with a multinational home-range and a need for regional management. *Journal of Applied Ecology* 46: 991–1002.

Farmer, C.J., Safi, K., Barber, D.R., Newton, I., Martell, M. & Bildstein, K.L. (2010) Efficacy of migration counts for monitoring continental populations of raptors: an example using the osprey (*Pandion haliaetus*). *The Auk* 127: 863–870.

Farnsworth, A. & Russell, R.W. (2007) Monitoring flight calls of migrating birds from an oil platform in the northern Gulf of Mexico. *Journal of Field Ornithology* 78: 279–289.

Fenton, M.B. & Griffin, D.R. (1997) High-altitude pursuit of insects by echolocating bats. *Journal of Mammalogy* 78: 247–250.

Fijn, R.C., Krijgsveld, K.L., Poot, M.J.M. & Dirksen, S. (2015) Bird movements at rotor heights measured continuously with vertical radar at a Dutch offshore wind farm. *Ibis* 157: 558–566.

Fleming, T.H. & Eby, P. (2003) Ecology of bat migration. In Kunz, T.H. & Fenton, M.B. (eds) *Bat Ecology*. Chicago, IL: University of Chicago Press. pp. 156–208.

Fortin, D., Liechti, F. & Bruderer, B. (1999) Variation in the nocturnal flight behaviour of migratory birds along the northwest coast of the Mediterranean Sea. *Ibis* 141: 480–488.

Fox, A.D., Desholm, M., Kahlert, J., Christensen, T.K. & Krag Petersen, I.B. (2006) Information needs to support environmental impact assessment of the effects of European marine offshore wind farms on birds. *Ibis* 148: 129–144.

Frey, K., Bach, L., Bach, P. & Brunken, H. (2012) Fledermauszug entlang der südlichen Nordseeküste. *Naturschutz und Biologische Vielfalt* 128: 185–204.

Frick, W.F., Baerwald, E.F., Pollock, J.F., Barclay, R.M.R., Szymanski, J.A., Weller, T.J., Russell, A.L., Loeb, S.C., Medellin, R.A. & McGuire, L.P. (2017) Fatalities at wind turbines may threaten population viability of a migratory bat. *Biological Conservation* 209: 172–177.

Fritzén, N.R. (2015) Kvarken Bats – nya resultat som stöder hypotesen om kvarkenöverskridande fladdermusmigration. *OA-Natur* 17: 14–27.

Gätke, H. (1895) *Heligoland as an Ornithological Observatory*. Edinburgh: Douglas.

Gauthreaux, S.A. & Belser, C.G. (2006) Effects of artificial night lighting on migrating birds. In Rich, C. & Longcore, T. (eds) *Ecological Consequences of Artificial Night Lighting*. London: Island Press. pp. 67–93.

Gauthreaux, S.A., Jr, Belser, C.G. & Welch, C.M. (2006) Atmospheric trajectories and spring bird migration across the Gulf of Mexico. *Journal of Ornithology* 147: 317–325.

GE (2018) Haliade-X Offshore Wind Turbine Platform. Retrieved 16 May 2018 from https://www.ge.com/renewableenergy/wind-energy/turbines/haliade-x-offshore-turbine

Gehring, J., Kerlinger, P. & Manville, A.M. II (2009) Communication towers, lights, and birds: successful methods of reducing the frequency of avian collisions. *Ecological Applications* 19: 505–514.

Germi, F. (2005) Raptor migration in east Bali, Indonesia: observations from a bottleneck watch site. *Forktail* 21: 93–98.

Global Wind Energy Council (GWEC) (2017) Offshore wind. Global Wind Report 2016: 58–65. Retrieved 13 June 2017 from http://www.gwec.net/wp-content/uploads/2017/05/Global-Offshore-2016-and-Beyond.pdf

Gnanadesikan, G.E., Pearse, W.D. & Shaw, A.K. (2017) Evolution of mammalian migrations for refuge, breeding, and food. *Ecology and Evolution* 7: 5891–5900.

Goodrich, L.J. & Smith, J.P. (2008) Raptor migration in North America. In Bildstein, K.L., Smith, J.P., Inzuna, E.R. & Veit, R.R. (eds) *State of North America's Birds of Prey*. Cambridge & Washington, DC: Nuttall Ornithological Club and American Ornithologist's Union. pp. 37–149.

Gosse, P.H. (1847) *Birds of Jamaica*. London: Van Voost.

Griffin, D.R. (1940) Migrations of New England bats. *Bulletin of the Museum of Comparative Zoology* 86: 217–246.

Griffin, L., Rees, E. & Hughes, B. (2011) Migration routes of Whooper Swans and geese in relation to wind farm footprints: Final report. Slimbridge: Wildfowl & Wetlands Trust. Retrieved 30 October 2017 from https://www.gov.uk/government/uploads/system/uploads/attachment_data/file/198201/OESEA2_Migration_Routes_WhooperSwans_Geese_Relation_to_Windfarms_v3.pdf

Gruber, S. & Nehls, G. (2003) Charakterisierung des offshore Vogelzuges vor Sylt mittels schiffsgestützter Radaruntersuchungen. *Vogelkundliche Berichte aus Niedersachsen* 35: 151–156.

Gundlach, J. (1861) Tabellarische Übersicht aller bisher auf Cuba beobachteten Vögel. *Journal für Ornithologie* 9: 321–349.

Hahn, S., Bauer, S. & Liechti, F. (2009) The natural link between Europe and Africa – 2.1 billion birds on migration. *Oikos* 118: 624–626.

Hansen, L. (1954) Birds killed at lights in Denmark 1886–1939. *Videnskabelige meddelelser, Dansk Naturhistorisk Forening I København* 116: 269–368.

Hatch, S.K. (2015) Behavior of migratory tree bats in the Western Basin of Lake Erie using telemetry and stable isotope analysis. MSc thesis, University of Akron, OH. Retrieved 13 October 2017 from https://etd.ohiolink.edu/

Hatch, S.K., Connelly, E.E., Divoll, T.J., Stenhouse, I.J. & Williams, K.A. (2013) Offshore observations of eastern red bats (*Lasiurus borealis*) in the mid-Atlantic United States using multiple survey methods. *PLoS ONE* 8 (12): e83803. doi: 10.1371/journal.pone.0083803

Haupt, H. (2009) The last one to turn on the lights. The consequences of illuminated high-rise buildings for nocturnal bird migration, as exemplified by the 'Post Tower' in Bonn. *Charadrius* 45: 1–19.

Hilgerloh, G. (2009) The desert at Zait Bay, Egypt: a bird migration bottleneck of global importance. *Bird Conservation International* 19: 338–352.

Hill, R., Hill, K., Aumüller, R., Boos, K. & Freienstein, S. (2014a) Testfeldforschung zum Vogelzug am Offshore-Piltopark *alpha ventus* und Auswertung der kontinuierlich auf FINO1 erhobenen Daten zum Vogelzug der Jahre 2008 bis 2012. Funded by the German Maritime and Hydrographic Agency. StUKplus Final Report. Retrieved 23 February 2017 from http://www.fino1.de/images/forschung/Schlussbericht%20Avitec%202014.pdf

Hill, R., Hill, K., Aumüller, R., Schulz, A., Dittmann, T., Kulemeyer, C. & Coppack, T. (2014b) Of birds, blades and barriers: detecting and analysing mass migration events at *alpha ventus*. In Federal Maritime and Hydrographic Agency & Federal Ministry for the Environment, Nature Conservation and Nuclear Safety (eds) *Ecological Research at the Offshore Windfarm Alpha Ventus. Challenges, results and perspectives*. Wiesbaden: Springer Spektrum. pp. 111–131.

Hope Jones, P. (1980) The effects on birds of a North Sea gas flare. *British Birds* 73: 547–555.

Horton, K.G., Van Doren, B.M., Stephanian, P.M., Hochachka, W.M., Farnsworth, A. & Kelly, J.F. (2016) Nocturnally migrating songbirds drift when they can and compensate when they must. *Scientific Reports* 6: 21249. doi: 10.1038/srep21249

Hüppop, O. & Hilgerloh, G. (2012) Flight call rates of migrating thrushes: effects of wind conditions, humidity and time of day at an illuminated offshore platform. *Journal of Avian Biology* 43: 85–90.

Hüppop, O. & Hill, R. (2016) Migration phenology and behaviour of bats at a research platform in the south-eastern North Sea. *Lutra* 59: 5–22.

Hüppop, O. & Hüppop, K. (2011) Bird migration on Helgoland: the yield from 100 years of research. *Journal of Ornithology* 152: S25–S40.

Hüppop, O., Dierschke, J. & Wendeln, H. (2004) Zugvögel und Offshore-Windkraftanlagen: Konflikte und Lösungen. *Berichte zum Vogelschutz* 41: 127–218.

Hüppop, O., Dierschke, J., Exo, K.-M., Fredrich, E. & Hill, R. (2006) Bird migration studies and potential collision risk with offshore wind turbines. *Ibis* 148: 90–109.

Hüppop, K., Dierschke, J., Dierschke, V., Hill, R., Jachmann, K.F. & Hüppop, O. (2010) Phenology of the 'visible bird migration' across the German Bight. *Vogelwarte* 48: 181–267.

Hüppop, K., Dierschke, J., Hill, R. & Hüppop, O. (2012) Jahres- und tageszeitliche Phänologie der Vogelrufaktivität über der Deutschen Bucht. *Vogelwarte* 50: 87–108.

Hüppop, O., Hüppop, K., Dierschke, J. & Hill, R. (2016) Bird collisions at an offshore platform in the North Sea. *Bird Study* 63: 73–82.

Hutterer, R., Ivanova, T., Meyer-Cords, C. & Rodrigues, L. (2005) Bat migrations in Europe. A review of banding data and literature. *Naturschutz und Biologische Vielfalt* 28: 69–162.

Jakobsen, B. (2008) *Fuglene ved Blåvandshuk: 1963–1992*. Dansk Ornitologisk Forening og Ribe Amt.

Jellmann, J. (1988) Leitlinienwirkung auf den nächtlichen Vogelzug im Bereich der Mündungen von Elbe und Weser nach Radarbeobachtungen am 8.8.1977. *Vogelwarte* 34: 208–215.

Jones, J. & Francis, C.M. (2003) The effects of light characteristics on avian mortality at lighthouses. *Journal of Avian Biology* 34: 328–333.

Kahlert, J., Leito, A., Laubek, B., Luigujoe, L., Kuresoo, A., Aaen, K. & Luud, A. (2012) Factors affecting the flight altitude of migrating waterbirds in Western Estonia. *Ornis Fennica* 89: 241–253.

Kemp, M.U., Shamoun-Baranes, J., Dokter, A.M., van Loon, E. & Bouten, W. (2013) The influence of weather on the flight altitude of nocturnal migrants in mid-latitudes. *Ibis* 155: 734–749.

Kerlinger, P. (1984) Flight behaviour of sharp-shinned hawks during migration. 2: Over water. *Animal Behaviour* 32: 1029–1034.

Kerlinger, P. (1985) Water-crossing behavior of raptors during migration. *Wilson Bulletin* 97: 109–113.

Kerlinger, P. & Moore, F.R. (1989) Atmospheric structure and avian migration. In Power, D.M. (ed.) *Current Ornithology 6.* New York: Plenum Press. pp. 109–142.

Kerlinger, P., Gehring, J.L., Erickson, W.P., Curry, R., Jain, A. & Guarnaccia, J. (2010) Night migrant fatalities and obstruction lighting at wind turbines in North America. *The Wilson Journal of Ornithology* 122: 744–754.

Kikuchi, R. (2008) Adverse impacts of wind power generation on collision behaviour of birds and anti-predator behaviour of squirrels. *Journal for Nature Conservation* 16: 44–55.

Krijgsveld, K.L., Fijn, R.C., Japink, M., Van Horssen, P.W., Heunks, C., Collier, M.P., Poot, M.J.M., Beuker, D. & Dirksen, S. (2011) Effect studies offshore wind farm Egmond aan Zee. Final report on fluxes, flight altitudes and behaviour of flying birds. *NoordzeeWind Report No.* 10-219/OWEZ_R_231_T1_20111114_flux&flight. Retrieved 12 July 2017 from https://www.researchgate.net/profile/Karen_Krijgsveld/publication/260591425_Effect_Studies_Offshore_Wind_Farm_Egmond_aan_Zee_Final_report_on_fluxes_flight_altitudes_and_behaviour_of_flying_birds/links/0deec531a16ba977ec000000/Effect-Studies-Offshore-Wind-Farm-Egmond-aan-Zee-Final-report-on-fluxes-flight-altitudes-and-behaviour-of-flying-birds.pdf

Krüger, T. & Garthe, S. (2001) Flight altitudes of coastal birds in relation to wind direction and speed. *Atlantic Seabirds* 3: 203–216.

Kunz, T.H., Arnett, E.B., Erickson, W.P., Hoar, A.R., Johnson, G.D., Larkin, R.P., Strickland, M.D., Thresher, R.W. & Tuttle, M.D. (2007) Ecological impacts of wind energy development on bats: questions, research needs and hypotheses. *Frontiers in Ecology and the Environment* 5: 315–324.

Lagerveld, S., Jonge Poerink, B., Verdaat, H. & Haselager, R. (2014) Bat activity in Dutch offshore wind farms in autumn 2012. *Lutra* 57: 61–69.

Lagerveld, S., Janssen, R., Manshanden, J., Haarsma, A.-J., de Vries, S., Brabant, R. & Scholl, M. (2017) Telemetry for migratory bats – a feasibility study. Wageningen Marine Research Report C011/17. Retrieved 26 June 2017 from https://www.noordzeeloket.nl/en/Images/Telemetry%20for%20migratory%20bats%20-%20a%20feasibility%20study%20-%20Wageningen%20University%20and%20Research%20Report%20C011-17_5357.pdf

Larkin, R.P. & Frase, B.A. (1988) Circular paths of birds flying near a broadcasting tower in cloud. *Journal of Comparative Psychology* 102: 90–93.

Liechti, F. (2006) Birds: blowin' by the wind? *Journal of Ornithology* 147: 202–211.

Lina, P.H.C. & Reinhold, J.O. (1997) Ruige dwergvleermuis *Pipistrellus nathusii* (Keyserling & Blasius, 1839). In Limpens, H., Mostert, K. & Bongers, W. (eds) *Atlas van de Nederlandse vleermuisen – Oenderzoek naar verspreiding en ecologie.* Utrecht: Stichting Uitgeverij Koninklijke Nederlandse Natuurhistorische Vereniging. pp. 164–171.

Longcore, T., Rich, C. & Gauthreaux, S.A., Jr (2008) Height, guy wires, and steady-burning lights increase hazard of communication towers to nocturnal migrants: a review and meta-analysis. *The Auk* 125: 485–492.

Longcore, T., Rich, C., Mineau, P., MacDonald, B., Bert, D.G., Sullivan, L.M., Mutrie, E., Gauthreaux, S.A., Jr, Avery, M.L., Crawford, R.L., Manville II, A.M., Travis, E.R. & Drake, D. (2012) An estimate of avian mortality at communication towers in the United States and Canada. *PLoS ONE* 7 (4): e34025. doi: 10.1371/journal.pone.0034025

Looft, V. & Busche, G. (1981) *Vogelwelt Schleswig-Holsteins. Band 2: Greifvögel.* Neumünster: Wachholtz.

Malmiga, G., Nilsson, C., Bäckman, J. & Alerstam, T. (2014) Interspecific comparison of the flight performance between sparrowhawks and common buzzards migrating at the Falsterbo peninsula: a radar study. *Current Zoology* 60: 670–679.

Martin, G.R. (1990) The visual problems of nocturnal migration. In Gwinner, E. (ed.) *Bird Migration: Physiology and ecophysiology.* Berlin: Springer. pp. 185–197.

Masden, E.A., Haydon, D.T., Fox, A.D., Furness, R.W., Bullman, R. & Desholm, M. (2009)

Barriers to movement: impacts of wind farms on migrating birds. *ICES Journal of Marine Science* 66: 746–753.

Masden, E.A., Reeve, R., Desholm, M., Fox, A.D., Furness, R.W. & Haydon, D.T. (2012) Assessing the impact of marine wind farms on birds through movement modelling. *Journal of the Royal Society Interface* 9: 2120–2130.

Mateos-Rodriguez, M. & Liechti, F. (2012) How do diurnal long-distance migrants select flight altitude in relation to wind? *Behavioral Ecology* 23: 403–409.

Mathews, F., Roche, N., Aughney, T., Jones, N., Day, J., Baker, J. & Langton, S. (2015) Barriers and benefits: implications of artificial night-lighting for the distribution of common bats in Britain and Ireland. *Philosophical Transactions of the Royal Society B, Biological Sciences* 370: 20140124.

McCracken, G.F., Gillam, E.H., Westbrook, J.K., Lee, Y.-F., Jensen, M.L. & Balsley, B.B. (2008) Brazilian free-tailed bats (*Tadarida brasiliensis*: Molossidae, Chiroptera) at high altitude: links to migratory insect populations. *Integrative and Comparative Biology* 48: 107–118.

McGuire, L.P., Guglielmo, C.G., Mackenzie, S.A. & Taylor, P.D. (2012) Migratory stopover in the long-distance migrant silver-haired bat, *Lasionycteris noctivagans*. *Journal of Animal Ecology* 81: 377–385.

Meyer, S.K., Spaar, R. & Bruderer, B. (2000) To cross the sea or to follow the coast? Flight directions and behaviour of migrating raptors approaching the Mediterranean Sea in autumn. *Behaviour* 137: 379–399.

Miller, R.A., Onrubia, A., Martín, B., Kaltenecker, G.S., Carlisle, J.D., Bechard, M.J. & Ferrer, M. (2016) Local and regional weather patterns influencing post-breeding migration counts of soaring birds at the Strait of Gibraltar, Spain. *Ibis* 158: 106–115.

Moore, R. (2000) A fallout of turkey vultures over Florida Bay with notes on water-crossing behavior. *Florida Field Naturalist* 28: 118–121.

Mote, W.R. (1969) Turkey vultures land on vessel in fog. *The Auk* 86: 766–767.

Muheim, R., Moore, F.R. & Phillips, J.B. (2006) Calibration of magnetic and celestial compass cues in migratory birds – a review of cue-conflict experiments. *Journal of Experimental Biology* 209: 2–17.

Müller, F., Taylor, P.D., Sjöberg, S., Muheim, R., Tsvey, A., Mackenzie, S.A. & Schmaljohann, H. (2016) Towards a conceptual framework for explaining variation in nocturnal departure time of songbird migrants. *Movement Ecology* 4: 24. doi: 10.1186/s40462-016-0089-2

Müller, H.H. (1981) Vogelschlag in einer starken Zugnacht auf der Offshore-Forschungsplattform 'Nordsee' im Oktober 1979. *Seevögel* 2: 33–37.

Myres, M.T. (1964) Dawn ascent and re-orientation of Scandinavian thrushes (*Turdus* spp.) migrating at night over the northeastern Atlantic Ocean in autumn. *Ibis* 106: 7–51.

Newton, I. (2008) *The Migration Ecology of Birds*. Amsterdam: Elsevier.

Newton, I. & Little, B. (2009) Assessment of wind-farm and other bird casualties from carcasses found on a Northumbrian beach over an 11-year period. *Bird Study* 56: 158–167.

Nichols, J.T. (1920) Red bat and spotted porpoise off the Carolinas. *Journal of Mammalogy* 1: 87.

Nilsson, C. & Sjöberg, S. (2016) Causes and characteristics of reverse bird migration: an analysis based on radar, radio tracking and ringing at Falsterbo, Sweden. *Journal of Avian Biology* 47: 354–362.

Norton, A.H. (1930) A red bat at sea. *Journal of Mammalogy* 11: 225–226.

Nourani, E. & Yamaguchi, N.M. (2017) The effects of atmospheric currents on the migratory behavior of soaring birds: a review. *Ornithological Science* 16: 5–15.

O'Shea, T.J., Cryan, P.M., Hayman, D.T.S., Plowright, R.K. & Streicker, D.G. (2016) Multiple mortality events in bats: a global review. *Mammal Review* 46: 175–190.

Oppel, S., Dobrev, V., Arkumarev, V., Saravia, V., Bounas, A., Kret, E., Velevski, M., Stoychev, S. & Nikolov, S.C. (2015) High juvenile mortality during migration in a declining population of a long-distance migratory raptor. *Ibis* 157: 545–557.

Pelletier, S.K., Omland, K., Watrous, K.S. & Peterson, T.S. (2013) Information synthesis on the potential for bat interactions with offshore wind facilities – final report. Herndon, VA: US Department of the Interior, Bureau of Ocean Energy Management. OCS Study BOEM 2013-01163. Retrieved 29 June 2017 from https://www.boem.gov/ESPIS/5/5289.pdf

Pereira, M.J.R., Salgueiro, P., Rodrigues, L., Coelho, M.M. & Palmeirim, J.M. (2009) Population structure of a cave-dwelling bat, *Miniopterus schreibersii*: does it reflect history and social organization? *Journal of Heredity* 100: 533–544.

Perrow, M.R. (2017) A synthesis of effects and impacts. In Perrow, M.R. (ed.) *Wildlife and Wind Farms, Conflicts and Solutions. Volume 1. Onshore: Potential Effects.* Exeter: Pelagic Publishing. pp. 241–276.

Petersen, I.K., Christensen, T.K., Kahlert, J., Desholm, M. & Fox, A.D. (2006) Final results of bird studies at the offshore wind farms at Nysted and Horns Rev, Denmark. Report Request. Commissioned by DONG Energy and Vattenfall A/S. National Environmental Research Institute, Ministry of the Environment, Denmark. Retrieved 14 June 2017 from http://www.folkecenter.net/mediafiles/folkecenter/pdf/final_results_of_bird_studies_at_the_offshore_wind_farms_at_nysted_and_horns_rev_denmark.pdf

Petersen, A., Jensen, J.-K., Jenkins, P., Bloch, D. & Ingimarsson, F. (2014) A review of the occurrence of bats (*Chrioptera*) on islands in the North East Atlantic and on North Sea installations. *Acta Chiropterologica* 16: 169–196.

Peterson, T., Pelletier, S., Wight, L. & Boyden, S. (2014): Tracking bats offshore using nanotags. Presented at Northeast Bat Working Group, Clinton, NJ, January 2014. Retrieved 16 May 2018 from http://nebwg.org/AnnualMeetings/2014/NEBWG2014Presentations/2T_1045_Peterson.pdf

Peterson, T.S., Pelletier, S.K. & Giovanni, M.D. (2016) Long-term monitoring on islands, offshore structures, and coastal sites in the Gulf of Maine, mid-Atlantic, and Great Lakes – Final Report. USDOE Office of Energy Efficiency and Renewable Energy (EERE), Wind and Water Technologies Office (EE-4W). DOE-Stantec-EE0005378. Retrieved 27 October 2017 from https://tethys.pnnl.gov/publications/long-term-bat-monitoring-islands-offshore-structures-and-coastal-sites-gulf-maine-mid

Pettersson, J. (2005) The impact of offshore wind farms on bird life in southern Kalmar Sound, Sweden: a final report based on studies 1999–2003. Swedish Energy Agency. Retrieved 26 October 2017 from https://tethys.pnnl.gov/sites/default/files/publications/The_Impact_of_Offshore_Wind_Farms_on_Bird_Life.pdf

Plonczkier, P. & Simms, I.C. (2012) Radar monitoring of migrating pink-footed geese: behavioural responses to offshore wind farm development. *Journal of Applied Ecology* 49: 1187–1194.

Poot, H., Ens, B.J., De Vries, H., Donners, M.A.H., Wernand, M.R. & Marquenie, J. (2008) Green light for nocturnally migrating birds. *Ecology and Society* 13 (2): 47. Retrieved 1 December 2016 from http://www.ecologyandsociety.org/vol13/iss2/art47/

Richardson, W.J. (1976) Autumn migration over Puerto Rico and the western Atlantic: a radar study. *Ibis* 118: 309–332.

Richardson, W.J. (1978a) Reorientation of nocturnal landbird migrants over the Atlantic Ocean near Nova Scotia in autumn. *The Auk* 95: 717–732.

Richardson, W.J. (1978b) Timing and amount of bird migration in relation to weather: a review. *Oikos* 30: 224–272.

Richardson, W.J. (1990) Timing of bird migration in relation to weather: updated review. In Gwinner, E. (ed.) *Bird Migration.* Berlin: Springer. pp. 78–101.

Robinson, T.R., Sargent, R.R. & Sargent, M.B. (1996) Ruby-throated hummingbird (*Archilochus colubris*). In Poole, A. (ed.) *Birds of North America Online.* Ithaca, NY: Cornell Lab of Ornithology. Retrieved 15 September 2017 from bna.birds.cornell.edu/bna/species/204

Ronconi, R.A., Allard, K.A. & Taylor, P.D. (2015) Bird interactions with offshore oil and gas platforms: review of impacts and monitoring techniques. *Journal of Environmental Management* 147: 34–45.

Russell, R.W. (2005) Interactions between migrating birds and offshore oil and gas platforms in the northern Gulf of Mexico: Final Report. New Orleans, LA: US Department of the Interior, Minerals Management Service, Gulf of Mexico OCS Region. OCS Study MMS 2005-009. dx.doi.org/10.3996/062015-JFWM-050. S6. Retrieved 26 October 2017 from http://www.fwspubs.org/doi/suppl/10.3996/062015-JFWM-050/suppl_file/10.3996_062015-jfwm-050.s6.pdf?code=ufws-site

Rydell, J., Bach, L., Dubourg-Savage, M.J., Green, M., Rodrigues, L. & Hedenström, A. (2010a) Bat mortality at wind turbines in northwestern Europe. *Acta Chiropterologica* 12: 261–274.

Rydell, J., Bach, L., Dubourg-Savage, M.J., Green, M., Rodrigues, L. & Hedenström, A. (2010b) Mortality of bats at wind turbines links to nocturnal insect migration? *European Journal of Wildlife Research* 56: 823–827.

Rydell, J., Bach, L., Bach, P., Guia-Diaz, L., Furmankiewicz, J., Hagner-Wahlsten, N., Kyheröinen, E.-M., Lilley, T., Masing, M., Meyer, M.M., Petersons, G., Šuba, J., Vasko, V., Vintulis, V.S. & Hedenström, A. (2014) Phenology of migratory bat activity across the Baltic Sea and the south-eastern North Sea. *Acta Chiropterologica* 16: 139–147.

Schmaljohann, H., Liechti, F. & Bruderer, B. (2009) Trans-Sahara migrants select flight altitudes to minimize energy costs rather than water loss. *Behavioral Ecology and Sociobiology* 63: 1609–1619.

Schmaljohann, H., Becker, P.J.J., Karaardic, H., Liechti, F., Naef-Daenzer, B. & Grande, C. (2011) Nocturnal exploratory flights, departure time, and direction in a migratory songbird. *Journal of Ornithology* 152: 439–452.

Schulz, A., Dittmann, T. & Coppack, T. (2014) Erfassung von Ausweichbewegungen von Zugvögeln mittels Pencil Beam Radar und Erfassung von Vogelkollisionen mit Hilfe des Systems VARS. Funded by the German Maritime and Hydrographic Agency. StUK-plus Final Report. Retrieved 10 July 2017 from http://www.bsh.de/de/Meeresnutzung/ Wirtschaft/Windparks/Windparks/Projekte/ Oekologische_Begleitforschung_alpha_ventus/ Abschlussberichte_StUKplus/Schlussber-icht_Vogelzug_Erfassung_von_Ausweichbe-wegungen_von_Zugvoegeln_mittels_Pencil_ Beam_Radar.pdf

Seebens, A., Fuß, A., Allgeyer, P., Pommeranz, H., Mähler, M., Matthes, H., Göttsche, M., Göttsche, M., Bach, L. & Paatsch, C. (2013) Fledermauszug im Bereich der deutschen Ostseeküste. Retrieved 30 October 2017 from http://www.energiewende-naturvertraeglich.de/fileadmin/Dateien/ Dokumente/themen/Windenergie_Offshore/ Artenschutz/2013Fledermauszug-Gutachten_F. pdf

Shamoun-Baranes, J., Liechti, F. & Vansteelant, W.M.G. (2017) Atmospheric conditions create freeways, detours and tailback for migrating birds. *Journal of Comparative Physiology A, Neuroethology, Sensory, Neural, and Behavioral Physiology* 203: 509–529.

Sjöberg, S. & Muheim, R. (2016) A new view on an old debate: type of cue-conflict manipulation and availability of stars can explain the discrepancies between cue-calibration experiments with migratory songbirds. *Frontiers in Behavioral Neuroscience* 10: 29. doi: 10.3389/fnbeh.2016.00029

Sjöberg, S., Alerstam, T., Åkesson, S., Schulz, A., Weidauer, A., Coppack, T. & Muheim, R. (2015) Weather and fuel reserves determine departure and flight decisions in passerines migrating across the Baltic Sea. *Animal Behaviour* 104: 59–68.

Skiba, R. (2007) Die Fledermäuse im Bereich der Deutschen Nordsee unter Berücksichtigung der Gefährdungen durch Windenergieanlagen. *Nyctalus* 12: 199–220.

Skov, H., Desholm, M., Heinänen, S., Kahlert, J.A., Laubek, B., Jensen, N.E., Žydelis, R. & Jensen, B.P.

(2016) Patterns of migrating soaring migrants indicate attraction to marine wind farms. *Biology Letters* 12: 20160804 http://dx.doi.org/10.1098/ rsbl.2016.0804

Smith, A.D. & McWilliams, S.R. (2016) Bat activity during autumn relates to atmospheric conditions: implications for coastal wind energy development. *Journal of Mammalogy* 97: 1565–1577.

Smolinsky, J.A., Diehl, R.H., Radzio, T.A., Delaney, D.K. & Moore, F.R. (2013) Factors influencing the movement biology of migrant songbirds confronted with an ecological barrier. *Behavioral Ecology and Sociobiology* 67: 2041–2051.

Stresemann, E. (1967) Vor- und Frühgeschichte der Vogelforschung auf Helgoland. *Journal für Ornithologie* 108: 377–429.

Šuba, J., Petersons, G. & Rydell, J. (2012) Fly-and-forage strategy in the bat *Pipistrellus nathusii* during autumn migration. *Acta Chiropterologica* 14: 379–385.

Taylor, P.D., Crewe, T.L., Mackenzie, S.A., Lepage, D., Aubry, Y., Crysler, Z., Finney, G., Francis, C.M., Guglielmo, C.G., Hamilton, D.J., Holberton, R.L., Loring, P.H., Mitchell, G.W., Norris, D., Paquet, J., Ronconi, R.A., Smetzer, J., Smith, P.A., Welch, L.J. & Woodworth, B.K. (2017) The Motus Wildlife Tracking System: a collaborative research network to enhance the understanding of wildlife movement. *Avian Conservation and Ecology* 12: 8.

Thaxter, C.B., Buchanan, G.M., Carr, J., Butchart, S.H.M., Newbold, T., Green, R.E., Tobias, J.A., Foden, W.B., O'Brien, S. & Pearce-Higgins, J.W. (2017) Bird and bat species' global vulnerability to collision mortality at wind farms revealed through a trait-based assessment. *Proceedings of the Royal Society B: Biological Sciences* 284 (1862): 20170829. doi: 10.1098/rspb.2017.0829

Thomas, O. (1921) Bats on migration. *Journal of Mammalogy* 2: 167.

Vaughan, R. (2009) *Wings and Rings: A History of Bird Migration Studies in Europe.* Penryn, Cornwall: Isabelline Books.

Voigt, C.C., Roeleke, M., Marggraf, L., Pētersons, G. & Voigt-Heucke, S.L. (2017) Migratory bats respond to artificial green light with positive phototaxis. *PLoS ONE* 12 (5): e0177748. doi: 10.1371/journal.pone.0177748

von Dalla Torre, K.W. (1889) Die Fauna von Helgoland. *Zoologisches Jahrbuch* 4, Supplement 11.

Walter, G. & Todeskino, D. (2005) Zur Richtung und Höhenverteilung des Vogelzuges im Bereich der Nordergründe (Wesermündung) auf Grundlage von Radaruntersuchungen. *Natur- und Umweltschutz (Zeitschrift Mellumrat)* 4: 29–35.

Weller, T.J., Castle, K.T., Liechti, F., Hein, C.D., Schirmacher, M.R. & Cryan, P.M. (2016) First direct evidence of long-distance seasonal movements and hibernation in a migratory bat. *Scientific Reports* 6: 34585. doi: 10.1038/srep34585

Wiederholt, R., López-Hoffman, L., Cline, J., Medellín, R.A., Cryan, P., Russell, A., McCracken, G., Diffendorfer, J. & Semmens, D. (2013) Moving across the border: modeling migratory bat populations. *Ecosphere* 4 (9): 1–16.

Wiese, F.K., Montevecchi, W.A., Davoren, G.K., Huettmann, F., Diamond, A.W. & Linke, J. (2001) Seabirds at risk around offshore oil platforms in the north-west Atlantic. *Marine Pollution Bulletin* 42: 1285–1290.

Williams, T.C. & Williams, J.M. (1990) Open ocean bird migration. *IEE Proceedings F (Radar and Signal Processing)* 137: 133–138.

Williams, T.C., Ireland, L.C. & Williams, J.M. (1973) High altitude flights of the Free-tailed Bat, *Tadarida brasiliensis*, observed with radar. *Journal of Mammalogy* 54: 807–821.

Williams, T.C., Williams, J.M., Ireland, L.C. & Teal, J.M. (1977) Autumnal bird migration over the western North Atlantic Ocean. *American Birds* 31: 251–267.

Yong, D.L., Liu, Y., Low, B.W., Española, C.P., Choi, C.-Y. & Kawakami, K. (2015) Migratory songbirds in the East Asian–Australasian flyway: a review from a conservation perspective. *Bird Conservation International* 25: 1–37.

Zimmerling, J.R. & Francis, C.M. (2016) Bat mortality due to wind turbines in Canada. *Journal of Wildlife Management* 80: 1360–1369.

CHAPTER 8

Seabirds: displacement

NICOLAS VANERMEN and ERIC W.M. STIENEN

Summary

The huge surface area assigned to current and future offshore wind farm (OWF) developments has raised concern over the impact of displacement and resultant habitat loss on seabird populations. The North and Baltic Seas host most of the world's OWFs, and numerous local studies have aimed to assess displacement effects on a range of seabirds. Extensive literature review demonstrates that divers, Northern Gannet, Common Guillemot and Razorbill show relatively consistent avoidance of areas occupied by turbines, although not all studies could detect displacement of these species. Spatial variation in observed displacement levels may be due to multiple local factors such as habitat quality, prey distribution, wind-farm configuration and location relative to the colony and/or feeding grounds. At the other end of the spectrum, Great Cormorants and Great Black-backed Gulls appear to be attracted to OWFs, where both species favour roosting on the foundations of the outermost turbines. Common Eider was the only species to show consistently indifferent behaviour among the available studies. Lastly, the response of a number of species, mainly gulls, was found to be inconsistent. Importantly, the ecological consequences of the observed changes in distribution remain poorly understood. Species avoiding wind farms are expected to expend more energy to find alternative habitat, where they may face increased competition with conspecifics, in turn affecting food intake rates. Otherwise, birds attracted to wind farms are likely to be subject to increased collision risk and additional mortality. A quantitative translation of these effects into a population impact, however, is very difficult. Individual-based modelling is a promising tool to estimate the impact of displacement by one or multiple wind farms on demographic parameters. Yet, a lack of knowledge on several crucial aspects of seabird ecology impedes a straightforward interpretation of individual-based model outcomes.

Introduction

The European Union's targets on renewable energy and spatial conflicts on land have inevitably led to the exploitation of offshore wind. What began with single and well-monitored pilot projects such as Horns Rev and Nysted in Denmark (DONG Energy & Vattenfall 2006) quickly evolved into large-scale industrial developments. More than 4,000 operational turbines have been installed in European waters, generating nearly 16 GW of electricity and the offshore wind industry is still growing rapidly (see Jameson *et al.*, Chapter 1). At the time of writing, the North and Baltic Seas host most of the world's wind turbines, alongside globally important seabird populations. Offshore wind farms (OWFs) need a lot of space, but so do seabirds. OWFs built close to large seabird colonies raise most concern as birds may need to fly through or around the turbines on a daily basis to access their foraging grounds, resulting in adverse effects such as increased collision risk or higher energy expenditure and longer foraging flights, potentially affecting chick provisioning rates. Avoidance behaviour or underwater physical and ecological changes (see Rees & Judd, Chapter 2; Broström *et al.*, Chapter 3; Dannheim *et al.*, Chapter 4; Gill & Wilhelmsson, Chapter 5) could imply that important foraging grounds are lost for seabirds. In addition, outside the breeding season, OWFs are likely to conflict with seabirds as a result of habitat loss in wintering areas or increased collision risk during migration or dispersal.

OWF-related seabird displacement could be defined as any change in temporal or spatial area use attributable to the presence of the wind farm (Hötker 2017). These changes can be due to a direct (visual) response or a more indirect stimulus because of a change in environmental conditions. A natural distrust of large, conspicuous rotating structures standing well over 100 m high in their usually wide open marine habitat may result in seabirds avoiding flying through wind farms when on migration or during commuting flights as a 'barrier effect', as well as in local birds avoiding swimming or flying into the wind-farm footprint area and its immediate surroundings. Avoidance thereby affects the favoured feeding distribution, resulting in habitat loss for seabirds. Increased vessel movements for maintenance activities may also result in some seabirds avoiding OWFs (Petersen *et al.* 2006; Fox *et al.* 2006; Drewitt & Langston 2006). Importantly, displacement levels are likely to be subject to spatiotemporal variation, as avoidance responses may change over time when birds habituate to the presence of turbines, and effects may differ between sites, resulting from differences in habitat quality of the wider area (Drewitt & Langston 2006).

Targeted monitoring programmes in and around European OWFs have indeed shown that local seabird distribution may change dramatically after wind-farm construction, and that the observed effects are highly species and site specific. Early results showed that while some species displayed avoidance, others seemed to be attracted towards OWFs (Petersen *et al.* 2006). As a result of effective habitat loss and increased competition in alternative foraging areas, Fox *et al.* (2006) suggested that seabirds avoiding OWFs could suffer from decreased food intake and increased energy expenditure, with potential consequences for their future survival or productivity. An increased presence of birds inside OWFs, in contrast, is often attributed to the newly created roosting possibilities or, alternatively, to enhanced food availability as a result of the 'artificial reef effect' (Drewitt & Langston 2006; Leonhard & Pedersen 2006; see Dannheim *et al.*, Chapter 4; Gill & Wilhelmsson, Chapter 5). On the downside, birds attracted to wind farms face increased collision risk.

A thorough understanding of the ecological drivers affecting spatiotemporal variation in seabird displacement levels is considered crucial for a reliable impact assessment of planned wind-farm projects, highlighting the need for wide-scale and long-term monitoring research. Equally important, however, is the assessment of the ecological consequences and cumulative impact of wide-scale offshore wind exploitation on seabird population levels, as quantifying this impact is the actual and ultimate goal of all past, present and future research on seabird–wind-farm interactions.

Scope

The aim in this chapter is to provide a thorough review of existing empirical knowledge on OWF-induced displacement of seabirds. As a starting point, the authors consulted the detailed review paper by Dierschke *et al.* (2016), bringing together the results of seabird displacement studies at 20 operational OWFs in north-west European waters. The methodology applied by Dierschke *et al.* (2016) was revisited to formulate species-specific conclusions on displacement. While the post-construction study results collected at the Luchterduinen and Greater Gabbard OWFs were added, some of the studies used by Dierschke *et al.* (2016) were not included because a few are not publicly available and, in the authors' view, others have too limited numerical information to allow a reliable interpretation or verification of the reported conclusions. The authors further consulted their own literature database, including peer-reviewed as well as grey literature, and performed a thorough internet search to look for possible knowledge gaps. The internet was scanned using Scholar Google and Google, using keywords such as 'seabird', 'displacement', 'avoidance', 'offshore wind' and 'cumulative impact' both singly and in combination alongside targeted species names. Geographically, this chapter focuses on results gathered in north-west European marine waters for the obvious reason that very little (if any) post-construction research has yet been conducted outside this region. Furthermore, the focus is on the effects of operational OWFs rather than on construction activities, because the latter are rather short term compared to a wind farm's operational phase. As will become clear as the chapter continues, most research has aimed at assessing the responses of seabirds to OWFs, measuring levels of avoidance or attraction, but very few have tried to quantify the actual consequences of the observed changes on seabird energetics or demographics. An important part of this chapter was therefore reserved for mapping the remaining uncertainties and knowledge gaps.

Themes

Survey design

In an ideal world, the scientific international community would have agreed on a best practice and affordable monitoring set-up long before the first offshore turbines were built, generating standardised study results and allowing reliable comparison between sites. In practice, however, there appears to be huge variation in the applied methodology and data-processing strategy. This variation is found in every step of the research process, which can be roughly divided into three parts: (1) recording and counting of birds, (2) monitoring set-up, and (3) data processing and modelling.

Recording and counting birds

In OWF-related seabird displacement studies, three methods are generally applied for counting birds: boat-based surveys with visual recording, and aerial surveys with either visual recording or digital imagery (Webb & Nehls, Chapter 3 in Volume 4 of this series). Each of these methods has its advantages and limitations. Boat-based surveys stand out when it comes to the detailed recording of birds to species level and detailed information on their behaviour (including that related to foraging and feeding) and flight direction. Boat-based surveys further allow simultaneous and real-time measurements of environmental variables such as water salinity and sea-surface temperature. The major drawbacks of boat-based surveys are the inevitable disturbance of divers (*Gavia* spp.) and seaducks such as scoters (*Melanitta* spp.), and the attraction of birds towards the vessel, as is the case for some gulls, and which may also apply to Northern Fulmar *Fulmaris glacialis* and Northern Gannet *Morus bassanus* in some circumstances. Most other bird species are less sensitive to disturbance and attraction, and at least allow a vessel to come close enough to allow visual detection with binoculars or are seemingly indifferent to the presence of vessels (Garthe & Hüppop 2004). Depending on flight height, aerial surveys suffer less from bias due to disturbance, and do not attract birds. Other advantages include the possibility to cover vast areas in a short period, providing a snapshot of wide-scale seabird distribution, as well as to cover shallow sea areas not accessible by vessels. On the downside, aerial surveys result in less species-specific data. Because of difficulties in identifying birds viewed from above, and considering the high speed of the survey platform in the case of visual recording, species often need to be lumped into species groups. Yet, similar-looking species within groups, such as 'auks' or 'gulls', do not necessarily occupy the same ecological niche and can certainly not be assumed to show the same response to OWFs. In practice, visual surveys have been almost entirely replaced by digital surveys (Webb & Nehls, Chapter 3 in Volume 4) as digital imagery allows the plane to fly higher, further reducing disturbance, and stored data allow for (re)checking and independent quality assurance; yet post-processing including species identification is very time consuming. In general, aerial survey results are highly sensitive to weather conditions (Camphuysen *et al.* 2004) and the snapshot character of aerial surveys may mean that potential diurnal (Schwemmer & Garthe 2005) and tidal (e.g. Embling *et al.* 2012) patterns in the presence of birds may not be adequately covered. The cost of multiple flights per day or over a few days may be prohibitive, however.

When aiming to monitor overall seabird responses to OWFs, one method is not undeniably better than the other. Nevertheless, when wanting to investigate the response of a specific bird species (group), it is easier to define the best way to go. For scoters and divers, aerial surveys undoubtedly supply better and less biased data; but when targeting Common Guillemot *Uria aalge*, for example, one will be better off with boat-based surveys. In the end, the chosen method is likely to be a compromise between the study goals and the budget and logistics available to reach those goals. Camphuysen *et al.* (2004) detail census techniques for both aerial and boat-based survey methods, and provide an extensive tabular overview of the suitability of either method depending on the monitoring objective. A full overview is also provided by Webb and Nehls (Chapter 3 in Volume 4 of this series).

Monitoring set-up

The monitoring set-up determines where and for how long monitoring should be performed, as well as the spatiotemporal resolution of the planned surveys. A commonly applied survey design in an OWF displacement framework has parallel and equally spaced transects in a wide study area with the wind farm located in the middle. Transects are typically spaced 1–4 km apart. When this kind of monitoring is performed before as well as after wind-farm construction, this is further referred to as a before–after gradient (BAG) design. In a BAG approach one assumes any impact-induced differences between the pre- and post-construction period to be a function of distance from the wind farm and that any effects would be roughly the same in all directions (Oedekoven *et al.* 2013). A significant before–after change in abundance centred on the impacted site and decaying with distance provides compelling evidence of a causal relationship between the impact and the observed change (Box 8.1).

Box 8.1   Comparing pre- and post-construction distributions of Long-tailed Ducks *Clangula hyemalis* in and around the Danish Nysted offshore wind farm

This case study presents the approach proposed by Petersen *et al.* (2011) to model abundance patterns of wintering seaducks in relation to the construction of the Nysted offshore wind farm (OWF), accounting for imperfect detection, local surface features and autocorrelation. The proposed method provides a quantification of distributional effects over a gradient in space and time, offering an alternative to before–after control-impact (BACI) designs. The study area of 1,350 km² included the OWF and a large reference area around it. The monitoring programme comprised three pre-construction years (2000–2002) and four post-construction years (2003–2007, except for 2006). In each of the seven years, three or four aerial surveys were conducted between January and early April along 26 north–south-oriented transects at 2 km intervals. The authors' goal was to model fine-scale abundance patterns of Long-tailed Ducks *Clangula hyemalis* (Figure 8.1) over time in relation to the construction of

Figure 8.1 Male Long-tailed Duck *Clangula hyemalis* in summer plumage. The Baltic Sea is an important wintering area, with around 1.5 million individuals in 2007–2009, representing a decline of around 65% compared to an earlier census in 1988–1993 (HELCOM 2017). The area around Nysted wind farm supports a relatively low density at less than five individuals per km². (Joris Everaert)

the wind farm, in a way that reflected environmental heterogeneity in the study area and accounted for important sources of uncertainty.

Using all survey data available, a multi-covariate distance analysis was performed to correct the counted numbers of Long-tailed Ducks for varying detectability. Based on the Akaike information criterion, 'survey', 'sea state', 'behaviour', 'group size' and 'observer' were all selected as covariates to model the detection function. This detection model was then used to estimate the number of seaducks present in each transect segment of 0.5 km in length and twice the truncation distance of 0.966 km in width.

The abundance estimates obtained after distance sampling were modelled using a spatially adaptive generalised additive model (GAM) (Hastie & Tibshirani 1990). The model included a smooth term for depth and a spatially adaptive two-dimensional smooth term for the spatial surface. These smoothers were permitted to differ before and after construction. As the collected data comprised repeated visits to transects over time, it was considered unlikely that the model residuals would be independent within transects. For this reason, generalised estimating equations (GEEs) were used to generate more realistic confidence intervals for model-based differences before and after construction. The 'panels' for the GEEs were chosen to be transect-days based on the inspection of autocorrelation functions. Residuals within

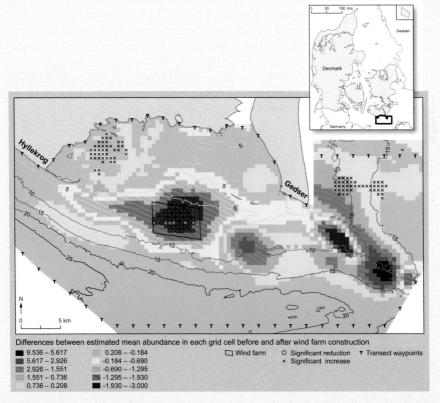

Differences between estimated mean abundance in each grid cell before and after wind farm construction

■ 9.536 – 5.617	0.208 – -0.184	▭ Wind farm ○ Significant reduction ▼ Transect waypoints
■ 5.617 – 2.926	-0.184 – -0.690	× Significant increase
■ 2.926 – 1.551	-0.690 – -1.295	
1.551 – 0.736	■ -1.295 – -1.930	
0.736 – 0.208	■ -1.930 – -3.000	

Figure 8.2 Map showing the estimated before–after changes in Long-tailed Duck *Clangula hyemalis* numbers across the Nysted offshore wind-farm study area in Danish waters (see inset), as derived from a spatially adaptive generalised additive model.

transects on a given day were thus permitted to be correlated, but independence was assumed between transects on a given day, as well as between days for a given transect. The uncertainty in the detection process and parameter uncertainty in the GAM-GEE model was combined using a two-step bootstrap approach, leading to 500,000 predictions for each 0.966 km^2 cell of a prediction grid covering the full survey area. For each set of predictions, the differences between estimated Long-tailed Duck numbers pre- and post-construction across the prediction grid were calculated and the 95% confidence intervals for these differences were assessed.

The model-based results were used to generate an 'after minus before' surface map, which indicates an estimated drop in Long-tailed Duck numbers in and around the wind farm, with an apparent displacement into neighbouring waters farther from the wind farm. There was evidence for significant decreases in Long-tailed Duck numbers only in survey grid cells between the turbines, despite the fact that the model was not informed about the OWF location, compared to significant increases in waters farther from the OWF (Figure 8.2).

Another widely adopted approach is the before–after control-impact (BACI) design. In OWF displacement studies, the extent of the survey area in a BACI set-up is typically smaller than in BAG studies. Samples are taken before and after wind-farm construction, in and around the OWF as well as in an undisturbed control area. The BACI approach assumes that naturally occurring changes will appear in both the control and impact sites (Oedekoven et al. 2013). In the case of no wind-farm effect, the trends in numbers of both areas are therefore expected to run parallel and any trend deviation between the impact and control areas is attributed to the presence of the OWF (Box 8.2). It is therefore very important for any underlying and explanatory variables driving the response variable to be similar, or at least to vary in an equal manner, in both areas. If not, these explanatory variables need to be accounted for during modelling (Stewart-Oaten & Bence 2001). Preferably, the control area will be close by and highly similar to the impact area in terms of species distribution and oceanographic variables, for any large-scale environmental changes to affect both areas, yet far enough away not to be influenced by the impact itself (Stewart-Oaten et al. 1986). As seabird numbers are generally subject to high temporal variation, replication in time is strongly advised and, ideally, samples in the control and impact areas are taken simultaneously (Bernstein & Zalinski 1983; Stewart-Oaten et al. 1986; Maclean et al. 2013). In both BAG and BACI, as many baseline years as possible should be sampled to account for natural year-to-year variation in bird abundance and distribution (Fox et al. 2006). For the same reason, a post-construction monitoring period of 3–5 years is commonly applied, and extending this time frame may strongly improve the statistical power to detect changes in seabird abundance (Maclean et al. 2013; Vanermen et al. 2015a).

Data processing and modelling

This step in the displacement assessment process is the one displaying most variation, some of this variation being the result of rapidly evolving insights and statistical techniques. Whereas earlier reports use basic calculations and/or statistical testing to look for evidence of displacement, most recent publications apply much more advanced statistical modelling techniques.

Box 8.2 Seabird displacement assessment at the Belgian Bligh Bank offshore wind farm within a before–after control-impact (BACI) framework

At the Belgian Bligh Bank, a before–after control-impact (BACI) monitoring programme was designed to detect seabird displacement following offshore wind farm (OWF) construction in 2010 (Vanermen *et al.* 2015b; 2016). Boat-based seabird counts were conducted according to the standardised and internationally applied European Seabirds at Sea (ESAS) method, combining a 'transect count' for birds on the water and repeated 'snapshot counts' for flying birds (Tasker *et al.* 1984; Camphuysen *et al.* 2004). The applied transect was 300 m wide and was counted along one side of the ship. The (future) OWF area was surrounded by a buffer zone of 3 km to define the 'impact area', where effects of the wind farm on the presence of seabirds could potentially occur. A nearby control area was also delineated, harbouring comparable numbers and similar species of seabirds before OWF construction, and showing a similar range in water depth and distance to the coast. The distance between the control and impact areas was kept small enough to be able to survey both on the same day by means of a 50 m research vessel typically travelling at 10 knots.

The Bligh Bank study area was studied intensively from April 2008 to April 2015, incorporating construction activities from September 2009 to September 2010. The survey design comprised four parallel south-west–north-east-oriented transect lines spaced 2.5 km apart. In addition, after OWF construction, four closely spaced transects (400 m) through the turbine corridors were surveyed to increase sample size and data reliability within the impact area (Figure 8.3). During the monitoring programme, both control and impact areas should have been visited monthly, but vessels were not always available and planned trips were sometimes cancelled owing

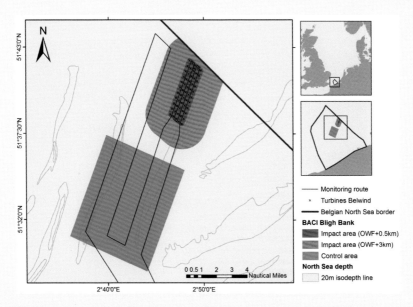

Figure 8.3 The before–after control-impact (BACI) monitoring set-up and post-construction seabird survey route in the Bligh Bank study area in Belgian waters.

to adverse weather conditions such as significant wave heights higher than 2 m and/ or poor visibility. Before 2008, the study area was also surveyed on an irregular basis, and surveys dating back to 1993 were included in the analysis provided that the control and impact areas were visited on the same day.

The numbers of seabirds observed on the water were corrected for decreasing detection probability with distance to the ship. As detection probability was further likely to depend on group size and observation conditions (Buckland *et al.* 2001; Marques & Buckland 2003), the following covariates were considered as explanatory variables for the species-specific detection functions: 'group size', 'log(group size)', 'wind force' and 'wave height'. The best detection model was selected according to the information-theoretic approach and Akaike information criterion (AIC) characteristics. This multi-covariate distance analysis thus resulted in species-specific detection probabilities varying with the selected covariates, and observed numbers were corrected accordingly.

For the BACI analysis, count data were aggregated for control and impact areas separately and per monitoring day, resulting in daily totals for both zones, thus avoiding autocorrelation between subsequent counts along transect lines and minimising overall variance. Only days on which both the control and impact areas were visited were used in the analysis, minimising variation resulting from short-term temporal changes in seabird abundance. Four different distributions, namely Poisson, negative binomial, zero-inflated Poisson and zero-inflated negative binomial, were considered. Explanatory factor variables included were 'area' (control/impact), 'period' (before/after), 'OWF' (present/absent) and 'fishery' (present/absent). For the last of these, only active fishing vessels observed within a distance of 3 km from the monitoring route were taken into account. To correct for varying monitoring effort, the area counted was included in the model as an offset variable. Lastly, 'month' was used as a continuous variable to model seasonal fluctuations by fitting a cyclic smoother or cyclic sine curve, the latter described by a linear sum of sine and cosine

Figure 8.4 Winter plumage Common Guillemot *Uria aalge:* the most abundant seabird in the Bligh Bank area during the autumn and winter months. (Hilbran Verstraete)

terms (Stewart-Oaten & Bence 2001). AIC was used to select the best-fitting model. The eventual and targeted output values were species-specific OWF coefficients estimating the effect of the wind farm on the different species abundances.

For Common Guillemot *Uria aalge* (Figure 8.4), analysis showed a significantly negative OWF coefficient of –1.39, corresponding to a decrease in numbers of 75% compared to the abundance in the control area and the period before impact (Figure 8.5). Other significant seabird displacement effects revealed in this study were avoidance by another common alcid auk, Razorbill *Alca torda* (–67%), as well as Northern Gannet *Morus bassanus* (–82%), in contrast to strong attraction of Lesser Black-backed *Larus fuscus* and Great Black-backed *Larus marinus* Gulls, which increased in number by a factor of 8.1 and 3.6, respectively.

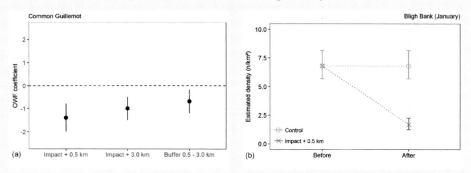

Figure 8.5 Modelling results for Common Guillemot *Uria aalge* in the Bligh Bank study area: (a) offshore wind farm (OWF) coefficients and their 95% confidence intervals; (b) before–after control-impact (BACI) density estimates for the month with maximum numbers.

When based on visual observations, raw data are likely to be subject to 'imperfect detection'. As distance to the subject increases, the chance of detection decreases according to a 'detection function' (Buckland *et al.* 2001). To be able to model such a detection function, the distance to the recorded birds needs to be assessed while surveying, which appears to be common practice. Performing a distance analysis results in species-specific detection probabilities, which, in turn, can be used to correct the counted numbers for birds missed during surveying. Detection probabilities tend to be lower for small, dark-coloured birds such as auks (Figure 8.4) than for large species such as Northern Gannet and large *Larus* spp. gulls. The detection probability may further vary with observation conditions, which can also be included in the modelling process by applying a multi-covariate distance analysis (Marques & Buckland 2003).

After performing distance correction, the analysis involves modelling the adjusted response variable as a function of explanatory covariates. Covariates to be considered may be oceanographic data that were collected simultaneously with the observations or remotely sensed data available online (Oedekoven *et al.* 2013). Alternatively, latitude and longitude are often included by means of a two-dimensional smoother as a proxy for underlying spatiotemporal covariates driving seabird distribution (Vallejo *et al.* 2017). Indeed, animal abundance at a specific location is likely to be the result of a complex mix of covariates, although many of these covariates are unavailable for inclusion in the model (Mackenzie *et al.* 2013). In an OWF displacement context, the development phase (before, during and after construction) or location (wind farm, buffer zone and control area) are generally included in the model as factor variables.

Correctly modelling seabird count data can be quite challenging for a number of reasons. First of all, considering the often patchy distribution of seabirds, count data tend to show large variance and/or a high proportion of zero values, leading to 'overdispersion', even after having included all relevant or available explanatory variables. Overdispersion can be accounted for using a negative binomial, or, alternatively, a zero-inflated distribution (Potts & Elith 2006; ver Hoef & Boveng 2007; Zeileis et al. 2008). The latter splits the model in two parts: a binomial part modelling the (excess in) zeros, and a count part modelling the positive values according to a Poisson or negative binomial distribution. Secondly, seabird count data may exhibit non-linear relationships with environmental or spatial covariates, which can be modelled through a generalised additive model or the spatially more flexible complex region spatial smoother (Oedekoven et al. 2013; Mackenzie et al. 2013). Lastly, owing to consecutive sampling along transect lines, collected data are expected to show some level of autocorrelation, implying that counts collected close to each other in either time or space are more similar than more distant counts, thereby violating the assumption of residual independence. This spatiotemporal correlation issue can be accounted for by including user-specified random effects in a generalised additive mixed model (Leopold et al. 2013; Vallejo et al. 2017) or by applying generalised estimating equations (Mackenzie et al. 2013). Better still, accounting for spatial correlation may be achieved through Bayesian modelling techniques, of which the integrated nested Laplace approximation approach is gaining popularity as it has proven to be a user-friendly alternative to the computationally very demanding Markov chain Monte Carlo technique (Rue et al. 2009).

Finally, there are several ways to illustrate OWF displacement evidence. In a BACI framework, the OWF effect is generally estimated by the interaction between the area (control/impact) and time factor (before/after). In a BAG model, the OWF location can also be included in the model as a factor variable, testing the level and significance of its effect on the response variable (Leopold et al. 2013). More often, though, the model structure does not incorporate wind-farm location, and instead a distribution map of pre- to post-impact differences in abundance is generated. Significant pre- to post-impact changes centred inside and around the OWF can then reliably be attributed to the OWF.

Avoidance and attraction

Review of displacement studies

Information on seabird displacement was compiled based on the results of monitoring studies in 16 different European OWFs. As already mentioned in the *Scope* above, very different methodologies have been applied, making consistent interpretation difficult and somewhat subjective. When different reports or publications on the same OWF were available, only the one including most recent data and applying the most up-to-date approach in terms of data processing and modelling was considered. In general, one key reference was considered per OWF, although in some cases different references were used for different species (Table 8.1). For example, at Robin Rigg OWF, Canning et al. (2013) report extensively on a range of species, while Vallejo et al. (2017) performed more advanced modelling on Common Guillemot in a further study. It should also be noted that not all species considered in the studies listed in Table 8.1 were included in this exercise. The preferred modelling strategy could not always be performed because of insufficient positive observations, and even if it was, the authors sometimes concluded that the results for certain species could be unreliable and should therefore be interpreted with care. Leopold et al. (2013), for instance, found the Prinses Amalia (PAWP) OWF to be largely out of the normal distribution range of divers, Great Crested Grebe *Podiceps cristatus* and

Table 8.1 List of offshore wind farms, their location, range of species assessed and key references included in this review.

Wind farm	Location	Key reference(s)	Key species
Robin Rigg	Irish Sea (Scotland)	Canning et al. 2013	Range of seven seabird species
		Vallejo et al. 2017	Common Guillemot
Kentish Flats	North Sea (England)	Percival 2014	Divers
London Array	North Sea (England)	APEM 2016	Auks and divers
Thanet	North Sea (England)	Percival 2013	Range of nine seabird species (groups)
Greater Gabbard	North Sea (England)	APEM 2014	Northern Gannet
Bligh Bank	North Sea (Belgium)	Vanermen et al. 2016	Range of 11 seabird species
Thornton Bank	North Sea (Belgium)	Vanermen et al. 2017	Range of 12 seabird species
Prinses Amalia (PAWP)	North Sea (Netherlands)	Leopold et al. 2013	Range of 12 seabird species (groups)
Egmond aan Zee (OWEZ)	North Sea (Netherlands)	Leopold et al. 2013	Range of 15 seabird species (groups)
Luchterduinen	North Sea (Netherlands)	Skov et al. 2016	Range of ten seabird species
Alpha Ventus	North Sea (Germany)	Welcker & Nehls 2016	Range of nine seabird species (groups)
Horns Rev 1	North Sea (Denmark)	Petersen et al. 2006	Divers
			Little Gull
			European Herring Gull
			Auks
			Common/Arctic Tern
		Petersen & Fox 2007	Common Scoter
Horns Rev 2	North Sea (Denmark)	Petersen et al. 2014	Divers
			Common Scoter
Nysted	Baltic Sea (Denmark)	Petersen et al. 2006	Divers
			Red-breasted Merganser
			Common Eider
			Common Scoter
			European Herring Gull
		Petersen et al. 2011	Long-tailed Duck

Table 8.1 – *continued*

Wind farm	Location	Key reference(s)	Key species
Tunø Knob	Baltic Sea (Denmark)	Guillemette *et al.* 1998	Common Eider and Common Scoter
		Guillemette *et al.* 1999	Common Eider and Common Scoter
Lillgrund	Baltic Sea (Sweden)	Nilsson & Green 2011	Great Cormorant
			Red-breasted Merganser
			Common Eider
			Long-tailed Duck
			European Herring Gull

Common Scoter *Melanitta nigra*, which were therefore not included in this exercise despite the reported displacement estimates.

Largely following Dierschke *et al.* (2016), the reported OWF-induced seabird displacement levels were classified in the following categories:

- **Category 1.** Strong avoidance – estimated decrease in numbers or negative effect of >50%.

- **Category 2.** Moderate avoidance – estimated decrease in numbers or negative effect of 20–50%.

- **Category 3.** (Rather) indifferent behaviour – estimated change in numbers or effect of <20%.

- **Category 4.** Moderate attraction – estimated increase in numbers or positive effect of 20–50%.

- **Category 5.** Strong attraction – estimated increase in numbers or positive effect of >50%.

To provide further information on the reliability and/or extent of these results, three symbols were added to the classification:

- * The change or effect is reported as statistically significant.

- d Distribution map shows an obvious gap, concentration or before–after change in and around the OWF.

- + The change or effect is noticeable up to more than 2 km from the OWF.

It should be stressed that even though an observed change is reported as significant it does not guarantee the observed change or effect to be true and/or undoubtedly attributable to the suspected source of impact. A change or effect is regarded as statistically significant when it is considered highly unlikely to occur by sheer coincidence, a consideration that is based on a number of assumptions. Most importantly, it assumes that all relevant sources of variation affecting the response variable, but also the data distribution of the response variable itself, are correctly accounted for. Not doing so may result in a poor estimate of the effect and/or its confidence intervals and falsely assigning an observed change in numbers to the presence of an OWF. In the same sense, it should be highlighted that a 'significant' effect should only be interpreted in the statistical sense of the word, and does not inform us about the actual extent or consequences of the observed effect. That is why, in the present analysis, a change or effect of more than 50% was classified as strong avoidance/attraction regardless of its statistical significance, as opposed to Dierschke *et al.* (2016), who increased this threshold to 80% for changes or effects without statistical 'evidence'. Significant results were thus not treated as more important or any different from results without statistical support. The overall score for a species' sensitivity to displacement was then calculated as the arithmetic mean of all classification scores of the included studies.

This exercise led to the results shown in Table 8.2, which was sorted according to the mean classification score. In the case of auks, summarising values are based on all classification scores recorded for unidentified auks as well as Common Guillemot and Razorbill *Alca torda*. Species with a mean score of 4.0 or more are coloured green and can be considered to be attracted to OWFs, while species with a score of 2.0 or less are coloured red and were generally found to avoid OWFs. This raw classification coincides with good consistency in the results, as all 'green' species show scores from 3 to 5, while all 'red' species show scores of 1 to 3. In between the green and red groups, there is a grey zone containing species that show inconsistency in displacement results, ranging from avoidance in some places to attraction in others, apart from the consistently indifferent Common Eider *Somateria mollissima*.

Species attracted to offshore wind farms

Great Cormorants *Phalacrocorax carbo* were found to be strongly attracted to OWFs in four out of five studies, with one study not finding a convincing effect (Figure 8.6). The attraction effects recorded at the three Dutch wind farms and the Scottish Robin Rigg wind farm all proved statistically significant in a BAG analysis (Canning *et al.* 2013; Leopold *et al.* 2013; Skov *et al.* 2016). In contrast, for the Swedish Lillgrund OWF, Nilsson and Green (2011) found Great Cormorant to decrease in numbers in the OWF footprint area, but to increase strongly in the 2 km buffer zone.

Figure 8.6 Great Cormorant *Phalacrocorax carbo* flying in to fish within a wind farm, a common sight in several wind farms in Dutch and UK waters. (Martin Perrow)

Table 8.2 Classification of the seabird displacement levels as found in 16 European offshore wind farms.

Species	Robin Rigg	Kentish Flats	London Array	Thanet	Greater Gabbard	Bligh Bank	Thornton Bank	Prinses Amalia	Egmond aan Zee	Luchterduin	Alpha Ventus	Horns Rev 1	Horns Rev 2	Nysted	Tuno Knob	Lillgrund	Mean	Minimum	Maximum	Number of studies
Sandwich Tern *Thalasseus sandvicensis*	5*(d)						5(+)										5.0	5	5	1
Great Cormorant *Phalacrocorax carbo*	3			3				5*(d)	5*(d)	5*(d)						3	4.6	3	5	5
Great Black-backed Gull *Larus marinus*						5*	5*	4	4	3	5*						4.0	3	5	8
Herring Gull *Larus argentatus*	5			3		5*	5	2	4	5	3	5(+)		3		3	3.9	2	5	11
Red-breasted Merganser *Mergus serrator*														5(+)		2	3.5	2	5	2
Common Gull *Larus canus*				3		5	1	3	3	5	3						3.3	1	5	7
Lesser Black-backed Gull *Larus fuscus*				3		5*(+)	4	2	2*	1	5*						3.1	1	5	7
Common Eider *Somateria mollissima*														3	3	3	3.0	3	3	3
Black-legged Kittiwake *Rissa tridactyla*	5			3		4	1*	1*	3	5*(d+)	2						3.0	1	5	8
Little Gull *Hydrocoloeus minutus*						1	1*	1*	5	1*(d)	1*	5(+)					2.1	1	5	7
Great Skua *Catharacta skua*	3					1	3										2.0	1	3	2
Common Guillemot *Uria aalge*				3		1*(+)	1*	2*	2*	1*(d)							1.9	1	3	7
Common Scoter *Melanitta nigra*												3	1*(d+)		3		1.8	1	3	5
Northern Gannet *Morus bassanus*	3			3	1(d)	1*	1*(+)	1*	1*	2	1*		1(+)				1.6	1	3	9
Auks *Uria aalge/Alca torda*			1*(d)								1*(+)	1					1.5	1	3	17
Divers *Gavia sp.*	3	1*(d)	3			1*		1*	1*	1*(d)	1*	1*(+)	1*(d+)	1(+)			1.4	1	3	9
Razorbill *Alca torda*	2			1*		1*	1	1*	3	1							1.4	1	3	7
Long-tailed Duck *Clangula hyemalis*														1*(d)		1(d)	1.0	1	1	2
Great-crested Grebe *Podiceps cristatus*									1*								1.0	1	1	1
Northern Fulmar *Fulmaris glacialis*						1*(+)	1(+)	1(d)	1(d)								1.0	1	1	4
Common/Arctic tern *Sterna hirundo/arctica*								1	1*			1					1.0	1	1	3

Attraction is shown by species shaded in green with mean category scores >4, with avoidance shown by species shaded in red with mean scores <2, compared to species shaded grey with scores >2 and <4 that are indifferent and/or inconsistent in their response. To further aid interpretation, species are also classified with (*) where results are reported as statistically significant, (d) where there is a clear change in distribution associated with the wind farm, and (+) where the effect is apparent up to 2 km from the wind farm. (Adapted from Dierschke *et al.* 2016)

Another species showing quite consistent attraction to OWFs is Great Black-backed Gull. Five of out eight studies report moderate to strong attraction, while three studies found no evidence of post-construction changes in the at-sea distribution of this species. Attraction of Great Black-backed Gulls was found over a wide regional scale, including Belgian, Dutch and German offshore waters (Leopold *et al.* 2013; Welcker & Nehls 2016; Vanermen *et al.* 2016; 2017).

As Great Cormorants need to dry their feathers in between foraging bouts, the presence of above-water constructions is a critical precondition to their being able to forage offshore for longer periods. As such, birds are now present in areas that were previously inaccessible, often resulting in very distinct increases and distributional concentrations (Leopold *et al.* 2013). At the Thornton Bank OWF, the majority of Great Cormorants was observed roosting on the turbine foundations (89%), with a clear preference for outer rather than inner turbines. Great Black-backed Gulls, by foraging on the wing and resting on the sea surface without waterlogging, do not suffer from the same ecological restriction as Great Cormorants. Nevertheless, they did show a strongly comparable preference for the foundations at Thornton Bank, with 79% of the birds observed roosting on turbine foundations and concentrating along the wind-farm edge (Figure 8.7). During operation of the Horns Rev OWF, all records of resting or perching cormorants and 91% of the gull records were made on turbines at the edge of the wind farm (Petersen *et al.* 2006). Apart from increased roosting possibilities, both species may also be attracted to OWFs because of improved food availability resulting from the attraction of fish to the turbine foundations and/or the exclusion of commercial fishing and, at least in the case of Great Black-backed Gull, possibly also because of benthic communities colonising the intertidal zones of the foundations (Leonhard & Pedersen 2006; Leonhard *et al.* 2011; Leopold *et al.* 2013; De Mesel *et al.* 2014; Reubens *et al.* 2014). Whatever the incentive, a concentration of birds staging at outer turbines points towards a 'collision' of the opposing forces of avoidance and attraction along the edge of the wind farm.

Figure 8.7 Great Black-backed Gulls *Larus marinus* (with one Lesser Black-backed Gull *Larus fuscus*) roosting, resting and preening on a turbine platform before the installation of the turbine. (Hilbran Verstraete)

Finally, a post-construction increase in Sandwich Tern *Thalasseus sandvicensis* has been reported at the Thornton Bank OWF, with this increase being strongest in the buffer area up to 3 km from the wind farm. Unfortunately, this is the only post-construction study reporting on displacement of Sandwich Terns, and general conclusions cannot be drawn from a single study. In contrast, Sandwich Terns have been shown to suffer from a barrier effect when approaching OWFs, with macro-avoidance rates of 28% and 30% reported by Krijgsveld *et al.* (2011) and Harwood *et al.* (2017), respectively. Clearly, in areas with a high proportion of commuting birds, a strong barrier effect may result in apparent displacement of Sandwich Terns from the wind-farm area. The Harwood *et al.* (2017) study of breeding birds during the construction period introduces further complexity as the rate of macro-avoidance was derived from the response of tracked birds (see Box 8.3), whereas the simultaneous boat-based surveys showed no clear reduction in density within the OWF, but a clear increase in density within 2 km of the wind farm. Attraction to the area around the OWF was linked to the intensive use of the eight navigation buoys around the OWF perimeter by resting and socialising Sandwich Terns.

Box 8.3   Individual tracking as a tool to monitor seabird–offshore wind farm interactions

Recent advances in, and better affordability of, Global Positioning System (GPS) tracking technologies have offered increasing possibilities to assess the likely impacts of offshore wind developments on seabirds (e.g. Thaxter *et al.* 2015; Thaxter & Perrow, Volume 4, Chapter 4) (Figure 8.8). While boat-based and aerial surveys are limited to specific time frames and conditions such as daylight and a favourable sea state, GPS tracking allows continuous monitoring of an individual bird's behaviour over the full tidal and diurnal cycle in all weather conditions. Owing to the detailed

Figure 8.8 Global Positioning System (GPS) tag with solar panel being fitted to a breeding European Herring Gull *Larus argentatus*. Rapid evolution of tag technology now allows a wide range of seabirds to be tagged, with the resulting data showing fine-scale details of the interaction of the individual birds with offshore wind farms. (Misjel Decleer, Flanders Marine Institute/Vlaams Instituut voor de Zee)

information on flight directions and heights, it allows monitoring of both vertical and horizontal responses of individual birds towards offshore turbines. On the other hand, the behavioural detail of foraging activity and subtle responsive changes in flight height and direction may be limiting when using GPS tracking. For highly detailed monitoring of behavioural responses, visual tracking of individual birds with a fast-moving rigid-hulled inflatable boat (Perrow *et al.* 2011; Harwood *et al.* 2017) or advanced camera systems installed inside OWFs have proven to be promising tools (Skov *et al.* 2018). GPS tracking is particularly effective in defining at-sea foraging movements of seabirds, which is fundamental to understanding their ecology and the potential impact of OWFs (Garthe *et al.* 2016; Thaxter & Perrow, Volume 4, Chapter 4). A key advantage of GPS tracking is that the birds involved are of known provenance and status as a result of capture, whereas boat-based or aerial surveys do not allow discrimination of breeding from failed and non-breeding birds (Perrow *et al.* 2015). The applicability of tracking data in wind-farm impact assessments is highlighted by their use in individual-based models to determine the effect of displacement on adult body condition and chick survival (Searle *et al.* 2014; Warwick-Evans *et al.* 2018). GPS tracking may also reveal habituation to wind farms as repeated tracking of the same individuals may help to identify changes in the birds' responses over time to existing wind farms (Garthe *et al.* 2016). A continuation of the widely applied tracking of seabirds may eventually also reveal general displacement levels by means of a pre–post construction comparison of distribution and habitat use. Thaxter *et al.* (2015) showed that changes in food availability may alter the behaviour of birds between years and even through the course of a single breeding season. As this may lead to the use of alternative foraging areas and different commuting patterns, the movements of birds need to be characterised over longer periods to fully appreciate the potential for variation in seabird–wind-farm interactions. For a full review of tracking seabirds in this context, see Thaxter & Perrow (Volume 4, Chapter 4).

Species avoiding offshore wind farms

This analysis included nine studies that aimed to assess displacement of divers following the construction of an OWF, resulting in a mean displacement score of 1.4. Except at Robin Rigg, where the analyses were conducted at species level for Red-throated Diver *Gavia stellata*, divers were mostly analysed as a species group owing to relatively high percentages of birds that remain undetermined. On the other hand, the percentage of Red-throated Divers among positively determined divers was often very high, for example 98% at Kentish Flats, 93% at Alpha Ventus and 91% at Horns Rev 2. Conclusions drawn on the displacement of divers can therefore be expected to hold true for Red-throated Diver. Seven out of nine studies showed strong avoidance of OWFs by divers, six of which were supported by statistical significance, while the two remaining studies could not find convincing signs of displacement. The change in diver abundance was often found to be strong, with reductions in numbers of 67% at Egmond aan Zee (OWEZ), 73% at Thanet, 94% at Kentish Flats and 90% at Alpha Ventus (Leopold *et al.* 2013; Percival 2013; 2014; Welcker & Nehls 2016).

Northern Gannets too seem to be sensitive to displacement by OWFs, with a mean score of 1.6 from the results of nine post-construction monitoring studies (Figure 8.9). Six studies reported strong avoidance, with five of these showing a significant change or effect,

Figure 8.9 A number of studies show that Northern Gannets *Morus bassanus* are displaced from offshore wind farms, at least in the autumn/winter months. Whether this pattern persists for breeding birds provisioning chicks will be tested with new wind farms planned relatively close to colonies in the UK. (Koen de Vos)

compared to one reporting moderate avoidance and two without apparent adverse effect. Where detected, displacement seemed to be strong, with reductions of 93% at PAWP, 74% at OWEZ, 95% at Greater Gabbard, 79% at Alpha Ventus, 82% at Bligh Bank and 97% at Thornton Bank (Leopold *et al.* 2013; APEM 2014; Welcker & Nehls 2016; Vanermen *et al.* 2016; 2017).

Common Guillemot, Razorbill and unidentified auks in general are another set of species (group) for which there is good consistency in displacement classification. Seventeen studies were inspected, three of which concluded indifference, whereas all the others illustrated avoidance of the wind farm under study. The resulting displacement score for the species group as a whole equals 1.5, with a score of 1.4 for Razorbill and 1.9 for Common Guillemot. Being common species with relatively homogeneously distributed numbers compared to gulls, for example, the changes in distribution and numbers reached significance quite often, even for moderate avoidance levels. Neither of the species or the generic group appeared to avoid wind farms completely, although generally high levels of change in abundance were found, such as a 75% reduction at Alpha Ventus (unidentified auks), 75% at Bligh Bank (Common Guillemot) and 80% at PAWP (Razorbill) (Leopold *et al.* 2013; Welcker & Nehls 2016; Vanermen *et al.* 2016). On the other hand, there is quite convincing evidence that in some places Common Guillemot did not avoid the OWF under study, as was the case for the Robin Rigg OWF (Vallejo *et al.* 2017). The lack of avoidance of Robin Rigg mainly involves breeding birds, in contrast to the non-breeding (dispersing and/or wintering) birds at all other sites, which are likely to be less constrained to use specific foraging areas and possibly more readily displaced.

Common Scoter reached a score of 1.8 based on five different post-construction studies, three of which reported strong avoidance with the other two showing indifference. At Tunø Knob, it was concluded that the changes in numbers and distribution of Common Scoter could not be attributed to the presence of the OWF and were more likely to be related to changes in food abundance (Guillemette *et al.* 1998; 1999). At Horns Rev 1, Common Scoters appeared to avoid the wind farm in the first 3 years after construction, but were found to occur in equal numbers in and outside the wind farm in 2007. Scoters did, however, appear to be displaced from the Nysted, Horns Rev 2 and OWEZ wind farms (Petersen *et al.* 2006; 2014; Petersen & Fox 2007; Leopold *et al.* 2013).

All four studies including Northern Fulmar showed strong avoidance by this species, although only one study claimed statistical significance. It is notable that these studies were conducted in Belgian and Dutch waters, where densities of Northern Fulmar are typically quite low, resulting in low confidence of the results obtained (Leopold *et al.* 2013; Vanermen *et al.* 2016; 2017). Post-construction monitoring in marine areas where OWFs overlap with high densities of Northern Fulmar is needed to enable better and more reliable assessment of this species' actual displacement response.

More species (groups) fall within the category of birds displaced by OWFs based on a mean displacement score below the value of 2.0, including Great Crested Grebe, Long-tailed Duck *Clangula hyemalis*, Great Skua *Stercorarius skua* and Common/Arctic Terns *Sterna hirundo/paradisaea*. However, these displacement scores all rely on only one to three post-construction studies and general conclusions are therefore difficult to draw.

Species with indifferent or inconsistent responses

For all but one of the gull species included within the indifferent/inconsistent group, responses ranged from strong avoidance to strong attraction. European Herring Gull *Larus argentatus* is the most consistent, but even so shows a response ranging from moderate avoidance in one study to indifference in four studies and moderate to strong attraction in six studies. The inconsistency found for gulls is not surprising as they are often strongly aggregated, especially in the vicinity of trawling activity. Even when present, whether or not large aggregations of gulls are encountered during standard monitoring procedures is largely coincidental, while the inclusion or exclusion of such large numbers in an impact study data set may strongly influence the eventual outcome.

Petersen *et al.* (2006) found numbers of Red-breasted Merganser *Mergus serrator* to increase strongly in and up to a distance of 3.5 km around Nysted OWF. At the Lillgrund OWF, however, built in an area hosting a wintering population of over 10,000 birds, the distribution of Red-breasted Merganser displayed moderate avoidance of the wind farm (Nilsson & Green 2011).

This leaves the Common Eider as the only species displaying consistent 'indifference', as this species' distribution and numbers were found to be unaffected by the presence of the Nysted, Tunø Knob and Lillgrund OWFs (Guillemette *et al.* 1998; 1999; Petersen *et al.* 2006; Nilsson & Green 2011). In contrast, Larsen and Guillemette (2007) showed strong behavioural responses of flying Common Eiders to the turbines at Tunø Knob, and hypothesised a possible reduction of available habitat within and around OWFs as a result.

Spatiotemporal variation in observed displacement effects

Offshore displacement studies showed quite good consistency in results, yet there still are obvious differences between sites. First of all, local factors such as food abundance, the location of the OWF relative to the colony or feeding grounds, and wind-farm configuration may be very important in explaining the observed response. When resources become limited, individuals may be prepared to take more risks or be more tolerant of the presence of wind farms (Hötker 2017). Therefore, displacement responses are likely to depend on the food availability in and outside OWFs, either within the foraging range from breeding colonies or in wintering areas.

In relation to wind-farm characteristics, all studies so far have investigated single and relatively small wind farms with fewer than 100 turbines, and how seabirds will respond towards much bigger clusters of turbines is still to be investigated. One can also easily imagine that wind farms with closely spaced turbines would provoke a stronger avoidance response, as argued by Leopold *et al.* (2013). In this study, the authors found Common Guillemot to be avoiding the PAWP more than the nearby OWEZ wind farm and suggested that the former, with its much higher turbine density (4.3 versus 1.3 turbines per km²), was less permeable to movement. Combining and analysing post-construction results from different sites may eventually lead to reliable statements on this matter.

Depending on the season, birds also may react differently towards wind farms. Migrating Common Eiders have been shown to change flight direction at distances of 1–2 km at Kalmar Sound, Sweden (Pettersson 2005), or 3–5 km at Nysted, Denmark (Petersen *et al.* 2006), while flying birds commuting between roosting and feeding grounds showed a strong behavioural response towards the Tunø Knob turbines (Larsen & Guillemette 2007). Yet when considering the general distribution patterns in their familiar winter staging areas, this same species seemed more or less indifferent towards OWFs. Likewise, the study site at Robin Rigg is the only one included in this review chapter where Common Guillemots are more abundant during the breeding season than during winter, perhaps coincident with this being one of the few studies not finding an avoidance response (Vallejo *et al.* 2017). In winter, Common Guillemots are known to range widely throughout the North Sea and may therefore more easily compensate for any habitat loss by moving elsewhere to forage (Dierschke *et al.* 2016). In contrast, being central-place foragers during summer, the individual Common Guillemots studied at the Robin Rigg study site are more constrained in terms of distribution and habitat choice, and may therefore be more tolerant of wind turbines. Ideally, seasonal variation would be assessed by studying the same bird species year-round at the same site. However, most study areas considered here are located in seabird wintering and/or migration areas rather than close to major seabird colonies. Therrefore, it has not yet been possible to investigate the full spectrum of seasonal variation in displacement level simply because the birds are not present during a large part of the year at the sites studied.

Lastly, OWFs are still a relatively new phenomenon, and it is often hypothesised that in time seabirds may habituate towards their presence (e.g. Drewitt & Langston 2006; Fox *et al.* 2006; Petersen & Fox 2007). Few studies have been conducted for long enough to support this hypothesis, yet the occurrence of Common Scoters in and around Horns Rev 2 provides a good example of potential habituation. Although Common Scoters were assumed to be displaced based on the three first post-construction years, Petersen and Fox (2007) reported these birds to occur in numbers higher than ever before, with a maximum of 4,624 birds within the wind-farm footprint in February 2007. In three out of four surveys carried out in 2007, there was no difference between mean encounter rates inside and immediately outside the wind farm, and thus no sign of avoidance during these surveys. The authors mention that this dramatic change in distribution consistent with habituation could well have been a reflection of a change in food abundance in the form of bivalve molluscs within the wind farm.

Ecological consequences

Importantly, a displacement effect is not synonymous with a displacement impact (Masden *et al.* 2010a). In an OWF context, the effect is often a decreased number of birds in the wind-farm footprint area due to an avoidance response, whereas the actual impact could be a change in population size resulting from a chain of consequences following the initial

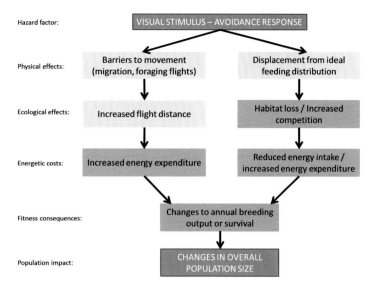

Figure 8.10 Flowchart summarising the ecological consequences resulting from the two main avoidance responses at offshore wind farms. The green boxes indicate potentially measurable effects while the red boxes indicate processes that need to be modelled. (Adapted from Fox *et al.* 2006)

avoidance. Figure 8.10 shows a schematic overview of the expected ecological consequences of displacement and their eventual impact resulting from the two possible avoidance responses: (1) barriers to movement and (2) displacement from an ideal foraging/feeding distribution.

Birds sensitive to displacement can be expected to respond towards the presence of an OWF by flying around rather than entering the wind farm and/or spending time searching for alternative foraging habitat, implying an increased energetic cost. This may result in decreased food intake when the alternative habitat proves to be of minor quality. Displacement from OWFs can further result in higher bird densities in other foraging areas, leading to increased interspecific and intraspecific competition, thus affecting the displaced birds as well as those already occupying the area (Fox *et al.* 2006; Petersen *et al.* 2014). All this may have adverse effects on body condition, potentially leading to an increased prospect of mortality and/or decreased productivity.

The sensitivity of a seabird species to displacement is likely to vary with specific stages of their annual life cycle (Busch & Garthe 2016). During the breeding season, seabirds become central-place foragers, commuting daily between breeding colonies and foraging sites to provide their chicks with food (Figure 8.11). As such, breeding seabirds could potentially interact with an OWF located within their foraging range several times a day, and are therefore suspected to be particularly vulnerable to displacement effects (Masden *et al.* 2010b). Adult birds suffering from poor body condition due to displacement may safeguard their own survival by abandoning ongoing breeding efforts or not attempting to breed in a particular year or for an even longer period (Furness 2013). Displacement from OWFs within the foraging range of a colony could result in longer foraging flights leading to decreased chick provisioning and/or increased foraging effort by both parents leading to increased chick predation; with both resulting in reduced chick survival (Searle *et al.* 2014). During the non-breeding season, seabirds are no longer attached to the colony and can be expected to suffer less from displacement. On the other hand, mortality of seabirds is often highest in winter, and displacement effects will be additive to other factors causing this annual survival bottleneck. As a result of carry-over effects, displacement during winter

Figure 8.11 Adult Common Guillemot *Uria aalge* with its chick at the colony. During the breeding season, seabirds become central-place foragers and may be particularly vulnerable to offshore wind farm impact. (Martin Perrow)

may lead to a poor body condition at the onset of the breeding season, in turn affecting productivity (Dierschke & Garthe 2006; Furness 2013).

The quantitative translation of a displacement effect into its ultimate impact in terms of reproductive success or chances of survival is highly challenging (Fox *et al*. 2006). Empirical evidence is needed to build reliable population models to estimate the impact of OWF displacement, but crucial parameters often lack such evidence (Green *et al*. 2016). In the case of seabirds, mechanisms of density dependence, for example, are poorly understood. Consequently, these mechanisms are either ignored in impact assessments or incorporated in a compensatory form that has little empirical validation. Compensatory density dependence is considered to offset a loss of individuals through increased recruitment of non-breeding adults or a lower age of first breeding (Horswill *et al*. 2016). The potential biological removal (PBR) approach that has been widely used in OWF impact assessments (O'Brien *et al*. 2017) implicitly assumes such compensatory density dependence. PBR is defined as the additional mortality that a population can theoretically sustain without causing its extinction. It ignores the fact that compensatory mechanisms may not be strong enough to offset the additive loss of individuals, or even worse, that density dependence might be depensatory. Depensation implies a decrease in survival and/or productivity rates as population size decreases, and thus a potential acceleration of population decline. Evidence of depensatory mechanisms has been found for several seabird species, such as terns, small gull species and auks (Horswill *et al*. 2016), and may result from the reduced prospect of colony defence from predators as the numbers of defenders reduces while the proportion of birds on the edge of the colony increases. O'Brien *et al*. (2017) indeed showed that an annual removal of a PBR-derived 'harvest' from a seabird population may be unsustainable over a 25-year projection, especially when populations were already in decline. The authors argue that a PBR approach is therefore unfit for use in Environmental Impact Assessments, while it can be useful for other purposes.

Another important source of uncertainty in the quantification of displacement impact is a lack of knowledge on the carrying capacity of seabird populations and the relative quality of available habitat. In a population close to carrying capacity, habitat loss due to displacement could lead to birds being unable to find alternative foraging habitat that is not already fully occupied. Related to this, seabird colony size is considered to be limited by intraspecific competition for food within foraging-range distance from colonies (Furness & Birkhead 1984), highlighting the potential ecological consequences of reducing the foraging habitat available to a specific colony. During the non-breeding season, seabirds are generally more widely dispersed. Considering their low at-sea densities during winter, Dierschke *et al.* (2017) found it difficult to imagine how interference with conspecifics could reduce the prey-intake rate of wintering Red-throated Divers, and consider a depletion of their prey due to increased bird numbers in non-impacted areas to be very unlikely. Thus, while there is evidence of density dependence (both compensatory and depensatory) affecting seabird populations during the breeding season, there is virtually no mention of such mechanisms occurring during the non-breeding season, when seabirds are much more difficult to study because of their wide-ranging distribution (Busch *et al.* 2015). In this respect, more insight into energy expenditure and time budgets of seabirds during the non-breeding season would be helpful. Birds spending a large proportion of their daily activity budget on foraging can be expected to struggle to find sufficient food, and are probably not able to buffer against any additional energetic costs of displacement effects (Dierschke *et al.* 2017).

The assessment of relative habitat quality is also key to assess displacement impacts on seabirds. Birds tend to occupy high-quality habitat first, and spill over into poor-quality habitat when all optimal habitat is fully occupied (Newton 1998). Displacement from high-quality habitat would therefore lead to redistribution of birds into poor-quality habitat, with implications for individual fitness and survival, compared to displacement from areas of poor-quality habitat, which is anticipated to have less impact (Furness 2013; Warwick-Evans *et al.* 2018).

A last important knowledge gap is the relationship between adult body mass and expected survival over the remainder of the year, which would help to quantify aforementioned carry-over effects. Accordingly, there is no empirical information on the effect of fledging chick mass on post-fledging survival (Searle *et al.* 2014).

While this theme has focused on the possible ecological impact of avoidance responses, it is important to note that seabird species that are attracted to OWFs will face increased collision risk. Whether resulting from displacement or collision, being long lived and late maturing, seabirds are highly sensitive to additional levels of adult mortality (Croxall & Rothery 1991; Sæther & Bakke 2000), highlighting the need to keep working towards a reliable impact assessment framework in the light of future large-scale OWF developments.

Cumulative impacts

While a single OWF generally affects only small numbers of birds relative to their population sizes, the cumulative impact of all current and future developments may be severe and potentially even greater than the sum of single wind-farm impacts (Masden *et al.* 2010a). As set out above (see *Spatiotemporal variation in observed displacement effects* and *Ecological consequences*), displacement affecting birds during the non-breeding season at a location several thousands of kilometres away from the colony may have fitness consequences that only become apparent a few months later during the breeding season or even in the years thereafter. Ideally, a cumulative impact assessment would therefore include all existing and new wind-farm developments within the year-round distribution range of the species or population under study. In practice, however, cumulative impact assessments are generally performed on smaller geographic scales coinciding with legislative units, for example to assess the impact of multiple wind-farm developments on bird populations within Special Protection Areas (SPAs) or national boundaries.

Given their strong displacement response and high conservation status, several cumulative impact assessments have been performed for wintering Red-throated Divers. Garthe and Mendel (2010) calculated that displacement by all current and proposed OWFs in German North Sea waters would affect 1,450 individuals. As OWFs and shipping exclude each other in marine spatial planning, disturbance by shipping traffic is expected to affect another 2,700 individuals, resulting in a total of 4,150 birds disturbed by the cumulative effect of wind exploitation and shipping. As the north-west European population size is estimated to be a minimum of 150,000 birds (Wetlands International 2017), this comprises almost 3% of the biogeographic population. Busch et al. (2013) performed a cumulative impact assessment by estimating the amount of OWF-induced habitat loss in English, German, Dutch and Belgian North Sea waters. Considering all operating, planned and approved OWFs, the authors concluded that 5.4% of potential diver habitat was endangered, with a 0.5% habitat loss in areas with very high diver abundance. The same exercise for Northern Gannet and Lesser Black-backed Gull resulted in an estimated habitat loss of 1.8 and 1.6%, respectively.

Topping and Petersen (2011) developed a more advanced individual-based model (IBM) to assess the impact of cumulative displacement effects on wintering Red-throated Divers. The authors compared the potential impact of three scenarios encompassing a range of possible wind-farm developments in the entire Baltic in combination with the eastern North Sea from the Netherlands in the south to mid-Norway in the north. The primary conclusion of this study was that in all scenarios there was a detectable, but small, impact resulting in a less than 2% decline in the diver population.

The use of such IBMs in combination with Global Positioning System (GPS) tracking data proves particularly promising. Searle et al. (2014) modelled the population consequences of displacement for five seabird species, namely Northern Gannet, Black-legged Kittiwake Rissa tridactyla, Common Guillemot, Razorbill and Atlantic Puffin Fratercula arctica breeding at four SPAs (Forth Islands, St. Abbs Head, Buchan Ness and Fowlsheugh) in the proximity of proposed OWF developments along the eastern coast of Scotland. Bird distribution was estimated based on the results generated by GPS loggers deployed on birds from all four SPAs during the chick-rearing phase. The impacts of displacement on population size were considered to operate via reduced body mass of adults leading to lower survival, specifically in the following winter, with the assumption that any current breeding attempt may be abandoned, and reduced survival of offspring during the breeding season. The model simulations indicated a possible decline in adult survival of almost 2% for the Black-legged Kittiwakes originating from the Forth Islands SPA, and a decline in breeding success of almost 5% for the Atlantic Puffins breeding in the Forth Islands SPA. For the latter, caution is needed because Atlantic Puffin distribution was inferred from GPS data of only seven birds, which on top of this may have suffered from tag effects (Harris et al. 2012). Searle et al. (2014) further stress that the model presented includes a number of assumptions that would benefit from parameterisation with local data, in particular prey distribution, behaviour of seabirds in response to OWFs, including habituation, and the relation between adult body mass and subsequent survival.

Warwick-Evans *et al.* (2018) too developed a spatially explicit IBM based on GPS tracking data to investigate the potential impacts of five OWF developments in the English Channel on the body mass, productivity and mortality of breeding Northern Gannets at Alderney. Model simulations indicated that the wind-farm installations under study should have little impact on the population, in relation to both avoidance behaviour and collision risk scenarios. To have population-level impacts, OWFs would need to be ten times larger or be located in more highly used areas.

All the modelling studies documented show that cumulative displacement effects can potentially have consequences for individual fitness, which may ultimately lead to seabird population declines. The models still rely on numerous assumptions on crucial aspects such as relative habitat quality and density-dependent mechanisms, and predicted changes should therefore be considered to be indicative of potential impacts rather than being seen as definitive predictions (Topping & Petersen 2011).

Concluding remarks

An overview of empirical evidence of seabird displacement from 16 monitoring studies conducted in European waters, including the North, Baltic and Irish Seas, classified all studied species based on the reported displacement responses in terms of abundance or distribution changes. Four species (groups) occupying several ecological niches that were studied in at least seven monitoring programmes showed a relatively consistent avoidance of turbine-built areas. These included divers and the alcid auks, Common Guillemot and Razorbill, which are all piscivorous diving species, and the large-bodied, piscivorous Northern Gannet that forages on the wing and typically plunge-dives from height. As not all studies could actually detect displacement of these species, the results were not unanimous. In contrast, Great Cormorants and Great Black-backed Gulls appeared to be attracted to OWFs, but again some studies failed to detect a corresponding change in abundance or distribution. These species are ecologically different, with one (Great Cormorant) a diving piscivorous species and the other (Great Black-backed Gull) a generalist predator and scavenger, although both may respond to foraging opportunities within wind farms. Common Eider, a diving seaduck specialising on benthic invertebrate prey, was the only species for which a consistent indifferent behaviour was found in all (three) included studies. Lastly, a number of species, mainly gulls, showed high inconsistency in their response. In the case of gulls, their tendency to aggregate in large numbers, especially in the vicinity of trawling activity, appears to be a key factor in this inconsistent response.

The overall variation in observed displacement levels is hypothesised to be due to multiple factors, such as habitat quality, prey distribution, wind-farm location relative to the colony and/or feeding grounds, and wind-farm configuration. Species-specific displacement responses are further suspected to vary with the different stages in the species' annual life cycle, including breeding, non-breeding and dispersal/migration periods. Unfortunately, there are few empirical data supporting these hypotheses. Yet, a thorough understanding of the ecological drivers affecting seabird displacement levels is considered crucial for a reliable impact assessment of single planned wind-farm projects and cumulative impact assessments in particular, highlighting the need for wide-scale and long-term monitoring research, applying standardised methods for data collection and analysis.

A displacement effect should not be considered synonymous with an impact (Masden *et al.* 2010a). But knowledge of a bird species' response towards OWFs is clearly an indispensable prerequisite ultimately to assess the resulting ecological consequences.

Birds subject to displacement can be expected to fly around the wind farm and/or spend time searching for alternative foraging habitat, implying an increased energetic cost. When the alternative habitat proves to be of minor quality or the displaced birds need to face increased competition, this may further lead to a decreased food intake rate. All this may have adverse effects on body condition, potentially leading to increased mortality and/or decreased productivity (Fox *et al.* 2006). The quantitative translation of displacement effects into their ultimate impact, however, is highly challenging. Nevertheless, individual-based modelling of a variety of species has begun to show that the fitness consequences of displacement from multiple wind farms could lead to population declines for a number of ecologically diverse species (Topping & Petersen 2011; Searle *et al.* 2004; Warwick-Evans *et al.* 2018). Meanwhile, these modelling exercises have highlighted that there is insufficient knowledge on a number of crucial input parameters, such as relative habitat quality, population carrying capacities, density-dependent mechanisms and the relation between adult body mass and subsequent survival rate. Gaining more knowledge on these key aspects of seabird ecology and demographics would support a more reliable assessment of the actual ecological consequences and ultimate cumulative impact of extensive OWF installation, and should be the primary goal of future research.

Acknowledgements

First of all, we would like to thank Martin Perrow for inviting us to write this book chapter, his excellent review and overall dedication to this book series. Of course, we would never have been asked to do this if we had not been given the chance to be involved in seabird displacement research ourselves. Therefore, we kindly thank the Belgian wind-farm concession holders for financing the environmental monitoring programme, and Robin Brabant and Steven Degraer of the Royal Belgian Institute of Natural Sciences (RBINS) for assigning the seabird monitoring to us. A special word of gratitude goes out to DAB Vloot and the Flanders Marine Institute/Vlaams Instituut voor de Zee (VLIZ) for providing ship time on the research vessels Zeeleeuw and Simon Stevin, as well as to RBINS and the Belgian Science Policy (BELSPO) for ship time on the research vessel Belgica. Lastly, we thank our colleagues Wouter Courtens, Marc Van de walle and Hilbran Verstraete for many hours of chilly seabird counting.

References

APEM (2014) Assessing northern gannet avoidance of offshore wind farms. APEM Report 512775. Stockport: APEM. Retrieved 1 September 2017 from https://tethys.pnnl.gov/sites/default/files/publications/Rehfisch-et-al-2014-APEM.pdf

APEM (2016) Assessment of displacement impacts of offshore windfarms and other human activities on red-throated divers and alcids. Natural England Commissioned Reports, No. 227. Stockport: APEM. Retrieved 1 September 2017 from http://publications.naturalengland.org.uk/file/6204240073064448

Bernstein, B.B. & Zalinski, J. (1983) An optimum sampling design and power tests for environmental biologists. *Journal of Environmental Management* 16: 35–43.

Buckland, S.T., Anderson, D.R., Burnham, K.P., Laake, J.L., Borchers, D.L. & Thomas, L. (2001) *Introduction to Distance Sampling – Estimating abundance of biological populations.* Oxford: Oxford University Press.

Busch, M. & Garthe, S. (2016) Approaching population thresholds in presence of uncertainty: assessing displacement of seabirds from offshore wind farms. *Environmental Impact Assessment Review* 56: 31–42.

Busch, M., Kannen, A., Garthe, S. & Jessopp, M. (2013) Consequences of a cumulative perspective on marine environmental impacts: offshore wind farming and seabirds at North Sea scale in context of the EU Marine Strategy Framework Directive. *Ocean & Coastal Management* 71: 213–224.

Busch, M., Buisson, R., Barrett, Z., Davies, S. & Rehfisch, M. (2015) Developing a habitat loss method for assessing displacement impacts from offshore wind farms. JNCC Report No. 551. Peterborough: Joint Nature Conservation Committee. Retrieved 1 September 2017 from http://jncc.defra.gov.uk/pdf/Report%20551_web2.pdf

Camphuysen, C.J., Fox, A.D., Leopold, M.F. & Petersen, I.K. (2004) Towards standardised seabirds at sea census techniques in connection with environmental impact assessments for offshore wind farms in the UK: a comparison of ship and aerial sampling methods for marine birds, and their applicability to offshore wind farm assessments. Report of the Royal Netherlands Institute for Sea Research, commissioned by COWRIE. London: The Crown Estate. Retrieved 1 September 2017 from https://www.thecrownestate.co.uk/media/6001/2004-04%20Towards%20standardised%20seabirds%20at%20sea%20census%20techniques%20in%20connection%20with%20environmental%20impact%20assessments%20for%20offshore%20wind%20farms%20in%20the%20UK.pdf

Canning, S., Lye, G. & Kerr, D. (2013) Analysis of marine ecology monitoring plan data from the Robin Rigg offshore wind farm, Scotland (Operational Year 3). Chapter 3: Birds. Dalry: Natural Power Consultants.

Croxall, J.P. & Rothery, P. (1991) Population regulation of seabirds: implications of their demography for conservation. In Perrins, C.M., Lebreton, J.-D. & Hirons, G.J.M. (eds) *Bird Population Studies, Relevance to Conservation and Management.* Oxford: Oxford University. pp. 272–296.

de Mesel, I., Kerckhof, F., Norro, A., Rumes, B. & Degraer, S. (2014) Succession and seasonal dynamics of the epifauna community on offshore wind farm foundations and their role as stepping stones for non-indigenous species. *Hydrobiologia* 756: 37–50.

Dierschke, V. & Garthe, S. (2006) Literature review of offshore wind farms with regard to seabirds. In Zucco, C., Wende, W., Merck, T., Köchling, I. & Köppel, J. (eds) *Ecological Research on Offshore Wind Farms: International exchange of experiences – Part B: Literature review of ecological impacts.* Bonn: Bundesamt für Naturschutz. pp. 131–197.

Dierschke, V., Furness, R.W. & Garthe, S. (2016) Seabirds and offshore wind farms in European waters: avoidance and attraction. *Biological Conservation* 202: 59–68.

Dierschke, V., Furness, R.W., Gray, C.E., Petersen, I.K., Schmutz, J., Zydelis, R. & Daunt, F. (2017) Possible behavioural, energetic and demographic effects of displacement of red-throated divers. JNCC Report No. 605. Peterborough: Joint Nature Conservation Committee. Retrieved 1 September 2017 from http://jncc.defra.gov.uk/pdf/Report_605_WEB.pdf

DONG Energy & Vattenfall A/S (2006) Review Report 2005: The Danish Offshore Wind Farm Demonstration Project: Horns Rev and Nysted Offshore Wind Farms Environmental Impact Assessment and Monitoring. Prepared for The Environmental Group of the Danish Offshore Wind Farm Demonstration Projects by DONG Energy and Vattenfall. Retrieved 1 September 2017 from https://corporate.vattenfall.dk/globalassets/danmark/om_os/horns_rev/review-report-2005.pdf

Drewitt, A.L. & Langston, R.H.W. (2006) Assessing the impact of wind farms on birds. *Ibis* 148: 29–42.

Embling, C.B., Illian, J., Armstrong, E., van der Kooij, J., Sharples, J., Camphuysen, C.J. & Scott, B.E. (2012) Investigating fine-scale spatio-temporal predator–prey patterns in dynamic marine ecosystems: a functional data analysis approach. *Journal of Applied Ecology* 49: 481–492.

Fox, A.D., Desholm, M., Kahlert, J., Christensen, T.K. & Petersen, I.K. (2006) Information needs to support environmental impact assessment of the effects of European marine offshore wind farms on birds. *Ibis* 148: 129–144.

Furness, R.W. (2013) Dogger Bank Creyke Beck Environmental Statement – Chapter 11 Appendix B – Extent of displacement and mortality implications of displacement of seabirds by offshore windfarms. Glasgow: MacArthur Green. Retrieved 1 September 2017 from http://www.forewind.co.uk/uploads/files/Creyke_Beck/Application_Documents/6.11.2_Chapter_11_Appendix_B_Seabird_Displacement_and_mortality_implications_-_Application_Submission_F-OFC-CH-011.pdf

Furness, R.W. & Birkhead, T.R. (1984) Seabird colony distributions suggest competition for food supplies during the breeding season. *Nature* 311: 655–656.

Garthe, S. & Hüppop, O. (2004) Scaling possible adverse effects of marine wind farms upon seabirds: developing and applying a vulnerability index. *Journal of Applied Ecology* 41: 724–734.

Garthe, S. & Mendel, B. (2010) Cumulative effects of anthropogenic activities on seabirds in the North Sea: methodology and first results. In Lange, M., Burkhard, B., Garthe, S., Gee, K., Kannen, A., Lenhart, H. & Windhorst, W. (eds) *Analyzing Coastal and Marine Changes: Offshore wind farming as a case study*. Geesthacht: GKSS Research Centre. pp. 86–90.

Garthe, S., Markones, N. & Corman, A.-M. (2016) Possible impacts of offshore wind farms on seabirds: a pilot study in Northern Gannets in the southern North Sea. *Journal of Ornithology* 158: 345–349.

Green, R.E., Langston, R.H.W., McCluskie, A., Sutherland, R. & Wilson, J.D. (2016) Lack of sound science in assessing wind farm impacts on seabirds. *Journal of Applied Ecology* 53: 1635–1641.

Guillemette, M., Larsen, J.K. & Clausager, I. (1998) Impact assessment of an offshore wind park on sea ducks. NERI Technical Report No 227. Denmark: National Environmental Research Institute. Retrieved 1 September 2017 from https://tethys.pnnl.gov/sites/default/files/publications/NERI_Impact_on_Sea_Ducks.pdf

Guillemette, M., Larsen, J.K. & Clausager, I. (1999) Assessing the impact of the Tunø Knob wind park on sea ducks: the influence of food resources. NERI Technical Report No 263. Denmark: National Environmental Research Institute. Retrieved 1 September 2017 from https://tethys.pnnl.gov/sites/default/files/publications/Guillemette-et-al-1999.pdf

Harris, M.P., Bogdanova, M.I., Daunt, F. & Wanless, S. (2012) Using GPS technology to assess feeding areas of Atlantic puffins *Fratercula arctica*. *Ringing & Migration* 27: 43–49.

Harwood, A.J.P., Perrow, M.R., Berridge, R.J., Tomlinson, M.L. & Skeate, E.R. (2017) Unforeseen responses of a breeding seabird to the construction of an offshore wind farm. In Köppel, J. (ed.) *Wind Energy and Wildlife Interactions*. Cham: Springer. pp. 19–41.

Hastie, T. & Tibshirani, R. (1990) *Generalised Additive Models*. London: Chapman & Hall.

Helsinki Commission (HELCOM) (2017) HELCOM Red List *Clangula hyemalis*. Retrieved 21 December 2017 from http://www.helcom.fi/Red%20List%20Species%20Information%20Sheet/HELCOM%20Red%20List%20Clangula%20hyemalis.pdf

Horswill, C., O'Brien, S.H. & Robinson, R.A. (2016) Density dependence and marine bird populations: are wind farm assessments precautionary? *Journal of Applied Ecology* 54: 1406–1414.

Hötker, H. (2017) Birds: displacement. In Perrow, M.R. (ed.) *Wildlife and Wind Farms, Conflicts and Solutions. Volume 1. Onshore: Potential Effects*. Exeter: Pelagic Publishing. pp. 119–154.

Krijgsveld, K.L., Fijn, R.C., Japink, M., van Horssen, P., Heunks, C., Collier, M., Poot, M., Beuker, D. & Dirksen, S. (2011) Effect studies Offshore Wind Farm Egmond aan Zee – Final report on fluxes, flight altitudes and behaviour of flying birds. Report commissioned by Noordzeewind. Culemborg: Bureau Waardenburg. Retrieved 1 September 2017 from https://www.buwa.nl/fileadmin/buwa_upload/Bureau_Waardenburg_rapporten/06-466_BW_research_OWEZ_cumulative_effects-web.pdf

Larsen, J.K. & Guillemette, M. (2007) Effects of wind turbines on flight behaviour of wintering common eiders: implications for habitat use and collision risk. *Journal of Applied Ecology* 44: 516–522.

Leonhard, S.B. & Pedersen, J. (2006) Benthic communities at Horns Rev before, during and after construction of Horns Rev offshore wind farm. Final Report – Annual Report 2005. Denmark: BioConsult AS. Retrieved 1 September 2017 from https://corporate.vattenfall.dk/globalassets/danmark/om_os/horns_rev/benthic-communities-at-horns.pdf

Leonhard, S.B., Stenberg, C. & Støttrup, J. (eds) (2011) Effect of the Horns Rev 1 offshore wind farm on fish communities – follow-up seven years after construction. Charlottenlund: DTU Aqua, National Institute of Aquatic Resources. Retrieved 1 September 2017 from http://orbit.dtu.dk/files/7615058/246_2011_effect_of_the_horns_rev_1_offshore_wind_farm_on_fish_communities.pdf

Leopold, M.F., van Bemmelen, R.S.A. & Zuur, A. (2013) Responses of local birds to the offshore wind farms PAWP and OWEZ off the Dutch mainland coast. Report C151/12. Report commissioned by Prinses Amaliawindpark. Texel: Imares. Retrieved 1 September 2017 from http://library.wur.nl/WebQuery/wurpubs/443769

Mackenzie, M.L., Scott-Hayward, L.A., Oedekoven, C.S., Skov, H., Humphreys, E. & Rexstad, E. (2013) Statistical modelling of seabird and cetacean data: guidance document. Report prepared for Marine Scotland. St Andrews: University of St Andrews, Centre for Research into Ecological and Environmental Modelling. Retrieved 1

September 2017 from https://creem2.st-andrews. ac.uk/download/mrsea-guidance/

Maclean, I.M.D., Rehfisch, M.M., Skov, H. & Thaxter, C.B. (2013) Evaluating the statistical power of detecting changes in the abundance of seabirds at sea. *Ibis* 155: 113–126.

Marques, F.F.C. & Buckland, S.T. (2003) Incorporating covariates into standard line transect analyses. *Biometrics* 59: 924–935.

Masden, E.A., Fox, A.D., Furness, R.W., Bullman, R. & Haydon, D.T. (2010a) Cumulative impact assessments and bird/wind farm interactions: developing a conceptual framework. *Environmental Impact Assessment Review* 30: 1–7.

Masden, E.A., Haydon, D.T., Fox, A.D. & Furness, R.W. (2010b) Barriers to movement: modelling energetic costs of avoiding marine wind farms among breeding seabirds. *Marine Pollution Bulletin* 60: 1085–1091.

Newton, I. (1998) *Population Limitation in Birds*. London: Academic Press.

Nilsson, L. & Green, M. (2011) Birds in southern Öresund in relation to the wind farm at Lillgrund. Final report of the monitoring program 2001–2011. Report commissioned by Vattenfall Vindkraft AB. Lund: Lunds Universitet. Retrieved 1 September 2017 from https:// corporate.vattenfall.se/globalassets/sverige/ om-vattenfall/om-oss/var-verksamhet/vind-kraft/lillgrund/birds_in_southern_oresund.pdf

O'Brien, S.H., Cook, A.S.C.P. & Robinson, R.A. (2017) Implicit assumptions underlying simple harvest models of marine bird populations can mislead environmental management decisions. *Journal of Environmental Management* 201: 163–171.

Oedekoven, C.S., Mackenzie, M.L., Scott-Hayward, L.A.S. & Rexstad, E. (2013) Statistical modelling of bird and cetacean distributions in offshore renewables development areas: literature review. Report prepared for Marine Scotland. St Andrews: University of St Andrews, Centre for Research into Ecological and Environmental Modelling. Retrieved 1 September 2017 from https://creem2.st-andrews.ac.uk/download/ mrsea-guidance/

Percival, S. (2013) Thanet Offshore Wind Farm – Ornithological Monitoring 2012–2013. Report commissioned by Thanet Offshore Wind Ltd. Peterborough: Royal Haskoning. Retrieved 1 September 2017 from http://ecologyconsult. co.uk/index_htm_files/Ornithological%20moni-toring%202012-13%20v3.pdf

Percival, S. (2014) Kentish Flats offshore wind farm: diver surveys 2011–12 and 2012–13. Report commissioned by Vattenfall Wind Power. Durham: Ecology Consulting. Retrieved 1 September 2017 from https://corporate. vattenfall.co.uk/globalassets/uk/projects/ redthroated-diver-2014.pdf

Perrow, M.R., Skeate, E.R. & Gilroy, J.J. (2011) Visual tracking from a rigid-hulled inflatable boat to determine foraging movements of breeding terns. *Journal of Field Ornithology* 82: 68–79.

Perrow, M.R., Harwood, A.J.P., Skeate, E.R., Praca, E. & Eglington, S.M. (2015) Use of multiple data sources and analytical approaches to derive a marine protected area for a breeding seabird. *Biological Conservation* 191: 729–738.

Petersen, I.K. & Fox, A.D. (2007) Changes in bird habitat utilization around the Horns Rev 1 offshore wind farm, with particular emphasis on Common Scoter. Report commissioned by Vattenfall A/S. Aarhus: National Environmental Research Institute, University of Aarhus. Retrieved 1 September 2017 from https://corpo-rate.vattenfall.dk/globalassets/danmark/om_os/ horns_rev/changes-in-bird-habitat.pdf

Petersen, I.K., Christensen, T.K., Kahlert, J., Desholm, M. & Fox, A.D. (2006) Final results of bird studies at the offshore wind farms at Nysted and Horns Rev, Denmark. NERI report commissioned by DONG Energy and Vattenfall A/S. Denmark: National Environmental Research Institute. Retrieved 1 September 2017 from https://corporate.vattenfall.dk/globalassets/ danmark/om_os/horns_rev/horns-rev-nysted-birds.pdf

Petersen, I.K., MacKenzie, M., Rexstad, E., Wisz, M.S. & Fox, A.D. (2011) Comparing pre- and post-construction distributions of long-tailed ducks *Clangula hyemalis* in and around the Nysted offshore wind farm, Denmark: a quasi-designed experiment accounting for imperfect detection, local surface features and autocorrelation. St Andrews: University of St Andrews, Centre for Research into Ecological and Environmental Modelling. Retrieved 1 September 2017 from https://research-repository.st-andrews. ac.uk/handle/10023/2008

Petersen, I.K., Nielsen, R.D. & Mackenzie, M.L. (2014) Post-construction evaluation of bird abundances and distributions in the Horns Rev 2 offshore wind farm area, 2011 and 2012. Report commissioned by DONG Energy. Aarhus: Aarhus University, Danish Centre for Environment and Energy. Retrieved 1 September 2017 from http://birdlife.se/1.0.1.0/1267/down-load_21126.php

Pettersson, J. (2005) The impact of offshore wind farms on bird life in Kalmar Sound, Sweden. A final report based on studies 1999–2003. Report prepared for the Swedish Energy Agency. Lund: Lunds Universitet. Retrieved 1 September 2017 from https://tethys.pnnl.gov/publications/impact-offshore-wind-farms-bird-life-southern-kalmar-sound-sweden

Potts, J.M. & Elith, J. (2006) Comparing species abundance models. *Ecological Modelling* 199: 153–163.

Reubens, J., Degraer, S. & Vincx, M. (2014) The ecology of benthopelagic fishes at offshore wind farms: a synthesis of 4 years of research. *Hydrobiologia* 727: 121–136.

Rue, H., Martino, S. & Chopin, N. (2009) Approximate Bayesian inference for latent Gaussian models by using integrated nested Laplace approximations. *Journal of the Royal Statistical Society – Series B (Statistical Methodology)* 71: 319–392.

Sæther, B.-E. & Bakke, Ø. (2000) Avian life history variation and contribution of demographic traits to the population growth rate. *Ecology* 81: 642–653.

Schwemmer, P. & Garthe, S. (2005) At-sea distribution and behaviour of a surface-feeding seabird, the lesser black-backed gull *Larus fuscus*, and its association with different prey. *Marine Ecology Progress Series* 285: 245–258.

Searle, K., Mobbs, D., Butler, A., Bogdanova, M., Freeman, S., Wanless, S. & Daunt, F. (2014) Population consequences of displacement from proposed offshore wind energy developments for seabirds breeding at Scottish SPAs (CR/2012/03). Scottish Marine and Freshwater Science Report Vol 5 No 13. Aberdeen: Marine Scotland Science. Retrieved 1 September from http://www.gov.scot/Publications/2014/11/6831/downloads

Skov, H., Heinänen, S., Lazcny, M. & Chudzinska, M. (2016) Offshore Windfarm Eneco Luchterduinen, Ecological monitoring of seabirds, T1 report. Report commissioned by Eneco. Denmark: DHI.

Skov, H., Heinänen, S., Norman, T., Ward, R., Méndez-Roldán, S. & Ellis, I. (2018) ORJIP bird collision and avoidance study. Final report – April 2018. London: Carbon Trust. Retrieved 20 June 2018 from https://www.carbontrust.com/media/675793/orjip-bird-collision-avoidance-study_april-2018.pdf

Stewart-Oaten, A. & Bence, J.R. (2001) Temporal and spatial variation in environmental impact assessment. *Ecological Monographs* 71: 305–339.

Stewart-Oaten, A., Murdoch, W.W. & Parker, K.R. (1986) Environmental impact assessment: 'pseudoreplication' in time? *Ecology* 67: 929–940.

Tasker, M.L., Jones, P.H., Dixon, T.J. & Blake, B.F. (1984) Counting seabirds at sea from ships: a review of methods employed and a suggestion for a standardised approach. *Auk* 101: 567–577.

Thaxter, C.B., Ross-Smith, V.H., Bouten, W., Clark, N.A., Conway, G.J., Rehfisch, M.M. & Burton, N.H.K. (2015) Seabird–wind farm interactions during the breeding season vary within and between years: a case study of lesser black-backed gull *Larus fuscus* in the UK. *Biological Conservation* 186: 347–358.

Topping, C. & Petersen, I.K (2011) Report on a red-throated diver agent-based model to assess the cumulative impact from offshore wind farms. Report commissioned by the Environmental Group. Aarhus: Aarhus University, Danish Centre for Environment and Energy. Retrieved 1 September 2017 from https://tethys.pnnl.gov/publications/report-red-throated-diver-agent-based-model-assess-cumulative-impact-offshore-wind

Vallejo, G.C., Grellier, K., Nelson, E.J., McGregor, R.M., Canning, S.J., Caryl, F.M. & McLean, N. (2017) Responses of two marine top predators to an offshore wind farm. *Ecology and Evolution* 2017: 8698–8708.

Vanermen, N., Onkelinx, T., Verschelde, P., Courtens, W., Van de walle, M., Verstraete, H. & Stienen, E.W.M. (2015a) Assessing seabird displacement at offshore wind farms: power ranges of a monitoring and data handling protocol. *Hydrobiologia* 756: 155–167.

Vanermen, N., Onkelinx, T., Courtens, W., Van de walle, M., Verstraete, H. & Stienen, E.W.M. (2015b) Seabird avoidance and attraction at an offshore wind farm in the Belgian part of the North Sea. *Hydrobiologia* 756: 51–61.

Vanermen, N., Courtens, W., Van de walle, M., Verstraete, H. & Stienen, E.W.M. (2016) Seabird monitoring at offshore wind farms in the Belgian part of the North Sea – updated results for the Bligh Bank & first results for the Thorntonbank. Report commissioned by the Management Unit of the North Sea Mathematical Models (MUMM). Brussels: Research Institute for Nature and Forest. Retrieved 1 September 2017 from https://pureportal.inbo.be/portal/files/12084979/Vanermen_etal_2016_SeabirdMonitoringAtOffshoreWindFarmsInTheBelgianPartOfTheNorthSea.pdf

Vanermen, N., Courtens, W., Van de walle, M., Verstraete, H. & Stienen, E.W.M. (2017) Seabird monitoring at the Thorntonbank wind farm – updated seabird displacement results & an explorative assessment of large gull behavior inside the wind farm area. Report commissioned by the Management Unit of the North Sea Mathematical Models (MUMM). Brussels: Research Institute for Nature and Forest. Retrieved 1 September 2017 from https://pureportal.inbo.be/portal/files/13271773/Vanermen_etal_2017_SeabirdMonitoringAt-TheThorntonbankOffshoreWindFarm.pdf

ver Hoef, J.M. & Boveng, P.L. (2007) Quasi-Poisson vs. negative binomial regression: how should we model overdispersed count data? *Ecology* 88: 2766–2772.

Warwick-Evans, V., Atkinson, P.W., Walkington, I. & Green, J.A. (2018) Predicting the impacts of wind farms on seabirds: an individual-based model. *Journal of Applied Ecology* 55: 503–515.

Welcker, J. & Nehls, G. (2016) Displacement of seabirds by an offshore wind farm in the North Sea. *Marine Ecology Progress Series* 554: 173–182.

Wetlands International (2017) Waterbird population estimates. Retrieved 1 October 2017 from http://wpe.wetlands.org/

Zeileis, A., Keibler, C. & Jackman, S. (2008) Regression models for count data in R. *Journal of Statistical Software* 27: 1–25.

CHAPTER 9

Seabirds: collision

SUE KING

Summary

The development of offshore wind energy is projected to increase in scale and distribution around the world although seabird collision is perceived as a major ecological risk. This chapter aims to review the existing evidence of seabird collision and avoidance at offshore wind farms (OWFs) and, where this is unavailable, at onshore and coastal turbines. It finds that few collision studies have been undertaken at constructed OWFs and empirical evidence of collision of seabirds is rather limited. However, both onshore and offshore, gull species are the most regularly reported collision victims followed by terns at coastal and onshore sites. A case study of tern mortality shows that this may be significant. In general, emerging evidence suggests that seabirds tend to avoid offshore wind farms at a range of scales with empirical avoidance rates now calculated at above 99%. This is well above the original default rates based on land birds at onshore wind farms. A review of the factors influencing collision risk, including species-specific and wind-farm specific factors and general location, showed that the percentage of time spent flying at collision risk height appears to be most important. In this case, as technology advances with larger turbines, the proportion of birds at risk height is likely to reduce. A review of the seabird sensitivity indices developed to understand the vulnerability of individual species to collision, and to identify collision risk in sea areas where development may be planned, concludes that they are most useful to identify species for impact assessment and as a strategic planning tool. They may also help to identify species and areas to take forward for further research and monitoring. However, their use of proxy measures of vulnerability and the limited empirical basis are limitations. Updating these resources with new evidence is recommended. In general, there is an urgent need for more research on the evidence of seabird behaviour and collision effects within and around constructed wind farms at sea to build on findings to date. This would enable planning decisions to be more evidence-based and offshore wind development to proceed within well-understood and ecologically sustainable limits.

Introduction

Offshore wind farms (OWFs) are a major intervention in the marine environment and are perceived as a potential collision risk to seabirds. This is particularly so if they are located close to breeding colonies or along a migration/dispersal flyway where effects from a number of wind farms may accumulate. The increasing number and size of consented and constructed OWFs, the increasing capacity of the turbines and their deployment farther from shore mean that the scale of potential collision mortality has become an important criterion in the planning consent process.

Additional mortality is a concern for long-lived species such as seabirds, many of which are of conservation importance and some of which are undergoing rapid declines (Paleczny *et al.* 2015; JNCC 2017). Their high adult survival rate, low productivity and delayed breeding may make them vulnerable to even relatively small increases in adult mortality with consequent effects on population stability (Tuck *et al.* 2001; Everaert & Stienen 2007; Igual *et al.* 2009; Reid *et al.* 2013). These same aspects of their life history and ecology mean they are at risk from other anthropogenic effects such as fisheries by-catch, disturbance or predation at the colony and climate change (Tuck *et al.* 2001; Croxall *et al.* 2012; Burthe *et al.* 2014). As a result, there are concerns that the effects of OWFs, including collision, may exacerbate the situation.

Most estimates of seabird collision are based on theory rather than empirical evidence. This is because the carcass collection technique and its statistical principles used onshore (see Huso *et al.* 2017) cannot be used offshore, where birds struck by the turbine blade would fall directly into the water. The location and scale of OWFs also make direct observation more difficult and remote sensing methods expensive as a range of techniques must often be employed to ensure 24-hour surveillance and identification of birds to species level (Collier *et al.* 2012; Mollis *et al.*, Volume 4, Chapter 5). The offshore environment also poses additional health and safety hazards and access difficulties for operatives who need to make observations and maintain equipment (Skov *et al.* 2018). As a result, consent conditions to monitor collision risk post-construction at individual projects are rare and, in the UK, have yielded relatively little collision information (MMO 2014).

The best evidence of bird behaviour within wind farms, including collision avoidance behaviour, has come from well-designed and often long-term studies at both onshore and offshore sites, for example in in Sweden, Denmark, the Netherlands, Belgium and Germany (e.g. Pettersson 2005; Petersen *et al.* 2006; Vanermen & Stienen 2009; Krijgsveld *et al.* 2011; Harwood *et al.* 2017). The industry and agency-funded Offshore Renewables Joint Industry Programme (ORJIP) study (Skov *et al.* 2018), specifically aimed at quantifying seabird collision and avoidance behaviour at OWFs in the UK is a significant recent addition to this body of work. Nonetheless, evidence of collision at OWFs remains sparse and, at the time of writing, the true scale of its impact on seabirds remains unknown.

Scope

The chapter begins with a review of the evidence of seabird collision and avoidance at operational OWFs. The topic is discussed primarily as it applies to installations in European waters, as 84% of current built capacity is installed off the European coast, mainly in the north and west. Outside Europe, consented development hotspots are located in the

People's Republic of China, northern Japan, South Korea, Taiwan and the eastern seaboard of North America, with upcoming markets in India and Vietnam (GWEC 2018).

The species at risk are identified, although with limited offshore evidence this also draws on previous reviews of collision fatalities at onshore and coastal wind farms such as Cook *et al.* (2014). Quantitative data such as collision rates per turbine are not reported, as methods of data collection and analysis are not standardised and may not translate to an offshore context. The evidence review is restricted to relatively few seabird species including gulls and terns (Laridae), gannets (Sulidae), skuas (Stercoraiidae) and petrels and shearwaters (Procellariidae), owing to the location of constructed OWFs. While divers (Gaviidae) and auks (Alcidae) may occasionally collide with turbines, they generally fly below rotor height and are more likely to experience population-level effects from displacement. Species such as ducks, swans, geese, waders and passerines, which may be exposed to collision risk while on migration, are discussed by Hüppop *et al.* in Chapter 7 of this volume.

The factors that may contribute to collision risk are then examined, including species-specific factors, locational factors and site-specific design (Marques *et al.* 2014). This is followed by an appraisal, in the context of existing evidence, of sensitivity indices that have been developed to identify collision vulnerability for individual seabird species and to identify sea areas where collision risk may be high.

The review draws on open-access internet search engines such as Google Scholar for both peer-reviewed and grey literature. Search terms used were 'offshore' 'wind' 'bird' and 'collision'. A targeted search for grey literature was also carried out on the websites of UK government departments dealing with planning consent and other sites that collate research on seabirds at OWFs.

Themes

Evidence of seabird collisions

Offshore, collision detection has been attempted using a range of methods (Desholm *et al.* 2006; Collier *et al.* 2012; Mollis *et al.*, Volume 4, Chapter 5). However, until 2017, the only published information about directly recorded bird collision with an offshore turbine, other than a single passerine detected using thermal observation (Petersen *et al.* 2006), was in Kalmar Sound, Sweden (Pettersson 2005). Here, four Common Eider *Somateria mollissima* in a large flock migrating through the Sound were observed colliding with a turbine blade, or its wake turbulence, and were knocked to the sea surface. Three birds recovered and quickly flew onwards. No other collisions were recorded during the observation of approximately 130,000 waterfowl on spring and autumn migration past the turbines. Based on this, and five occasions when avoidance was observed, a collision rate of one bird per turbine per year was estimated. Although Common Eider is a member of the duck family (Anatidae) rather than one of the seabird families identified in the scope of this chapter, this study is notable for good design and the early use of direct observation combined with other technologies, including an optical rangefinder and radar, to calculate an offshore collision rate per turbine.

Subsequently, the only other reported collisions at an OWF have been recorded at Thanet, UK, as part of an industry-level study on bird avoidance behaviour (ORJIP). This study recorded six collisions using video camera linked to a tracking radar, the species being Black-legged Kittiwake *Rissa tridactyla*, Lesser/Great Black-backed Gull *Larus*

Table 9.1 Seabird species recorded as collision victims at studies of onshore and coastal wind farms and Thanet offshore wind farm.

Wind farm	Country	No. of turbines	Collision fatalities	Data source(s)
Altamont	USA	685	Gull species	Thelander et al. 2003[a]
Avonmouth	UK	3	Black-headed Gull	The Landmark Practice 2013[a,b]
Bluff Point	Tasmania, Australia	37	Common Diving Petrel, White-faced Storm Petrel, Wilson's Storm Petrel, Short-tailed Shearwater, Fairy Prion	Hull et al. 2013
Blyth Harbour	UK	9	Black-headed Gull, Herring Gull, Great Black-backed Gull, Fulmar, Gannet	Newton & Little 2009; Painter et al. 1999; Lawrence et al. 2007[a]
Boudewijnkanaal	Belgium	14	Black-headed Gull, Herring Gull	Everaert 2002; Everaert & Kuikjen 2007; Everaert 2014[a,b]
Bouin	France	8	Black-headed Gull, Yellow-legged Gull, Mediterranean Gull	Dulac 2008[a,b]
Bremerhaven	Germany	2	Black-headed Gull, Herring Gull, Common Gull	Dierschke & Garthe 2006
Buffalo Ridge	USA	143	Herring Gull	Johnson et al. 2000[a]
Burgar Hill	UK	3	Black-headed Gull	Meek et al. 1993
De Put, Nieuwkapelle	Belgium	2	Black-headed Gull, Common Gull	Everaert & Kuikjen 2007; Everaert 2008[a,b]
Friedrich-Wilhelm-Lubke Koog	Germany	13	Black-headed Gull, Common Gull	Dierschke & Garthe 2006
Gneizdzewo	Poland	19	Black-headed Gull	Zielinski et al. 2008; 2010; 2011; 2012[a,b]
Groettocht	Netherlands	7	Herring Gull, Common Gull	Krijgsveld et al. 2009[a]
Hellrigg	UK	4	Herring Gull	Percival 2012; 2013[a,b]
Helgoland	Germany	1	Herring Gull	Dierschke & Garthe 2006
Hooksiel	Germany	1	Black-headed Gull, Herring Gull	Dierschke & Garthe 2006
Keewaunee County	USA	31	Herring Gull	Howe et al. 2002[a]
Kessingland	UK	2	Black-headed Gull, Herring Gull, Common Gull	Wild Frontier Ecology 2013[a,b]
Klarwerk, Westerland	Germany	1	Black-headed Gull, Herring Gull	Dierschke & Garthe 2006
Kleine Pathoekeweg	Belgium	7	Gulls, including Black-headed Gull, Herring Gull, Lesser Black-backed Gull	Everaert & Kuikjen 2007[a]; Everaert 2014

Table 9.1 – continued

Wind farm	Country	No. of turbines	Collision fatalities	Data source(s)
Kluisendok, Gent	Belgium	11	Black-headed Gull	Everaert 2014
Kreekrak	Netherlands	5	Gull species, including Black-headed Gull, Herring Gull, Little Gull	Musters *et al.* 1996
Marienkoog	Germany	15	Black-headed Gull	Dierschke & Garthe 2006
Nordholz	Germany	25	Common Gull	Dierschke & Garthe 2006
Oosterbierum	Netherlands	18	Gull species, including Black Headed Gull and Herring Gull	Winkelman 1992[a,b]
Project West Wind	New Zealand	62	Southern Black-backed Gull, Sooty Shearwater, Fairy Prion	Bull *et al.* 2013
Reussenkoge	Germany	17	Herring Gull	Dierschke & Garthe 2006
Simonsberger Koog	Germany	13	Black-headed Gull, Common Gull, Black Tern	Dierschke & Garthe 2006
Studland Bay	Tasmania, Australia	25	Grey-backed Storm Petrel, Short-tailed Shearwater, Australian Gannet	Hull *et al.* 2013
Thanet	UK	100	Black-legged Kittiwake, Great/Lesser Black-backed Gull, unidentified gull	Carbon Trust 2016
Tjaereborg	Denmark	1	Gull species	Pedersen & Poulsen 1991, in Percival 2003
Urk	Netherlands	25	Gull species, including Black-headed Gull, Herring Gull and Common Gull	Winkelman 1989
Waterkaaptocht	Netherlands	8	Black-headed Gull	Krijgsveld *et al.* 2009
Westkuste, Dithmarschen	Germany	32	Black-headed Gull, Common Gull	Dierschke & Garthe 2006
Wilmhelmshaven, Jadewindpark	Germany	3	Herring Gull	Dierschke & Garthe 2006
Zeebrugge	Belgium	25	Black-headed Gull, Herring Gull, Lesser Black-backed Gull, Black-legged Kittiwake, Sandwich Tern, Common Tern, Little Tern	Everaert 2002; 2008; Everaert & Stienen 2007[b]

[a]Reported in Cook *et al.* (2014); [b]used to calculate empirical within-wind-farm avoidance rates by Cook *et al.* (2014, 2018).

Figure 9.1 Black-headed Gulls *Chroicocephalus ridibundus*, here in winter plumage, are the most frequently recorded member of any seabird family noted as victims of collision with coastal and onshore turbines. (Martin Perrow)

fuscus/marinus, two large gulls and two unidentified gulls (Skov *et al.* 2018). This was from a total of approximately 12,000 bird tracks logged over a twenty two month period. All collisions were recorded during daylight and it was noted that night time videos recorded from the same turbine positions recorded only between 2% and 3% of the daytime activity levels, with many of the birds recorded sitting on the water, suggesting that for these species, at this site, nocturnal collision risk was low.

Collisions of seabirds at onshore and coastal wind farms have been summarised in reviews by a number of authors, including Percival (2003), Dierschke and Garthe (2006), MacArthur Green *et al.* (2013) and Cook *et al.* (2014, 2018), as summarised in Table 9.1. In northern Europe and the USA, gulls are the most commonly recorded collision victims, the most frequent being Black-headed Gull *Chroicocephalus ridibundus* (Figure 9.1), European Herring Gull *Larus argentatus* and Common Gull *Larus canus*. Other gull species included Lesser Black-backed Gull, Great Black-backed Gull, Black-legged Kittiwake, Yellow-legged Gull *Larus michahellis* and Mediterranean Gull *Ichthyaetus melanocephalus*. Other seabirds were Northern Gannet *Morus bassanus*, Sandwich Tern *Thalasseus sandvicensis*, Common Tern *Sterna hirundo* and Little Tern *Sternula albifrons* (Painter *et al.* unpublished data; Everaert & Stienen 2007; Everaert 2008; 2014). Collisions are generally assumed to have been with the turbine blades, based on the injuries recorded. However, a single collision of a Northern Fulmar *Fulmarus glacialis* with a turbine tower was observed at the coastal wind farm at Blyth Harbour, UK, in conditions of poor visibility (Newton & Little 2009). The list clearly indicates that gulls and terns are the main species group affected by collisions at onshore and coastal wind farms in this region.

In the southern hemisphere, where petrels and shearwaters nest in high numbers near coastal wind farms, they have regularly been recorded as collision victims (Hull *et al.* 2013; Bull *et al.* 2013). Fatalities have included Kelp Gull *Larus dominicanus*, Fairy Prion *Pachyptila turtur*, Sooty Shearwater *Ardenna grisea*, Short-tailed Shearwater *Ardenna tenuirostris*, Common Diving Petrel *Pelecanoides urinatrix*, White-faced Storm Petrel *Pelagodroma marina* and Wilson's Storm Petrel *Oceanites oceanicus*. Although pelagic species such as petrels and shearwaters have not been recorded as fatalities at coastal wind farms in northern Europe and the USA, this is most likely because their ranges are rarely likely to overlap with terrestrial wind farms.

The caveat to the above is that the list of species is likely to be incomplete as not all onshore and coastal wind farms collect data and, if they do, it may be in inaccessible grey literature.

Evidence of seabird avoidance

Evidence of collision avoidance is potentially easier to record than collision itself as events should, according to theory, be more frequent. However, avoidance may occur at a range of scales. These were initially defined by Cook *et al.* (2014) with modifications added by Skov *et al.* (2018) as: (1) macro-avoidance, involving total avoidance of the wind-farm footprint and in some cases a buffer around its perimeter of up to 3 km; (2) meso-avoidance, describing any within wind-farm responses to the turbines, such as flying between rows, sometimes with a specific buffer around the rotor–swept zone e.g. 10 m; and (3) micro-avoidance, being a 'last-minute' action to avoid collision with the blades within any defined buffer. An overall avoidance rate would combine all these factors.

Where seabirds avoid wind farms during both construction and operation they cannot collide with the infrastructure. Studies in the Dutch North Sea estimated that 28% of all birds avoided the wind farm entirely (Desholm & Kahlert 2005; Petersen *et al.* 2006). Vanermen and Stienen discuss avoidance and displacement from the wind-farm footprint in detail in Chapter 8 in this volume. However, in terms of collision risk, it is important to understand both the scale and rate of avoidance so it can be incorporated into the calculation of avoidance rates for collision risk modelling (Cook & Masden, Volume 4, Chapter 6).

Consistently high macro-avoidance rates have been recorded for Northern Gannet (Garthe *et al.* 2017) estimated at 64% and 85% in Dutch and Belgian wind farms respectively (Krijgsveld *et al.* 2011; Vanermen *et al.* 2015) and Skov *et al.* (2018) calculated an empirical macro-avoidance rate of 79.7% at Thanet, UK. Gull species at Horns Rev and Nysted, Denmark, avoided the wind farm at rates between 56% and 61% (Skov *et al.* 2012), although at some wind farms they still provide the bulk of local flight activity (Fijn *et al.* 2015) but not necessarily at pre-construction densities (Vanermen *et al.* 2017). At Thanet, there was variation in empirical avoidance rates between species with European Herring Gull and Great Black-backed Gull at 44.2% and 46.4% respectively and Lesser Black-backed Gull at 61.9%. Black-legged Kittiwake was intermediate at 56.6% (Skov *et al.* 2018). These relatively low rates meant that macro-avoidance contributed least to overall collision risk. During the construction of Sheringham Shoal, visual tracking showed that 30% of Sandwich Terns avoided the wind farm completely, compared to pre-construction, and although post-construction utilisation increased, birds tended to transit the site rather than forage within it (Harwood *et al.* 2017). Macro-avoidance also occurs at night, particularly by birds on migration (see Hüppop *et al.*, Chapter 7), although different wind farms have recorded different behaviours, some seabirds avoiding by lesser and some by greater distance during darkness (Krijgsveld *et al.* 2011). At Thanet, night time activity within the wind farm was less than 3% of that recorded during the day (Skov *et al.* 2018) although whether this represents macro-avoidance or simply low levels of nocturnal activity is not known.

Within-wind-farm avoidance (meso-avoidance) was demonstrated in early radar studies which showed that seabirds entering the wind farm were likely to reorientate their flight direction to fly between turbine rows, often remaining equidistant from them (Desholm & Kahlert 2005; Pettersson 2005; Petersen *et al.* 2006) and to avoid operational turbines at a greater distance than stationary ones (Krijgsveld *et al.* 2011; Hill *et al.* 2014). This behaviour has been quantified at Thanet with radar and video data showing that 99.4% of individuals flew between rows rather than avoiding the turbine by flying below or, even more rarely, above the rotor. Empirical meso-avoidance rates were calculated ranging between 84.2% for Great Black-backed Gull and 92.1% for Northern Gannet with European Herring Gull, Lesser Black-backed Gull, and Black-legged Kittiwake at intermediate values (Skov *et al.* 2018).

Until recently, micro-avoidance behaviour has rarely been observed offshore but at Egmond aan Zee radar backed up by visual observation showed that very few birds approached within 50 m of the rotating turbine and, of those, only 7% entered the rotor-swept area (Krijgsveld *et al.* 2011). Similarly, remote three-dimensional camera-tracking on a single turbine at Sheringham Shoal, UK, showed that no birds approached within 60 m of the turbine blades (M. Mellor & K. Hawkins, unpublished report 2013). The use of video at Thanet, also suggested that only a small proportion of birds entered the zone within 10 m of the rotor blade. For all those seabirds that did, the empirical micro-avoidance rate was 95%, with birds often making flight adjustments to fly parallel to the rotor. The overall empirical avoidance rate was most sensitive to this element of avoidance behaviour. There was a high risk attached to those flights which crossed perpendicular to the rotor blade as of the 15 occasions on which this was recorded, six resulted in collision (Skov *et al.* 2018).

Overall avoidance rates can only be back-calculated from collision risk models where flight movement, including flight height, and fatality data are collected simultaneously and the differing contributions of macro-, meso- and micro-avoidance are well understood (Cook *et al.* 2014). Avoidance rates calculated previously for seabirds and reviewed by MacArthur Green *et al.* (2013) have not always met these requirement. For example, data on the micro-avoidance rates for tern and gull species at Zeebrugge (Everaert & Stienen 2006; Everaert 2014) were generated using passage rates collected over a shorter period than fatalities, and the authors themselves noted that the results should be treated with caution. However, if treated as micro-avoidance rates they are in line with those calculated at Thanet (Skov *et al.* 2018).

Cook *et al.* (2014, 2018) identified nine sites (Table 9.1) where mortality and bird movement data had been collected with sufficient spatial overlap to derive empirical micro-avoidance rates and used standard methods to calculate them using different options of the Band collision risk model (Band 2012). Where appropriate, these were combined with macro- and meso-avoidance rates calculated from other wind farms to produce an overall avoidance rate. Rates were calculated for individual gull species and gulls as a species group. In the UK, all statutory advisers have agreed the use of the avoidance rates from this study with the Band collision risk model (Band 2012) in relation to European Herring Gull, Lesser Black-backed Gull and Great Black-backed Gull (99.5% for the basic model and 98.9% for the extended model excepting Herring Gull at 99%). Rates for Black-legged Kittiwake and Northern Gannet have also been agreed based on this work, although empirical rates could not be calculated (JNCC *et al.* 2014). This has been helpful in reducing conflict between developers, conservation advisers and the regulator.

The key question about the above study is whether rates derived from evidence collected at terrestrial wind farms reflect reality offshore. In fact they appear to be conservative. Empirical avoidance rates calculated at Thanet (Skov *et al.* 2018) by combining macro-, meso- and micro-avoidance avoidance rates derived from bird tracks within the operational wind farm concluded the following rates which could be applied to all versions of the Band (2012) collision risk model: Northern Gannet, 99.9%; Black-legged Kittiwake, 99.8%; European Herring Gull, 99.9%; Great Black-backed Gull, 99.6% and Lesser Black-backed Gull, 99.8%. All rates were estimated plus or minus two Standard Deviations. This study represents a major step forward in terms of understanding empirical rates of offshore avoidance and collision and, whilst the data have some limitations because they are from a single wind farm, collected mainly during daylight hours and the birds present are not necessarily breeding birds, they suggest that the avoidance rates used until now have been conservative. However, they have yet to be reviewed by the UK's statutory nature conservation advisers.

Species-specific factors

Marques *et al.* (2014) propose that three interconnected elements influence onshore collision risk: (1) species-specific factors covering morphology, ecology and behaviour; (2) wind-farm specific factors and (3) general locational factors. There is no reason to suppose that these factors differ offshore although they are less well understood. The available evidence for seabird collision offshore is therefore reviewed under similar categories below.

Flight height and behaviour

The number of flights, the time spent at risk height and flight behaviours that increase bird encounters with turbines are key indicators of seabird collision risk at onshore wind farms. For these reasons, terns experienced the highest collision rates at Zeebrugge (Everaert & Stienen 2007) (Box 9.1). At the same site, Common Tern collision was sex biased, with more males colliding during the egg-laying, incubation and early chick-rearing period when they undertook a greater number of provisioning flights (Everaert & Kuijken 2007; Everaert & Stienen 2007; Everaert 2008; Stienen *et al.* 2008). In general, Common Tern collided more frequently than Sandwich Tern and this was attributed to the species' more sinuous, less direct flights, which exposed them to risk for longer periods. By contrast, flights of seabirds at Thanet OWF were often non-linear within the wind-farm compared to the straight flight lines assumed by the Band collision risk model. This would effectively reduce flux rate and thus collision estimates (Skov *et al.* 2018).

Box 9.1   Fatalities of terns at turbines along a breakwater at Zeebrugge in Belgium

Joris Everaert & Eric W.M. Stienen

During the 1980s, a new outer harbour of Zeebrugge was constructed on the coast of Flanders, northern Belgium. Two main breakwaters were built and parts of the area between them became very attractive for breeding seabirds, especially gulls and terns. As a result, colonies of international importance (>1% of the biogeographic population) have existed since 1989 for Common Tern *Sterna hirundo* and since 1991 for Sandwich Tern *Thalasseus sandvicensis* and Little Tern *Sternula albifrons*, although the last two species have shown strong fluctuations in population size.

Initially, most terns nested in the western port area. However, further industrialisation of this area and the extension of a 'tern peninsula' along the eastern breakwater, specifically constructed as new breeding habitat, gradually displaced all the terns by 2004 towards the tern peninsula (Figure 9.2). Because of the breeding terns, parts of the harbour area became protected as a Natura 2000 site.

At that time, along the eastern breakwater, a row of 25 (mostly) 400 kW wind turbines was already present. As a result, during the breeding season many terns

Figure 9.2 The previous and current location of turbines in relation to the tern colony on the eastern port breakwater in Zeebrugge, Belgium.

Figure 9.3 Collided Common Tern *Sterna hirundo* at the peninsula in Zeebrugge (Eric Stienen)

Figure 9.4 Examples of collision fatalities of (a) Little Tern *Sternula albifrons*, (b) Sandwich Tern *Thalasseus sandvicensis*, and (c) Common Tern *Sterna hirundo* at the peninsula in Zeebrugge. (Joris Everaert)

Table 9.2 Impact of wind turbines at Zeebrugge on the breeding populations of terns in Flanders, Belgium.

	Number of collision fatalities of adult birds after corrections for available search area, scavenging and search efficiency			Number of breeding adults on the peninsula in Zeebrugge, next to the wind turbines			Total number[a] of breeding adults in Flanders		
	Little Tern	Sandwich Tern	Common Tern	Little Tern	Sandwich Tern	Common Tern	Little Tern	Sandwich Tern	Common Tern
2004	5	54	109	276	8,134	3,664	350	8,134	6,500
2005	2	30	129	30	5,076	2,950	138	5,076	4,900
2006	4	9	156	168	4,124	4,086	202	4,124	5,428
2007	12	7	114	156	2,254	5,582	242	2,254	6,030
2008	2	0	32	250	498	4,006	355	498	4,474
2009	0	0	0	38	8	250	86	8	1,582
2010	0	0	5	60	0	2,500	60	0	3,218
2011	2	0	34	204	108	2,708	204	108	3,140
2012	0	0	5	170	2	1,708	170	2	2,110
2013	0	0	3	164	294	1,346	164	294	1,926
2014	0	0	0	8	2	464	8	2	1,034

The new turbines were installed in early 2009, as shown by the black line. In the period from 2009 to 2014, a Red Fox *Vulpes vulpes* was present on the peninsula, with implications of predation and removal of collided terns The shaded parts of the table indicate estimated mean annual mortality effects >1% of the regional population as calculated from known mortality rates in north-western Europe.
[a]During the last years, the data is probably not 100% complete for Flanders.

collided with the turbines as they undertook regular foraging flights to and from the sea, thereby crossing the row of turbines (Figure 9.3 and Figure 9.4). Between 2004 and 2008, these collisions resulted in a significant impact on the regional breeding population in Flanders (Table 9.2) (Everaert & Stienen 2007; Everaert 2014).

In early 2009, fewer (14) and larger 850 kW turbines, at greater spacing and with more free space under the blades, replaced the small turbines. The reason for this repowering was two-fold: (1) to achieve higher electricity production and (2) to reduce tern collision fatalities, because most terns fly at relatively low altitude when crossing the breakwater. This 'win–win' situation was one of the conservation objectives in this Natura 2000 site.

However, in the period 2009–2014, the colony was significantly disturbed by the presence of a Red Fox *Vulpes vulpes* and Brown Rats *Rattus norvegicus*, resulting in a large decrease in the number of breeding pairs and decreased reproductive output. This made it difficult to make a reliable comparison with previous years and to separate the effect of the repowering from disturbance and predation. Nonetheless, compared with the period before 2009, a clear reduction in the number of collision fatalities, both absolute and proportional to the number of breeding pairs, was noted (Table 9.2). However, the exact rate of change is still to be determined, mainly because of other confounding impact factors such as the presence of the Red Fox. Further study during the coming years is recommended to clarify the actual impact.

At OWFs, most gull species have been shown to maintain consistent flight heights before and after construction (e.g. Mendel *et al.* 2014), sometimes increasing height within the wind farm and closer to the turbines (Krijgsveld *et al.* 2011; Skov *et al.* 2012, 2018). This height increase may contribute to the number of gull fatalities recorded at coastal wind farms and could potentially lead to higher numbers of collisions than pre-construction estimates. By contrast, Northern Gannet, Black-headed Gull and Black-legged Kittiwake flew at reduced heights inside the operational wind farm at Egmond aan Zee compared to outside (Krijgsveld *et al.* 2011), suggesting that behavioural adaptation may occur (Petersen *et al.* 2006; May 2015) and that actual collisions would be lower than predicted. As evidence emerges, these general patterns may be shown to vary, for example at Thanet both Northern Gannet and Black-legged Kittiwake flew slightly higher within the wind farm than outside (Skov *et al.* 2018). However, in general, the percentage of birds at rotor height is expected to be a relatively good predictor of collision risk (Band 2012; Cook *et al.* 2012; Johnston *et al.* 2014) and this percentage is used in collision risk modelling derived either from site-based surveys or from a generic database.

In more detail, Johnston *et al.* (2014) pooled flight height data from 32 pre-construction surveys of UK OWFs to estimate continuous flight height distributions for 25 seabird species. Data were collected mainly by boat-based surveys with the addition of four studies from either onshore observers or fixed offshore platforms. Species that forage on the wing and subsequently dip to the sea surface to seize prey such as gulls and terns, or dive from a height to take prey from below the surface, such as Northern Gannet, have a tendency to fly higher than species that dive from the sea surface, such as auks and shearwaters (Figure 9.5 and Figure 9.6). This basic pattern may then be modified by how the bird flies (see *Morphology* below) and the reason for flight, such as whether the bird is foraging, commuting or migrating and may explain why gannet flights appear to be bi-modal (Skov *et al.* 2017, Cleasby *et al.* 2015a). The analysis showed that a variety of gull species and

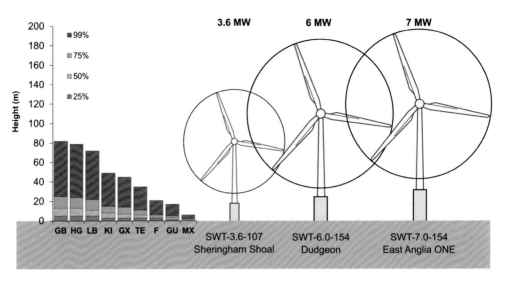

Figure 9.5 Flight heights of different seabird species relative to the rotor-swept zone of three turbine models (3.6 MW, 6 MW and 7 MW) as installed (3.6 MW at Sheringham Shoal and 6 MW at Dudgeon) or planned (7 MW at East Anglia ONE) at three UK offshore wind farms. Heights at which 25%, 50%, 75% and 99% of flights are expected according to the modelled distributions of Johnston *et al.* (2014) are shown. Species are shown in order of declining flight heights and are denoted by their respective British Trust for Ornithology codes. GB: Great Black-backed Gull *Larus marinus*; HG: European Herring Gull *Larus argentatus*; LB: Lesser Black-backed Gull *Larus fuscus*; KI: Black-legged Kittiwake *Rissa tridactlya*; GX: Northern Gannet *Morus bassanus*; TE: Sandwich Tern *Thalasseus sandvicensis*; F: Northern Fulmar *Fulmarus glacialis*; GU: Common Guillemot *Uria aalge*; MX: Manx Shearwater *Puffinus puffinus*. The species selected represent key seabird families in relation to collision risk, including Laridae (GB, HG, LB, KI and TE), Sulidae (GX), Alcidae (GU) and Procellariidae (F and MX). (Andrew Harwood)

Figure 9.6 A number of species within several common groups of seabirds, including petrels, shearwaters and this group of alcid auks comprising Razorbills *Alca torda* and Common Guillemots *Uria aalge* in winter plumage, typically fly close to the sea surface and rarely reach the rotor-swept zone. As such, they are at little risk of collision with turbine blades. (Martin Perrow)

Northern Gannet were among the species that flew within the rotor-swept area (>20 m), although flights were skewed towards its lower reaches (Figure 9.5). It was estimated that overall, these birds flew at risk height between 10% and 35% of the time (Jongbloed 2016).

While such data are a useful way of standardising impact assessments across wind farms at the planning stage, generic data collected pre-construction may not accurately reflect bird behaviour within constructed wind farms. In addition, seabirds may exhibit different flight behaviours at different sites, meaning that generic data may either underestimate or overestimate collision.

For example, flight height is also known to vary according to whether the bird is commuting or foraging (Cleasby *et al.* 2015a) and may also be inconsistent between species. Similarly, although foraging strategies which include aerial searching and pursuit have been shown to be significant predictors of collision at onshore wind farms (Hull *et al.* 2013) and may be the case for Northern Gannet, where foraging birds had a greater median flight height than commuting birds (Cleasby *et al.* 2015a), by contrast, Global Positioning System (GPS)-tagged Lesser Black-backed Gulls flew lower on more tortuous, potentially foraging flights than when in transit (Corman & Garthe 2014), potentially reducing collision. This suggests that although foraging strategy may be an important indicator of risk (Scott *et al.* 2015), impacts may differ between species. Indirect flight patterns may also mean that the direct flight speeds (e.g. from Pennycuick 1997; Alerstam *et al.* 2007) used in collision risk modelling to calculate passage rate may be inappropriate as the actual track speed of the bird, as measured by the straight line distance between its start and end point, is relatively slow. Speeds measured at Thanet OWF, based on large sample sizes suggest that for some species, such as Black-legged Kittiwake, previous estimates were too high and recommends that for collision risk calculations, track speed should be used to calculate passage rate and flight speed as the rate through the actual rotor (Skov *et al.* 2018).

Other factors, such as wind speed and direction (Spear & Ainley 1997; Ainley *et al.* 2015), season and time of day, can also influence flight altitude. For example, Skov *et al.* (2018) found that seabirds flew slightly lower in head-winds within Thanet offshore wind farm and Ross-Smith *et al.* (2016) found that GPS-tagged Lesser Black-backed Gulls flew lower at night than by day when 34% occurred at nominal rotor height (22–250 m) and tended to fly lower nearshore than offshore although the latter data were not recorded within constructed wind farms. Behaviour may also vary between years (Garthe *et al.* 2011; Thaxter *et al.* 2015).

In general, it seems difficult to make generic predictions about how seabird flight behaviour in open water would be modified within a constructed wind farm, as it appears to be species, site and circumstance specific. Post-construction monitoring is essential to investigate differences and to see whether generic behaviours can be established.

Morphology

The morphology that makes gliding birds predisposed to collision onshore, including wingspan, aspect ratio and wing loading (Barrios & Rodriguez 2004; de Lucas *et al.* 2008), may act to reduce risk offshore as pelagic seabirds with a high aspect ratio and lower wing loading, such as petrels and shearwaters, utilise wind shear and lift associated with waves (Gibb *et al.* 2017) and rarely fly at rotor height (Johnston *et al.* 2014), although the probability of higher flight is greater in stronger winds (Ainley *et al.* 2015). However, petrels and shearwaters have been recorded as collision victims at onshore wind farms in Tasmania and New Zealand (Hull *et al.* 2013; Bull *et al.* 2013). This suggests that where these species traverse land their morphology remains a risk factor owing to their lack of manoeuvrability. Hull *et al.* (2013) noted that shearwaters were not recorded during daytime utilisation studies so must have been colliding during crepuscular periods or at night when visibility was reduced. Other causal factors could include: (1) an increase in flight height into the rotor-swept zone due to increased wind strength (Spear & Ainley 1997; Ainley *et al.* 2015), such as at coastal cliffs and slopes; (2) obligatory downwind flight (Ainley *et al.* 2015), such as when accessing nesting colonies; or (3) turbulence effects caused by topography (Stumpf *et al.* 2011); all of which reduce a gliding bird's ability to control its flight.

In theory, species such as gulls, which utilise more flapping flight and have greater manoeuvrability, should be less prone to collision. However, the fact that they are regularly recorded as collision fatalities (Table 9.1) suggests that for species in this group, flight height behaviour is a better collision predictor than morphology.

Avoidance behaviour

Avoidance is a key factor in the theoretical prediction of collision risk as small changes in avoidance rate have a big influence on fatality estimates (Cook & Masden, Volume 4, Chapter 6). Empirically calculated avoidance rates such as those derived by Cook *et al.* (2014, 2018) are generally higher than the default rates used previously in the UK, which were initially as low as 95% before being revised to 98% for most species based on evidence from onshore wind farms. MacArthur Green *et al.* (2013) made the first well-argued case that the 98% generic rate was too precautionary for seabirds. This was primarily because terrestrial survey methods covered only a relatively small area of the wind farm site and data could only represent micro-avoidance rates. Any macro-avoidance as shown by some seabirds at sea would therefore not be incorporated. Furthermore, unlike offshore surveys, terrestrial surveys did not correct for imperfect detection to derive density and thus potential passage rates. The study went on to cite or re-estimate micro-avoidance rates from a number of both onshore and offshore wind farms of 99.25–100% for gulls and 99.83–100% for terns, and recommended an overall rate of 99.8% for Northern Gannet. Based on this, MacArthur Green *et al.* (2013) argued that the seabird default rate should be raised from 98%. Subsequently, Cook *et al.* (2014) reanalysed a number of studies using standard methods and more closely prescribed criteria as described in the previous section. This confirmed that empirically derived seabird avoidance rates were indeed closer to 99% and sometimes higher. The most recent study at the time of writing (Skov *et al.* 2018) supports the conclusion that previous rates have been precautionary and confirms that for the target species studied: Northern Gannet, Black-legged Kittiwake, European Herring Gull, Lesser Black-backed Gull and Great Black-backed Gull, empirical avoidance rates exceed 99.6%.

In contrast to avoidance, a few species are attracted to OWFs, specifically Great Cormorant *Phalacrocorax carbo* and European Shag *Phalacrocorax aristotelis*, which use parts of the turbine structures as a perch for wing-drying and roosting (Dierschke & Garthe 2006; Krijgsveld *et al.* 2011; Dierschke *et al.* 2016; Vanermen & Stienen, Chapter 8). In theory, this could increase collision risk by bringing birds more regularly into contact with the rotor-swept area of the turbines. Outside wind farms, the flight height of Great Cormorant has been shown to be variable (Cook *et al.* 2012) but within wind farms 50% of cormorants were recorded at rotor height (Krijgsveld *et al.* 2011).

Large gulls (*Larus* spp.) have been shown to be attracted to some Belgian wind farms (Vanermen *et al.* 2015). However, the pattern is inconsistent as GPS tracking of large gulls at Thorntonbank showed that although birds were present within the wind farm, roosting on the turbine jackets and occasionally feeding on mussels growing in the intertidal sections, densities were lower inside than outside the wind farm and birds were more common on the outer rows than farther in and spent less time in flight. This suggests either macro-avoidance or a barrier effect (Vanermen *et al.* 2017). Similar results were recorded at Thanet, where gulls were seen to roost on the transformer module, but densities were lower within the wind farm than outside and Herring Gull rarely penetrated beyond the two outermost turbine rows (Skov *et al.* 2018).

Phenological period

Collision risk may be associated with phenological period as seabird abundance is often highly seasonal (Krijgsveld *et al.* 2011). If so, the scale and timing of effects will depend on whether the wind farm is located close to a breeding colony, foraging area, migration route or wintering location.

For breeding GPS-tagged Lesser Black-backed Gulls, pre-construction data showed peak use of offshore foraging areas, including potential wind-farm locations, during the pre-breeding season. Use decreased while birds were incubating and increased again during early chick rearing. However, use varied between years, sexes and individuals (Thaxter *et al.* 2015; Thaxter & Perrow, Volume 4, Chapter 4). Similar data for Northern Gannet showed differences in foraging patterns between colonies and sexual segregation in the use of foraging habitat during the breeding season, although birds tracked in German waters tended to avoid the wind farms (Corman 2015). In Scotland, before the construction of any wind farms, males foraged in more mixed coastal waters with females foraging farther offshore (Cleasby *et al.* 2015b). In theory, this has the potential to lead to a sex bias in collisions, depending on wind-farm location, for different reasons than those described previously for Common Tern (Stienen *et al.* 2008).

For breeding seabirds, which are central-place foragers, their reproductive status may drive them to take greater risks during the chick-rearing period, leading them to fly through, rather than avoid wind farms, potentially increasing collisions. At Zeebrugge, high collision rates of terns during the breeding season were attributed to specific turbines located between a large colony and the birds' feeding grounds at sea (Everaert & Stienen 2007; Everaert 2014).

In winter, risk may be increased where wind farms lie on flight paths between feeding grounds and winter roosts. Such commuting flights made up the bulk of local bird movements recorded during the summer and winter periods using both radar and visual observations at Egmond aan Zee wind farm in the Dutch North Zee (Fijn *et al.* 2015).

Vision

Other species-specific factors may influence collision risk but evidence of how they apply to seabirds at OWFs is not yet available. These include sensory perception, for example the way in which seabird visual perception appears to have evolved to detect highest detail in the lateral rather than frontal field of view for life in an uncluttered airspace (Martin 2011). Other related factors, such as perception distance, the visibility of moving blades (Hodos 2003) and habituation/spatial memory that could lead to evasion (May 2015), may all have a role in collision risk. These factors are described more fully in de Lucas and Perrow (2017).

Wind-farm specific factors

Turbines

Theoretical considerations suggest that larger offshore turbines will reduce collision risk owing to their lower rotational speeds (Krijgsveld *et al.* 2009; Johnston *et al.* 2014) and increased tip clearance above highest astronomical tide where bird flight height distribution is skewed towards sea level (Johnston *et al.* 2014). However, a theoretical reduction in collision risk may be offset in part by the larger rotor-swept area of larger turbines although, for such turbines the greatest proportional increase in swept area is above the majority of seabird flight heights. Any contrast in colour between the transition piece and the tower could reduce collisions, as demonstrated onshore (de Lucas & Perrow 2017; May 2017). Conversely, jacket structures may also increase collision risk as they provide more perching and roosting opportunities than monopile structures (Vanermen *et al.* 2017), perhaps encouraging more birds to come into close proximity with the turbine rotors. However, any effect of offshore turbine type on bird collision has rarely been investigated and none of the effects mentioned here has been demonstrated in practice.

Configuration

Wind-farm configuration may affect collision risk in a number of ways. To illustrate, radar (Desholm & Kahlert 2005) and visual tracking (Harwood *et al.* 2017) have shown that birds which enter the wind farm often fly between turbine rows, and Krijgsveld *et al.* (2011) noted higher levels of flight activity in parts of the wind farm where turbines were more widely spaced, although a later comparative study by the same author could not confirm that turbine density affected avoidance rates (Krijgsveld, K. personal communication, 2014). Standard offshore turbine separation is increasing as turbine capacity increases because greater distances improve wind recovery (Archer *et al.* 2013) and reduce turbine loading owing to wake interaction, which may cause effects up to 10 rotor diameters (D) in some circumstances (DNV GL 2016). In the UK, offshore turbine separations between 6 and 8 D are common, often giving inter-turbine spacing of greater than 1 km. This may permit higher rates of meso-avoidance.

Where turbine strings are isolated from the main body of the wind farm, birds have been seen to fly closer to the machines, undertaking more risky behaviour (Krijgsveld *et al.* 2011). Similar effects incurring higher collision rates have been demonstrated in the Waddensee at a terrestrial wind farm located on a coastal headland where bird flights are concentrated (Brenninkmeijer & Klop 2017). At Thanet, macro-avoidance at the corner of the wind farm was lower than elsewhere although this may have been influenced by factors such as the proximity of the transformer module used by roosting gulls or the presence of fishing vessels making estimates conservative (Skov *et al.* 2018).

Lighting

The lighting of OWFs is mandatory for safe air and sea navigation. Although Danish studies found no evidence that large migrating birds such as seabirds were attracted to OWFs at night (Petersen *et al.* 2006), bright lighting at offshore structures such as oil and gas platforms is known to attract migrants, particularly in poor weather, resulting in high collision numbers for some species, particularly passerines (Ronconi *et al.* 2015; Hüppop *et al.* 2016; Hüppop *et al.*, Chapter 7). Experimental use of directional lighting and white lights low in the red spectrum has mitigated such effects at some gas platforms (Poot *et al.* 2008), although intermittent rather than continuous red lights appeared to make no difference (Kerlinger *et al.* 2010). These findings may be relevant to both turbines and associated wind-farm infrastructure such as substation platforms, although there are no mandatory requirements in relation to birds.

Locational factors

Weather

Wind speed and direction are known to influence seabird flight behaviour (Spear & Ainley 1997; Ainley *et al.* 2015). The flux of migrants through Egmond aan Zee wind farm was greater with tailwinds and flight height was higher than during headwinds (Krijgsveld *et al.* 2011), potentially increasing collision risk. Northern Gannets showed a similar pattern at Horns Rev (Skov *et al.* 2012) as did seabirds in general at Thanet (Skov *et al.* 2018). However, in the Krijgsveld *et al.* (2011) study, large gulls flew lower with tailwinds and no effects could be demonstrated for small gulls. In a separate study on Lesser Black-backed Gull, no effects of wind speed on flight height were observed (Corman & Garthe 2014). These variable results suggest that responses to wind speed and direction may be species and/or location specific. Moreover, a key factor in relation to wind direction may simply be whether this orientates the turbines perpendicular to the predominant flight path of the birds concerned and thus occupies a greater proportion of the air space. At Zeebrugge, more collisions occurred in this scenario (Everaert & Stienen 2007).

Good weather, defined as decreasing humidity and increasing barometric pressure, may also influence flight height, as demonstrated at Horns Rev for both Northern Gannet and terns, which flew higher under such conditions, potentially increasing their collision risk (Skov *et al.* 2012). The same study also stated that humidity was inversely correlated with visibility, implying that in poor visibility, seabird flight height would decrease, although this is not explicitly stated. At Egmond aan Zee, seabirds were found to avoid even stationary turbines and the wind farm as a whole at a greater distance in conditions of poor visibility (Krijgsveld *et al.* 2011).

In general, the influence of bad weather on seabird flight height is not well understood as many methods of offshore bird survey, particularly visual boat-based or digital aerial surveys, are curtailed in poor weather, and data quality for radar may be compromised by wave height or rain (Fijn *et al.* 2015). Heavy rain is thought to reduce flight height, particularly of migrating birds (Hüppop *et al.* 2006), which may bring them down from high altitude and into the risk zone of rotating turbines.

Risk indices

Where empirical evidence is thin, as it is for most effects on seabirds from OWFs, proxy measures are required to assess the sensitivity of a site or species to wind-farm development for the purposes of Environmental Impact Assessment (EIA). As a result, a number of sensitivity indices have been developed. On a species level, their use is primarily for screening whether the species should be taken forward for assessment, in this case for collision risk. At a site level, risk indices are ideally used strategically to identify whether an area is suitable for offshore wind development or whether the putative levels of collision (or some other effect) make it unsuitable.

The first species index was developed for the German waters of the southern North Sea (Garthe & Hüppop 2004). Sensitivity was scored on a scale of 1–5 for nine factors based on existing knowledge and expert judgement and combined to form a risk index incorporating both collision and displacement. The attributes used as collision risk indicators were flight altitude, manoeuvrability, percentage of time flying during the day and a nocturnal activity factor. These were combined with the species' biogeographic population size, adult survival rate and conservation status. The indices were combined with density distributions of seabirds in German waters to map areas of high wind-farm sensitivity on an annual and a seasonal basis. Overall, seabirds were assessed as most vulnerable to wind farms located closest to the German coast, where bird densities were consistently higher. Subsequently, developers avoided those sensitive areas. A similar approach used for seabirds in the Dutch North Sea reached similar conclusions (Leopold & Dijkman 2010).

Furness et al. (2013) refined this model by calculating separate indices for collision and disturbance. While the collision predictors proposed by Garthe and Hüppop (2004) were considered robust, the percentage of birds at rotor height was used rather than a 1–5 score and the calculation was weighted to reflect the strong influence of this factor on collision risk (Band 2012). Overall, the large Larus gulls were considered to be at greatest risk of collision, with Great Black-backed Gull (Figure 9.7), European Herring Gull and Lesser Black-backed Gull given estimated percentages of time at blade height of the earlier generation of turbines, such as 3.6 MW machines (Figure 9.4), of 35%, 31% and 27% respectively. These values multiplied by one-third (manoeuvrability score + percentage of time flying score + nocturnal flight score) multiplied by conservation score (ranked by index value) took these three species to the top of the rankings of concern of collision risk. See Furness et al. (2013) for details of the basis for these calculations.

Work on sensitivity indices was extended by Wade et al. (2016) to reflect uncertainty around some of the factors used for particular species. Although the Furness et al. (2013) vulnerability rankings remained broadly similar to those of Garthe and Hüppop (2004), the uncertainty indices identified species for which data were more robust. As such, they also inform qualitative discussion of any assessment results, particularly for those species where uncertainty around their vulnerability is highest. In addition, uncertainty can be used to identify areas for strategic monitoring and research.

Bradbury et al. (2014) built on the methods of Furness et al. (2013) to develop a sensitivity-mapping tool for seabirds in English waters (SeaMAST). Sensitivity in this case is governed by areas of higher use by different seabirds, which in turn is likely to be linked to the distribution of breeding colonies in the breeding season, as well as illuminating key overwintering areas or those concentrating seabirds in the passage period, i.e. their ecological importance. This publicly available, geographic information system (GIS)-based package combines species sensitivity information on collision and/or displacement with density surface models in a high-resolution grid (3 km × 3 km). Although a useful tool, it was developed after the award of many UK OWF development zones, some of which were already located in potentially sensitive areas with high levels of consenting risk. However, it has since been used by to identify sea areas with potential for future development based on the biological 'headroom' that remains after taking into account the predicted collision risk for wind farms already in planning, consented or constructed (MacArthur Green, unpublished report, 2017). The approach assumes that predicted levels of collision are likely to be precautionary and remain acceptable, which is a view not held by all stakeholders.

If developed ahead of planning decisions, risk and sensitivity indices are a useful strategic tool and similar indices identifying areas of high wind-farm sensitivity and/

Figure 9.7 Great Black-backed Gull *Larus marinus*; the species ranked by Furness *et al.* (2013) as the most vulnerable to collision with turbines of all 38 seabirds or species strongly associated with the sea, occurring in Scottish waters. The list included White-tailed Eagle *Haliaeetus albicilla*, a species known to be particularly vulnerable to collision with turbines onshore (de Lucas & Perrow 2017), which was ranked fourth. (Martin Perrow)

or conservation priority have also been or are being produced for seabirds along the Norwegian coast (Gressetvold 2013), the Pacific coast of North America (e.g. Adams *et al.* 2016) and the US Atlantic Outer Continental Shelf (OCS) (e.g. Winiarski *et al.* 2013). Zipkin *et al.* (2015) advanced this work by identifying seabird 'hot' and 'cold' spots on the US OCS and making recommendations on the design of survey campaigns with sufficient power to detect changes in seabird abundance at the relevant scale.

While recognising the importance of such decision-making tools, particularly the seminal work of Garthe and Hüppop (2004), they are clearly less useful where their development lags behind marine spatial planning decisions (Maclean *et al.* 2014). Certain *et al.* (2015) also critiqued some basic assumptions of such models and their underlying mathematics, suggesting that refinements would permit better management decisions. Recommendations included separate assessment of pressures such as displacement and collision as per Furness *at al.* (2013) and Bradbury *et al.* (2014) and, more importantly, the separation of vulnerability between primary and secondary factors in order to weight them differently. In the collision index, for example, the percentage of flights at rotor height and time spent in flight were identified as primary factors, whereas manoeuvrability and nocturnal activity were classed as secondary, 'aggravating', factors, being relevant only if a species flies at risk height. Finally, Certain *et al.* (2015) also argue that the multiplication of factors such as vulnerability and abundance can result in information loss, potentially making the risk maps such as those in Bradbury *et al.* (2014) less useful than they might be. Generic modifications to the formulae are suggested.

Another common feature of the sensitivity index approach is that factors lacking empirical evidence depend on expert opinion (Bailey *et al.* 2014). However, there is consistent identification of gulls as the group most sensitive to collision risk, and inshore/coastal areas and some shallow banks as the most sensitive locations for wind-farm development. Large concentrations of breeding seabirds, particularly gulls, have been identified as a significant influence in this respect (Leopold & Dijkman 2010; Gressetvold 2013; Bradbury *et al.* 2014). While the vulnerability of gulls to collision is borne out by the limited evidence available to date at onshore and offshore wind farms, theoretical risk predicted in EIAs has not been an accurate predictor of mortality at onshore wind farms (Ferrer *et al.* 2012), as abundance may not equate to collision risk (de Lucas *et al.* 2008; Hull *et al.* 2013). To illustrate the point, Northern Gannet features high in terms of collision risk using the revised methods of Certain *et al.* (2015), but building evidence suggests that Northern Gannet is generally displaced from wind farms (Krijgsveld *et al.* 2011; Dierschke *et al.* 2016; Vanermen & Stienen, Chapter 8). This illustrates the importance of refining these factors based on within-wind-farm evidence as it accrues, such as from Skov *et al.* (2018).

Finally, indices cannot always readily identify species that require special assessment under the EU Birds (2009/147/EC) and Habitats Directives (92/43/EEC), even though a measure of conservation value may be introduced as a factor, as was the case in Garthe and Hüppop (2004). In general, whether particular protected species require special attention in relation to a wind farm is likely to depend on the proximity of Special Protection Areas (SPAs) designated under these Directives and the predicted connection or connectivity between the SPA and the birds within the wind-farm footprint.

Population-scale effects

Perhaps the most important question about collision mortality is how it affects species at the population scale, be that local, regional, national or international. Such effects are rarely likely to be caused by individual wind farms as the planning process is designed to identify and exclude such projects. For example, in the UK, the proposed Docking Shoal wind farm was refused consent because of its potential effects on internationally important colonies of Sandwich Terns contained within a single SPA (DECC 2012).

Cumulative effects across a region or along a migration flyway are much more likely to occur but harder to quantify. Calculations require a knowledge of the area used by a species at all stages of its life history, a realistic knowledge of the effects from OWFs across this area, and an understanding of the population size, trend and demographic factors of the species in question so that the effects can be modelled. Having undertaken the modelling, agreement is then needed about the threshold at which impacts become unacceptable. As many of the factors described are either unknown or uncertain, theoretical results should be viewed with caution. However, if models converge in their identification of particular species at risk, it would be logical to make them a priority for monitoring.

In the Dutch North Sea, Poot *et al.* (2011) modelled cumulative effects on seabirds to consider alternative scenarios for future wind-farm development in the region. While no significant population-level effects were predicted, the authors noted that where populations such as European Herring Gull were already declining, additional wind-farm mortality could exacerbate this effect. More recently, theoretical modelling for the southern North Sea (Leopold *et al.* 2014) identified potentially worrying levels of effect for Lesser Black-backed Gull, Great Black-backed Gull, Black-legged Kittiwake, European Herring Gull and Northern Gannet. Similar species, plus Common Gull, were also identified in a cumulative study of potential effects in Belgian waters, extrapolated to the wider North Sea (Brabant *et al.* 2015). Looking at cumulative effects on all ecological receptors, Plateeuw *et al.* (2017) developed an overarching assessment framework to understand how Dutch offshore wind targets could be accommodated without endangering protected species.

In the Scottish North Sea, a study of cumulative effects on 27 seabird species concluded that in general, collisions would not cause population-level effects, although for European Herring Gull and Great Black-backed Gull, certain modelling scenarios indicated cause for concern (WWT & MacArthur Green 2014). The cumulative effects of collisions from UK North Sea wind farms have since been recalculated using the final structure of the project as it is built out, rather than the results of the worst case scenario reported in the projects' environmental statements. The study concluded that the originally predicted effects would be likely to be reduced in reality, although once again, in the North Sea, large *Larus* gulls were the species group predicted to be closest to allowable thresholds of mortality (MacArthur Green, unpublished report, 2017). It should be stressed that these studies are, by necessity, theoretical owing to the lack of evidence, but are a useful tool for the early identification of potential issues. Cook & Masden (Volume 4, Chapter 6) provide further details of the modelling process and the issues that surround it.

Concluding remarks

This review shows that evidence for bird collision with truly offshore turbines has been slow to emerge and remains a major ecological concern (Marques *et al.* 2014). Currently, the best available evidence on collision and avoidance comes from the study at Thanet OWF (Skov *et al.* 2018). This confirms findings from onshore and coastal wind farms, where they overlap with seabird distribution, that gulls as a species group are the most regular fatalities, with the main predictor apparently being the time spent in flight at rotor height. At onshore and coastal wind farms, terns are also susceptible to collision and the case study shows that, at some sites, this may be significant.

Seabird avoidance rates, calculated from land-based data, have proved higher than previously anticipated at around 99% or above. The minimum empirical avoidance rate calculated at Thanet OWF was 99.6%, confirming this direction of travel and indicating that rates are higher than previously permitted. Revised guidance may therefore be necessary. Nonetheless, if these values are to be verified and calculated for a greater range of species there is a need for further studies at OWFs in which fatality, passage rate and flight height data are collected simultaneously (Cook *et al.* 2014, 2018; Skov *et al.* 2018).

In relation to the factors which influence collision risk such as time spent flying at rotor height, flight speed and nocturnal activity, many of the data come from birds at sea in general and, while useful in understanding seabird ecology and behaviour, more information from behaviour within wind farms is clearly needed. The Thanet study has provided a useful indication of at least some of the species and factors which should be monitored offshore in future although there is still little evidence of how general wind-farm location and site-specific factors affect collision. The trend towards fewer, larger, more widely separated turbines with a greater rotor clearance above sea levels will, at least theoretically, mitigate risk, perhaps even to a significant extent (Figure 9.5).

The development of seabird sensitivity indices in relation to collision risk has been most useful in respect of marine spatial planning decisions at a strategic level and for identifying particular species at risk. They converge in their identification of gulls as the main species group at risk of collision. However, they include many proxy factors or those based on or bird pre-construction behaviours and distributions. While this is a pragmatic approach based on the available data, post-construction monitoring is a vital requirement to understand the real risks to species and whether generalisations can be applied (May 2015). Such indices should be updated as evidence emerges.

In the absence of more empirical data, uncertainty about the true scale of collision and other effects on seabirds will remain, potentially making the planning and consenting of OWFs more extended, complex and risky (Masden *et al.* 2015). However, collision or avoidance monitoring is rarely required as a consent condition for individual wind farms, at least in the UK, mainly because of the difficulty of recording rare events and the expense of the technology required (although see Harwood *et al.* 2017). Rather, collision and avoidance monitoring is regarded as an industry-level issue. To date, it has arguably been best addressed by well-designed and managed, long-term projects run with the buy-in of government agencies and advisers and funded from a range of sources including industry, such as examples in Denmark (DONG Energy *et al.* 2006), the Netherlands (Krijgsveld *et al.* 2011), Belgium (Vanermen *et al.* 2015) and the ORJIP project in the UK (Skov *et al.* 2018). Further studies of this type would be beneficial.

Finally, a crucial factor in project reporting is that it is timely and widely disseminated at conferences, as peer-reviewed articles or in textbooks such as Köppel (2017), in a manner which makes it accessible to all interested parties. As the cost of offshore wind energy continues to decrease (CRMF 2017; Jameson *et al.*, Chapter 1) and its development around the world continues to grow (GWEC 2018; Jameson *et al.*, Chapter 1), this evidence will be vital in understanding the available capacity in keeping with the sustainable development of this form of renewable energy while minimising wildlife impacts.

Acknowledgements

I am grateful to Liz Masden of the University of the Highlands and Islands for useful comments on an earlier draft of this chapter. Many thanks to Joris Everaert and Eric Stienen for making their influential work at Zeebrugge available as a case study. Thanks also to Andrew Harwood of ECON Ecological Consultancy Ltd for drawing Figure 9.5 and to Martin Perrow of the same, for photographs and constructive restructuring of the chapter content.

References

Adams, J., Kelsey, E.C., Felis, J.J. & Peretska, J. (2016) Collision and displacement vulnerability among marine birds of the California Current System associated with offshore wind energy infrastructure. Report in cooperation with the Bureau of Ocean Energy Management (BOEM). Open-File Report 2016-1154. Reston, VA: US Geological Survey. Retrieved 21 May 2018 from https://pubs.er.usgs.gov/publication/ofr20161154.

Ainley, D.G., Porzig, E., Zajanc, D. & Spear, L.B. (2015) Seabird flight behavior and height in response to altered wind strength and direction. *Marine Ornithology* 43: 25–36.

Alerstam, T., Rosen, M., Backman, J., Ericson, P.G., & Hellgren, O. (2007) Flight speeds among bird species: allometric and phylogenetic effects. PLoS Biology, 5 (8): e197. https://doi.org/10.1371/journal.pbio.0050197

Archer, C.L., Mirzaeisefat, S. & Lee, S. (2013) Quantifying the sensitivity of wind farm performance to array layout options using large-eddy simulation. *Geophysical Research Letters* 40: 4963–4970.

Bailey, H., Brookes, K.L. & Thompson, P.M. (2014) Assessing environmental impacts of offshore wind farms: lessons learned and recommendations for the future. *Aquatic Biosystems* 10: 8. doi: 10.1186/2046-9063-10-8.

Band, B. (2012) Using a collision risk model to assess bird collision risk for offshore windfarms. Strategic Ornithological Support Services (SOSS), Project SOSS-02. Thetford: British Trust for Ornithology. Retrieved 21 May 2018 from https://www.bto.org/sites/default/files/u28/downloads/Projects/Final_Report_SOSS02_Band1ModelGuidance.pdf

Barrios, L. & Rodríguez, A. (2004) Behavioural and environmental correlates of soaring-bird mortality at on-shore wind turbines. *Journal of Applied Ecology* 41: 72–81.

Brabant, R., Vanermen, N., Stienen, E.W.M. & Degraer, S. (2015) Towards a cumulative collision risk assessment of local and migrating birds in North Sea offshore wind farms. *Hydrobiologia* 756: 63–74.

Bradbury, G., Trinder, M., Furness, B., Banks, A.N., Caldow, R.W.G. & Hume, D. (2014) Mapping seabird sensitivity to offshore wind farms. *PLoS ONE* 9 (9): e106366. doi: 10.1361/journal/pone/0106366.

Brenninkmeijer, A. & Klop, E. (2017) Bird mortality in two Dutch wind farms: effects of location, spatial design and interactions with powerlines. In Köppel, J. (ed.) *Wind Energy and Wildlife Interactions: Presentations from the CWW2015 Conference.* Cham: Springer. pp. 99–116.

Bull, L.S., Fuller, S. & Sim, D. (2013) Post-construction avian mortality monitoring at Project West Wind. *New Zealand Journal of Zoology* 40: 28–46.

Burthe, S.J., Wanless, S., Newell, M.A., Butler, A. & Daunt, F. (2014) Assessing the vulnerability of the marine bird community in the western North Sea to climate change and other anthropogenic impacts. *Marine Ecology Progress Series* 507: 277–295.

Certain, G., Jørgensen, L.L., Christel, I., Planque, B. & Bretagnolle, V. (2015) Mapping the vulnerability of animal community to pressure in marine systems: disentangling pressure types and integrating their impact from the individual to the community level. *ICES Journal of Marine Science* 72: 1470–1482.

Cleasby, I.R., Wakefield, E.D., Bearhop, S., Bodey, T.W., Votier, S.C. & Hamer, K.C. (2015a) Three dimensional tracking of a wide-ranging marine predator: flight heights and vulnerability to offshore wind farms. *Journal of Applied Ecology* 52: 1474–1482.

Cleasby, I.R., Wakefield, E.D., Bodey, T.W., Davies, R.D. Patrick, S.C., Newton, J., Votier, S.C., Bearhop, S. & Hamer, K.C. (2015b) Sexual segregation in a wide-ranging marine predator is a consequence of habitat selection. *Marine Ecology Progress Series* 518: 1–12.

Collier, M., Dirksen, S. & Krijgsveld, K. (2012) A review of methods to monitor collisions or microavoidance of birds with offshore wind turbines. Report for The Crown Estate. Strategic Ornithological Support Services Project SOSS-03a. Retrieved 21 May 2018 from https://www.bto.org/sites/default/files/u28/downloads/Projects/Final_Report_SOSS03A_Part1.pdf

Cook, A.S.C.P., Johnston, A., Wright, L.J. & Burton, N.H.K. (2012) A review of flight heights and avoidance rates of birds in relation to offshore wind farms. Report to The Crown Estate. Strategic Ornithological Support Services Project SOSS-02. Retrieved 21 May 2018 from https://www.bto.org/sites/default/files/u28/downloads/Projects/Final_Report_SOSS02_BTOReview.pdf.

Cook, A.S.C.P., Humphreys, E.M., Masden, E.A. & Burton, N.H.K. (2014) The avoidance rates of collision between birds and offshore turbines. BTO Research Report 656. Thetford: British Trust for Ornithology. Retrieved 21 May 2018 from http://www.gov.scot/Resource/0046/00464979.pdf

Cook, A.C.S.P., Humphreys, E.M., Bennet, F., Masden, E.A. & Burton, N.H.K. (2018) Quantifying avian avoidance of offshore wind trubines: Current evidence and key knowledge gaps. *Marine Environmental Research* https://doi.org/10.1016/j.marenvres.2018.06.017.

Corman, A.-M. (2015) Flight and foraging patterns of lesser black-backed gulls and northern gannets in the southern North Sea. Dissertation. Christian-Albrechts University, Kiel. Retrieved 21 May 2018 from https://macau.uni-kiel.de/servlets/MCRFileNodeServlet/dissertation_derivate_00006292/diss_corman_seabirds_north_sea.pdf

Corman, A.-M & Garthe, S. (2014) What flight heights tell us about foraging and potential conflicts with wind farms: a case study in Lesser Black-backed Gulls (*Larus fuscus*). *Journal of Ornithology* 155: 1037–1043.

Cost Reduction Monitoring Framework (CRMF) (2017) Cost Reduction Monitoring Framework 2016. Summary report on behalf of the Offshore Wind Programme Board & Offshore Renewable Energy Catapult. Retrieved 21 May 2018 from http://crmfreport.com/wp-content/uploads/2017/01/crmf-report-2016.pdf

Croxall, J., Butchart, S., Lascelles, B., Stattersfield, A., Sullivan, B., Symes, A. & Taylor, P. (2012) Seabird conservation status, threats and priority actions: a global assessment. *Bird Conservation International* 22 (1): 1–34. doi: 10.1017/S0959270912000020.

de Lucas, M. & Perrow, M.R. (2017) Birds: collision. In Perrow, M.R. (ed.) *Wildlife and Wind Farms, Conflicts and Solutions. Volume 1. Onshore: Potential effects.* Exeter: Pelagic Publishing. pp. 155–190.

de Lucas, M., Janss, G.F.E., Whitfield, D.P. & Ferrer, M. (2008) Collision fatality of raptors in wind farms does not depend on raptor abundance. *Journal of Applied Ecology* 45: 1695–1703.

Department of Energy & Climate Change (DECC) (2012) Record of the Appropriate Assessment Undertaken for Applications Under Section 36 of the Electricity Act 1989. Projects: Docking Shoal Offshore Wind Farm (as amended); Race Bank Offshore Wind Farm (as amended) and Dudgeon Offshore Wind Farm (December 2011, updated June 2012). Retrieved 21 May 2018 from https://www.og.decc.gov.uk/EIP/pages/projects/AAGreaterWash.pdf

Desholm, M. & Kahlert, J. (2005) Avian collision risk at an offshore wind farm. *Biology Letters* 1: 296–298.

Desholm, M., Fox, A.D., Beasley, P. & Kahlert, J. (2006) Remote techniques for counting and estimating the number of bird–wind turbine collisions at sea: a review. *Ibis* 148: 76–89.

Dierschke, V. & Garthe, S. (2006) Literature review of offshore wind farms with regard to seabirds. In Zucco, C., Wende, W., Merck, T., Kochling, I. & Köppel, J. (eds) *Ecological Research on Offshore Wind Farms – International exchange of experience.* Part B. Bonn: Federal Agency for Nature Conservation. pp. 193–198.

Dierschke, V., Furness, R.W. & Garthe, S. (2016) Seabirds and offshore wind farms in European waters: avoidance and attraction. *Biological Conservation* 202: 59–68.

DNV GL (2016) Standard DNVGL-ST 0437. Load and site conditions for wind turbines. DNV GL AS. November 2016. Retrieved 21 May 2018 from https://rules.dnvgl.com/docs/pdf/DNVGL/ST/2016-11/DNVGL-ST-0437.pdf

DONG Energy; Vattenfall AB; Danish Energy Authority; Danish Forest and Nature Agency (2006). Danish Offshore Wind: Key Environmental Issues. Report by Danish Energy Agency, Danish Nature Agency, DONG Energy, and Vattenfall. pp 144. Retrieved 22 May 2018 from https://tethys.pnnl.gov/sites/default/files/publications/Danish_Offshore_Wind_Key_Environmental_Issues.pdf

Everaert, J. (2008) Effecten van windturbines op de fauna in Vlaanderen. Onderzoeks resultaten, discussie en aanbevelingen. Brussels: Research Institute for Nature and Forest. Report No. INBO.R.2008.44. Retrieved 21 May 2018 from https://www.vlaanderen.be/nl/publicaties/detail/effecten-van-windturbines-op-de-fauna-in-vlaanderen-onderzoeksresultaten-discussie-en-aanbevelingen-effects-of-wind-turbines-on

Everaert, J. (2014) Collision risk and micro-avoidance rates of birds with wind turbines in Flanders. *Bird Study* 61: 220–230.

Everaert, J. & Kuijken, E. (2007) Wind turbines and birds in Flanders (Belgium). Preliminary summary of the mortality research results. Brussels: Research Institute for Nature and Forest. Retrieved 21 May 2018 from http://www.wattenrat.de/files/everaert_kuijken_2007_preliminary.pdf.

Everaert, J. & Stienen, E.W.M. (2007) Impact of wind turbines on birds in Zeebrugge (Belgium): significant effect on breeding tern colony due to collisions. *Biodiversity and Conservation* 16: 3345–3359.

Ferrer, M., de Lucas, M., Janss, G.F.E., Casado, E., Muñoz, A.R., Bechard, M.J. & Calabuig, C.P. (2012) Weak relationship between risk assessment studies and recorded mortality in wind farms. *Journal of Applied Ecology* 49: 38–46.

Fijn, R., Krijgsveld, K., Poot, M.J.M. & Dirksen, S. (2015) Bird movements at rotor height measured continuously with vertical radar at a Dutch offshore wind farm. *Ibis* 157: 558–566.

Furness, R.W., Wade, H.M. & Masden, E.A. (2013) Assessing vulnerability of marine bird populations to offshore wind farms. *Journal of Environmental Management* 119: 56–66.

Garthe, S. & Hüppop, O. (2004) Scaling possible adverse effects of marine wind farms on seabirds: developing and applying a vulnerability index. *Journal of Applied Ecology* 41: 724–734.

Garthe, S., Montevecchi, W.A. & Davoren, G.K (2011) Inter-annual changes in prey fields trigger different foraging tactics in a large marine predator. *Limnology & Oceanography* 56: 802–812.

Garthe, S., Markones, N. & Corman, A.-M. (2017) Possible impacts of offshore wind farms on seabirds: a pilot study in Northern Gannets in the southern North Sea. *Journal of Ornithology* 158: 345–349.

Gibb, R., Shoji, A., Fayet, A.l., Perrins, C.M., Guilford, T. & Freeman, R. (2017) Remotely sensed wind speed predicts soaring behaviour in a wide-ranging pelagic seabird. *Journal of the Royal Society Interface* 14 (132): pii: 20170262. doi: 10.1098/rsif.2017.0262.

Gressetvold, M. (2013) Identifying conservative criteria for (environmental) assessments of vulnerability: seabirds and offshore wind farms as a case study. Unpublished MSc thesis, Norwegian University of Science and Technology, Trondheim, Norway. Retrieved 17 April 2015 from http://www.diva-portal.org/smash/get/diva2:659652/FULLTEXT01.pdf.

Global Wind Energy Council (GWEC) (2018) Global Wind Report – Annual Market Update 2017. Retrieved 21 May 2018 from http://gwec.net/publications/global-wind-report-2/

Harwood, A.J.P., Perrow, M.R., Berridge, R.J., Tomlinson, M.L. & Skeate, E.R. (2017) Unforeseen responses of a breeding seabird to the construction of an offshore wind farm. In Köppel, J. (ed.) *Wind Energy and Wildlife Interactions.* Cham: Springer. pp. 19–41.

Hill, R., Hill, K., Aumüller, R., Schulz, A., Dittman, T., Kulemeyer, C. & Coppack, T. (2014) Of birds, blades and barriers: detecting and analysing mass migration events at *alpha ventus*. In BSH & BMU (eds) *Ecological Research at the Offshore Windfarm Alpha Ventus: Challenges, results and perspectives*. Weisbaden: Springer Spektrum. pp. 111–131.

Hodos, W. (2003) Minimization of motion smear: reducing avian collisions with wind turbines. Report No. NREL/SR 500-33249, August 2003. Golden, CO: National Renewable Energy Laboratory. Retrieved 21 May 2018 from https://www.nrel.gov/docs/fy03osti/33249.pdf.

Hull, C.L., Stark, E.M., Peruzzo, S. & Sims, C.C. (2013) Avian collisions at two wind farms in Tasmania, Australia: taxonomic and ecological characteristics of colliders versus non-colliders. *New Zealand Journal of Zoology* 40: 47–62.

Hüppop, O., Dierschke, J., Exo, K.-M., Fredrich, E. & Hill, R. (2006) Bird migration studies and potential collision risk with offshore wind turbines. *Ibis* 148: 90–109.

Hüppop, O., Hüppop, K., Dierschke, J. & Hill, R. (2016) Bird collisions at an offshore platform in the North Sea. *Bird Study* 63: 73–82.

Huso, M., Dalthorp, D. & Korner-Nievergelt, F. (2017) Statistical principles of post-construction fatality monitoring. In Perrow, M.R. (ed.) *Wildlife and Wind Farms, Conflicts and Solutions. Volume 2. Onshore: Monitoring and mitigation*. Exeter: Pelagic Publishing. pp. 84-102.

Igual, J.M., Tavecchia, G., Jenouvrier, S., Forero, M.G. & Oro, D. (2009) Buying years to extinction: is compensatory mitigation for marine bycatch a sufficient conservation measure for long-lived seabirds? *PLoS ONE* 4 (3): e4826. doi: 10.1371/journal.pone.0004826.

Joint Nature Conservation Committee (JNCC) (2016) Seabird population trends and causes of change: 1986–2015 report. Joint Nature Conservation Committee. Updated September 2016. Retrieved 4 October 2017 from http://jncc.defra.gov.uk/page-3201

Joint Nature Conservation Committee (JNCC), Natural England (NE), Northern Ireland Environment Agency (NIEA), Natural Resource Wales (NRW) & Scottish Natural Heritage (SNH) (2014) Joint response from the statutory nature conservation bodies to the Marine Scotland Science Avoidance Rate Review. 25 November 2014. Retrieved 21 May 2018 from https://www.nature.scot/sites/default/files/2018-02/SNCB%20Position%20Note%20on%20avoidance%20rates%20for%20use%20in%20collision%20risk%20modelling.pdf

Johnston, A., Cook, A.S.C.P., Wright, J.W., Humphreys, E.M. & Burton, N.H.K. (2014) Modelling flight heights of marine birds to more accurately assess collision risk with offshore wind turbines. *Journal of Applied Ecology* 51: 31–41.

Jongbloed, R.H. (2016) Flight heights of seabirds: a literature study. Institute for Marine Resources and Ecosystem Studies (IMARES), Wageningen UR. Report C024/16. Retrieved 21 May 2018 from http://edepot.wur.nl/378293

Kerlinger, P., Gehring, J.L., Erickson, W.P., Curry, R., Jain, A. & Guarnaccia, J. (2010) Night migrant fatalities and obstruction lighting at wind turbines in North America. *Wilson Journal of Ornithology* 122: 744–754.

Köppel, J. (ed.) (2017) *Wind Energy and Wildlife Interactions: Presentations from the CWW2015 Conference*. Cham: Springer.

Krijgsveld, K.L. (2014). *Avoidance behaviour of birds around offshore wind farms. Overview of knowledge including effects of configuration*. Bureau Waardenburg bv. Retrieved 23 May 2018 from https://tethys.pnnl.gov/sites/default/files/publications/Krijgsveld-2014.pdf

Krijgsveld, K.L., Akershoek, K., Schenk, F., Dijk, F. & Dirksen, S. (2009) Collision risk of birds with modern large wind turbines. *Ardea* 97: 357–366.

Krijgsveld, K.L., Fijn, R.C., Japink, M., van Horssen, P.W., Heunks, C., Collier, M.P., Poot, M.J.M., Beukers, D. & Dirksen, S. (2011) Effect studies Offshore Wind Farm Egmond aan Zee. Flux, flight altitude and behaviour of flying birds. Bureau Waardenburg Report 10–219. Culemborg: Bureau Waardenburg. Retrieved 21 May 2018 from https://tethys.pnnl.gov/publications/effect-studies-offshore-wind-farm-egmond-aan-zee-final-report-fluxes-flight-altitudes.

Leopold, M.F.L. & Dijkman, E.M. (2010) Offshore wind farms and seabirds in the Dutch sector of the North Sea: what are the best and worst locations for future developments? Institute for Marine Resources and Ecosystem Studies (IMARES), Wageningen UR. Report C134/10. Retrieved 21 May 2018 from https://www.researchgate.net/publication/283339247_Offshore_wind_farms_and_seabirds_in_the_Dutch_Sector_of_the_North_Sea

Leopold, M.F., Boonman, M., Collier, M.P., Davaasuren, N., Fijn, R.C., Gyimesi, A., de Jong, J., Jongbloed, R.H., Jonge Poerink, B., Kleyheeg-Hartman, J.C., Krijgsveld, K.L., Lagerveld, S.,

Lensink, R., Poot, M.J.M., van der Wal, J.T. & Scholl, M. (2014) A first approach to deal with cumulative effects on birds and bats of offshore wind farms and other human activities in the southern North Sea. Institute for Marine Resources and Ecosystem Studies (IMARES), Wageningen UR. Report C166/14. Retrieved 21 May 2018 from https://tethys.pnnl.gov/institution/imares-wageningen-ur

MacArthur Green, British Trust for Ornithology, NIRAS & Royal Haskoning DHV (2013) Review of avoidance rates in seabirds at offshore wind farms and applicability of use in the Band collision risk model. Report for SMart-Wind & Forewind. Retrieved 21 May 2018 from https://infrastructure.planninginspectorate.gov.uk/wp-content/ipc/uploads/projects/EN010053/EN010053-001025-Appendix%20Z_Review%20of%20Avoidance%20Rates%20-%20December%202013.pdf

Maclean, I.M.D., Inger, R., Benson, D., Booth, C.G., Embling, C.B., Grecian, W.J., Heymans, J.J. Plummer, K.E., Shackshaft, M., Sparling, C., Wilson, B., Wright, L.J., Bradbury, G., Christen, N., Godley, B.J., Jackson, A., McCluskie, A., Nichols-Lee, R. & Bearhop, S. (2014) Resolving issues with environmental impact assessment of marine renewable energy installations *Frontiers in Marine Science* 75 (1): 1–5. doi: 10.3389/fmars.2014.00075

Marine Management Organisation (MMO) (2014) Review of environmental data associated with post-construction monitoring of licence conditions of offshore wind farms. MMO Project No. 1031. April 2014. Retrieved 21 May 2018 from https://assets.publishing.service.gov.uk/government/uploads/system/uploads/attachment_data/file/317787/1031.pdf

Marques, A.T., Batalha, H., Rodrigues, S., Costa, H., Pereira, M.J.R., Fonseca, C., Mascarenhas, M. & Bernardino, J. (2014) Understanding bird collisions at wind farms: an updated review on the causes and possible mitigation strategies. *Biological Conservation* 179: 40–52.

Martin, G. (2011) Understanding bird collisions with man-made objects: a sensory ecology approach. *Ibis* 153: 239–254.

Masden, E.A., McCluskie, A., Owen, E. & Langston, R.H.W. (2015) Renewable energy developments in an uncertain world: the case of offshore wind and birds in the UK. *Marine Policy* 51: 169–172.

May, R.F. (2015) A unifying framework for the underlying mechanisms of avian avoidance of wind turbines. *Biological Conservation* 190: 179–187.

May, R. (2017) Mitigation for birds. In Perrow, M.R. (ed.) *Wildlife and Wind Farms, Conflicts and Solutions. Volume 2. Onshore: Monitoring and mitigation.* Exeter: Pelagic Publishing. pp. 124–144.

Meek, E.R., Ribbands, J.B., Christer, W.G., Davy, P.R. & Higginson, I. (1993) The effects of aerogenerators on the moorland bird populations in the Orkney Islands, Scotland. *Bird Study* 40: 140–143.

Mendel, B., Kotzerka, J., Sommerfeld, J., Schwemmer, H., Sonntag, N. & Garthe, S. (2014) Effects of the alpha ventus offshore test site on distribution patterns, behaviour and flight heights of seabirds. In BSH & BMU (eds) *Ecological Research at the Offshore Windfarm Alpha Ventus – Challenges, results and perspectives.* Weisbaden: Springer Spektrum. pp. 95–110.

Musters, C.J.M., Noordervliet, M.A.W. & Keurs, W.J.T. (1996) Bird casualties caused by a wind energy project in an estuary. *Bird Study* 43: 124–127.

Newton, I. & Little, B. (2009) Assessment of wind farm and other bird casualties from carcasses found on a Northumbrian beach over an 11-year period. *Bird Study* 56: 158–167.

Paleczny, M., Hammill, E., Karpouzi, V. & Pauly, D. (2015) Population trend of the world's monitored seabirds, 1950–2010. *PLoS ONE* 10 (6): e0129342. doi: 10.1371/journal.pone.0129342.

Pennycuick, C.J. (1997) Actual and 'optimum' flight speeds: field data reassessed. *Journal of Experimental Biology* 200, 2355-2361.

Percival, S. (2003) Birds and wind farms in Ireland: a review of potential issues and impact assessment. Durham: Ecology Consulting. Retrieved 21 May 2018 from https://tethys.pnnl.gov/sites/default/files/publications/Percival_2003.pdf

Petersen, I.K., Christensen, T.K., Kahlert, J., Desholm, M. & Fox, A.D. (2006) Final results of bird studies at the offshore wind farms at Nysted and Horns Rev, Denmark. NERI report commissioned by DONG Energy and Vattenfall A/S. Denmark: National Environmental Research Institute. Retrieved 21 May 2018 from https://corporate.vattenfall.dk/globalassets/danmark/om_os/horns_rev/horns-rev-nysted-birds.pdf

Pettersson, J. (2005) The impact of offshore wind farms on bird life in Kalmar Sound, Sweden. A final report based on studies 1999–2003. Report prepared for the Swedish Energy Agency. Lund: Lunds Universitet. Retrieved 21 May 2018 from https://tethys.pnnl.gov/publications/impact-offshore-wind-farms-bird-life-southern-kalmar-sound-sweden

Platteeuw, M., Bakker, J., van den Bosch, I., Erkman, A., Graafland, M., Lubbe, S. & Warnas, M. (2017) A framework for assessing ecological and cumulative effects (FAECE) of offshore wind farms on birds, bats and marine mammals in the southern North Sea. In Koppel, J. (ed.) *Wind Energy and Wildlife Interactions: Presentations from the CWW2015 Conference.* Cham: Springer. pp. 219–237.

Poot, H., Ens, B.J., de Vries, H., Donners, M.A.H., Wernand, M.R. & Marquenie, J.M. (2008) Green light for nocturnally migrating birds. *Ecology and Society* 13 (2): 47. http://www.ecologyandsociety. org/vol13/iss2/art47/

Poot, M.J.M., van Horssen, P.W., Collier, M.P., Lensink, R. & Dirksen, S. (2011) Effect studies Offshore Wind Egmond Aan Zee: cumulative effects on seabirds, a modelling approach to estimate effects on populations levels on seabirds. Report commissioned by Noordzeewind. Retrieved 21 May from from https:// www.buwa.nl/fileadmin/buwa_upload/ Bureau_Waardenburg_rapporten/06-466_BW_ research_OWEZ_cumulative_effects-web.pdf

Reid, T., Hindell, M., Lavers, J.L. & Wilcox, C. (2013) Re-examining mortality sources and population trends in a declining seabird: using Bayesian methods to incorporate existing information and new data. *PLoS ONE* 8 (4): e58230. doi: 10.1371/ journal.pone.0058230.

Ronconi, R.A., Allard, K.A. & Taylor, P.D. (2015) Bird interactions with offshore oil and gas platforms: review of impacts and monitoring techniques. *Journal of Environmental Management* 147: 34–45.

Ross-Smith, V.H., Thaxter, C.B., Masden, E., Shamoun-Baranes, J., Burton, N.H.J., Wright, L.J., Rehfisch, M.M. & Johnston, A. (2016) Modelling flight heights of lesser black-backed gulls and great skuas from GPS: a Bayesian approach. *Journal of Applied Ecology* 53: 1676–1685.

Scott, B., Langton, R., Philpott, E. & Waggitt, J.J. (2015) Seabirds and marine renewables: are we asking the right questions? In Shields, M.A. & Payne, A.I.L. (eds) *Marine Renewable Energy Technology and Environmental Interactions. Humanity and the sea.* Dordrecht: Springer. pp. 81–92.

Skov, H., Heinänen, S., Norman, T., Ward, R., Méndez-Roldán, S. & Ellis, I. (2018) ORJIP bird collision and avoidance study. Final report – April 2018. London: Carbon Trust. Retrieved 20 June 2018 from https://www.carbontrust.com/ media/675793/orjip-bird-collision-avoidances- tudy_april-2018.pdf

Skov, H., Leonhard, S.B., Heinänen, S., Zydelis, R., Jensen, N.E., Durinck, J., Johansen, T.W., Jensen, B.P., Hansen, B.L., Piper, W. & Grøn, P.N. (2012) Horns Rev 2 monitoring 2010–2012. Migrating birds. Orbicon, DHI, Marine Observers and Biola. Report commissioned by DONG Energy. Retrieved 11 December 2017 from https://tethys. pnnl.gov/sites/default/files/publications/Horns_ Rev_2_Migrating_Birds_Monitoring_2012.pdf.

Spear, L.B. & Ainley, D.G. (1997) Flight speed of seabirds in relation to wind speed and direction. *Ibis* 139: 234–251.

Stienen, E.W.M., Courtens, W., Everaert, J. & Van de Valle, M. (2008) Sex-biased mortality of common terns in wind farm collisions. *The Condor* 110: 154–157.

Stumpf, J.P., Denis, N., Hamer, T.E., Johnson, G. & Verschuyl, J. (2011) Flight height distribution and collision risk of the marbled murrelet *Brachyramphus marmoratus*: methodology and preliminary results. *Marine Ornithology* 39: 123–128.

Thaxter, C.B., Ross-Smith, V.H., Bouten, W., Clark, N.A., Conway, G.J., Rehfisch, M.M. & Burton, N.H.K. (2015) Seabird–wind farm interactions during the breeding season vary within and between years: A case study of lesser black-backed gull *Larus fuscus* in the UK. *Biological Conservation* 186: 347–358.

Tuck, G.N., Polacheck, T., Croxall, J.P. & Weimerskirch, H. (2001) Modelling the impact of fishery by-catches on albatross populations. *Journal of Applied Ecology* 38: 1192–1196.

Vanermen, N. & Stienen, E.W.M. (2009) Seabirds & offshore wind farms: monitoring results 2008. In Degraer, S. & Brabant, R. (eds) *Offshore Wind Farms in the Belgian Part of the North Sea: State of the art after two years of environmental monitoring.* Report by Ghent University, Institute for Agricultural and Fisheries Research (ILVO), Management Unit of the North Sea Mathematical Models (MUMM), Research Institute for Nature and Forest (INBO), and Royal Belgian Institute of Natural Sciences (RBINS). pp. 151–221. Retrieved 21 May 2018 from https://tethys.pnnl.gov/publications/ offshore-wind-farms-belgian-part-north-sea- state-art-after-two-years-environmental.

Vanermen, N., Onkelinx, T., Courtens, W., van de Walle, M., Verstraete, H. & Stienen, E.W.M. (2015) Seabird avoidance and attraction at an offshore wind farm in the Belgian part of the North Sea. *Hydrobiologia* 756: 51–61.

Vanermen, N., Courtens, W., van der Walle, M., Verstraete, H. & Stienen, E.W.M. (2017) Large gull behaviour in an offshore wind farm. In

Abstracts of CWW Conference, Lisbon, 2017. Retrieved 21 May 2018 from http://cww2017. pt/images/Congresso/conference-materials/ cww-final-programme_05-09-2017.pdf

Wade, H.M., Masden, E.A., Jackson, A.C. & Furness, R.W. (2016) Incorporating data uncertainty when estimating potential vulnerability of Scottish seabirds to marine renewable energy developments. *Marine Policy* 70: 108–113.

Wildfowl & Wetlands Trust (WWT) & MacArthur Green (2014) Migratory species collision risk modelling assessments. Report for Marine Scotland. Slimbridge: Wildfowl & Wetlands Trust.

Retrieved 4 October 2017 from http://www.gov. scot/Resource/0044/00443170.pdf

Winiarski, K.J., Miller, D.L., Paton, P.W.C. & McWilliams, S.R. (2014) A spatial conservation prioritization approach for protecting marine birds given proposed offshore wind energy development. *Biological Conservation* 169: 79–88.

Zipkin, E.F., Kinlan, B.P., Sussman, A., Rypkema, D., Wimer, M. & O'Connell, A.F. (2015) Statistical guidelines for assessing marine avian hotspots and coldspots: a case study on wind energy development in the U.S. Atlantic Ocean. *Biological Conservation* 191: 216–223.

CHAPTER 10

A synthesis of effects and impacts

MARTIN R. PERROW

Summary

This chapter aims to present a unifying synthesis of the effects and impacts of offshore wind farms upon the different trophic levels of marine life, from seabed communities and fish to marine mammals and seabirds, as well as migratory birds and birds passing over the sea, derived in the previous chapters in this volume. Most information originates from north-western European waters, the current focus of offshore wind. The installation of turbines and scour protection affects local coastal processes and ocean dynamics, with potential for the wind-wake effect to generate upwelling over considerable areas, thereby affecting biological productivity. During construction, noisy pile driving may displace sensitive fish and marine mammals and some seabirds. Return may be rapid, although the fish prey base of a breeding seabird was affected in one case. After construction, rapid colonisation of hard surfaces by species such as Blue Mussels occurs. The 'reef effect', with aggregation of fish including large predators, thereby attracting marine mammals and some seabirds, is generally seen as positive. The displacement of seabirds from operational wind farms is species specific, with divers and Northern Gannet the most sensitive, and has implications for individual fitness. Collision of birds such as gulls is rarely recorded, although sampling such events is technically difficult. Further evidence is required to address concern regarding possible collision of migratory birds and bats. Cumulative effects have barely begun to be addressed and little targeted research has yet been conducted to detect population-scale impacts. In essence, wind farms are likely to induce a number of top–down or bottom–up cascading effects through the food web via diverse mechanisms, which can only be understood through integrated ecosystem-based research. Until knowledge gaps are filled, uncertainty should be embraced by avoiding possible conflict through careful site selection to achieve the ultimate goal of a 'win–win' for wind energy and wildlife.

Introduction

Offshore wind is of increasing importance for countries with a coastline on our fundamentally blue planet, mainly covered in sea, which offers a huge resource of potential renewable energy. Recent technological advances, especially with regard to larger turbines, ease of installation and the advent of floating wind, coupled with the reduction of costs, mean that offshore wind "has come of age … and is now a mainstream source of low-carbon electricity, at least in Europe" (Jameson *et al.*, Chapter 1). Thus, offshore wind is an increasingly important component in the strategy to curb climate change, which is widely perceived to be a key threat to many ecosystems and the wildlife that they support (IPCC 2014).

At the time of writing, it has been 27 years since the installation of the first offshore wind farm (OWF), the 11-turbine 4.95 MW Vindeby wind farm in the Danish part of the Baltic Sea, in 1991. The growth of offshore wind was slow at first, with a few small sites in Denmark and Sweden followed by the Netherlands, and then the first sites in Round 1 of development in the UK during the early years of the current millennium (Jameson *et al.*, Chapter 1). Sufficient time has passed that some early generation sites, including Vindeby, have now been decommissioned (Offshore Wind 2017). Onshore, there was a lag in understanding the effects of wind farms on wildlife until Orloff and Flannery (1992) described the potential scale of impacts upon birds of prey at Altamont Pass, California. Perhaps learning from these experiences, early intensive studies at OWFs focused on migratory waterbirds and seabirds (Desholm & Kahlert 2005; Pettersson 2005; Petersen *et al.* 2006). In the UK, early studies included fundamental research into the potential effect of wind farms on the distribution of Common Scoter *Melanitta nigra*, governed through their habitat preferences and bivalve mollusc prey resource (Kaiser *et al.* 2006), and the interactions of breeding Little Tern *Sternula albifrons* and their prey with Scroby Sands wind farm (Perrow *et al.* 2006; 2011a). Scroby Sands also saw the first impact study on seals in the UK (Skeate *et al.* 2012), mirroring earlier studies in Denmark (Teilmann *et al.* 2006). In European waters, the potential for construction noise to have far-reaching impacts upon cetaceans then became the focus of research (Tougaard *et al.* 2009; Thompson *et al.* 2010; Scheidat *et al.* 2011).

Technical challenges shape the nature of the data gathered, and the offshore wind industry has been obliged to be at the 'cutting edge' in order to develop appropriate and meaningful survey, assessment and monitoring methods. Digital (camera) aerial (aeroplane) surveys for birds (and marine mammals) have replaced visual surveys (Buckland *et al.* 2012; Webb & Nehls, Volume 4, Chapter 3), and sophisticated techniques such as radar, thermal devices, video and integrated systems have been implemented to monitor collisions passively (Desholm *et al.* 2006; BMU & BSH 2014; Skov *et al.* 2018; Molis *et al.*, Volume 4, Chapter 5). In theory, this removes the need for intense human effort under challenging conditions, as illustrated by Pettersson (2005), who lived in the Utgrunden lighthouse to monitor the response of migrating waterfowl. Nonetheless, the data produced by remote techniques still generally demand human verification and analysis.

There is a certain irony that extensive survey programmes, especially for birds, expanded what was often poor knowledge of the abundance and distribution of birds likely to be present in areas proposed for development. For example, surveys across Liverpool Bay in the west of the UK crystallised the importance of this area for wintering Common Scoter and Red-throated Diver *Gavia stellata* (Anonymous 2009), ultimately leading to the establishment of the internationally important Liverpool Bay Special Protection Area (SPA)

(JNCC 2017) and the initial relocation of the Shell Flat site, which was ultimately rejected over concerns over interference with defence radar and low-flying aviation (4C Offshore 2018). Similarly, surveys for the London Array led to the realisation of the importance of the Thames estuary for wintering Red-throated Divers (Goodship *et al.* 2015). While the first phase of the London Array was built (until recently the largest operational OWF in the world), its expansion was cancelled owing to uncertainty over whether monitoring would reveal a negative effect upon divers within the SPA (Harvey 2014). This followed the earlier rejection of Docking Shoal over projected cumulative effects on the North Norfolk Coast SPA population of Sandwich Tern *Thalasseus sandvicensis*, again following an extensive survey programme (DECC 2012).

Site-based Environmental Impact Assessment (EIA) and especially Habitat Regulations Assessment (HRA), where European protected sites were thought likely to be affected by development, have often relied on outputs from predictive assessment tools such as modelling of noise, collision risk and population trends (DECC 2012; Thompson *et al.* 2013; Cook & Masden, Volume 4, Chapter 6). From the developers' perspective, multi-billion pound investment has thus often hinged on uncertain predictions, the need to demonstrate a lack of reasonable scientific doubt in relation to any conclusions reached and fitting within the precautionary principle. This last factor, in particular, is difficult to define and the level of precaution adopted may vary according to the perceived sensitivity of the receptor species and the context of the site.

Any uncertainty and relative lack of information has not constrained development, however, as after the slow start described above, development has been rapid in recent years. By the end of 2017, a total of 4,149 turbines had been connected to the grid in 92 OWFs in the North and Baltic Seas (Wind Europe 2018). Clearly, many national regulators in north-western Europe have been sure that consented sites do not represent a significant risk to any wildlife either in isolation or in a cumulative capacity. The fact that very few wind farms have been rejected as a result of potential impacts upon wildlife, invariably on birds, could be testament to good planning in the first place (debated by Köppel *et al.*, Volume 4, Chapter 9) or could mask a more fundamental lack of understanding of the actual effects of OWFs. In this context, Green *et al.* (2016) argued that scientific knowledge of the likely population effects of relatively well-studied seabirds is still lacking, whereas others, such as Cleasby *et al.* (2015), question the accuracy of the underpinning data and how this may underestimate impacts. In addition, Scott *et al.* (2013, 2014) consider whether researchers are even asking the right questions in relation to the potential effects/impacts on seabirds.

So, fundamentally, have we accumulated enough knowledge to avoid significant impacts and achieve the ultimate objective of a 'win–win' for wind power and wildlife (Kiesecker *et al.* 2011)? Knowledge lies at the core of achieving that goal, the expansion of which has been the fundamental aim of this book series. Within that wider context, this chapter aims to describe and assess the current state of the science in relation to the effects and impacts of offshore wind upon wildlife.

Scope

This final chapter in Volume 3 attempts to synthesise the findings of the preceding chapters that describe effects and impacts (Box 10.1) in order to provide a platform for Volume 4, which describes the monitoring tools available and outlines the means of mitigating the most likely impacts. This synthesis fundamentally relies on the material presented and conclusions reached by the expert authors of the chapters; although some attempt

Box 10.1 Definitions of effects and impacts

In this book series as a whole, the distinction between effects and impacts is as defined in the *Oxford English Dictionary*, with an impact being a 'marked effect or influence'. As such, an effect becomes an impact once it leads to a definable change in an important parameter or measure of the species or group concerned, such as a significant decline or increase in numbers or density, or where a population parameter such as productivity is significantly affected

is also made to explore the ramifications of important points as well as the nature of any effects and impacts described. The latter includes both where effects and impacts have been shown and therefore realised , but also where they have the potential to occur from a consideration of ecological principles. Moreover, whereas the authors have generally utilised peer-reviewed contributions "as the gold standard for veracity" (Lovich & Ennen 2017), this chapter also uses supporting information from the 'grey' literature in reports or websites. It is important to state that this chapter is a personal perspective and while I have faithfully tried to represent the findings of the respective authors, my interpretation of the information they provide may differ, and readers are encouraged to also refer to the original work.

The intended scope of all chapters in this volume was global, but information was invariably heavily biased towards experiences in the North and Baltic Seas in north-west Europe as the current epicentre of the industry. The intention was to learn from these experiences so they could be applied, or where necessary the standard of research and monitoring could be improved, within those rapidly emerging markets in Asia, including China, Japan and Taiwan; in the USA; and in southern Europe, including France, Portugal and Spain; among others. The different species and marine communities in other parts of the world, especially those in warm waters with different trophic dynamics and structure and function, may demand a different focus and approach for investigation of effects and impacts.

To help structure the *Themes* of this chapter, the information provided in each chapter on the six 'groups' considered in this volume, namely seabed communities, fish, marine mammals, bats, migratory non-seabirds and seabirds, was summarised according to the evidence for effects and impacts (Table 10.1). As in the equivalent chapter in the onshore Volume 1 (Perrow 2017), the *Themes* classified the main effects across faunal groups under the following headings: *Disturbance and displacement, Colonisation and attraction*, and *Collision with turbines*. In addition, it was important to consider *Food-web and indirect effects* under a specific heading, because these invariably occur to a greater or lesser extent as offshore wind farms alter physical processes and ocean dynamics, and as communities and ecological functioning change following the installation of new hard substrates available for colonisation. The evidence for *In-combination and cumulative effects*, as the sum of the effects from different factors and wind farms, and *Population-scale impacts*, as the sum of effects upon individuals from all factors and sites within the range of a species' population, was then summarised.

Finally, every attempt was made to identify key knowledge gaps in order to make recommendations on the best targets of future research efforts in the *Concluding remarks*.

Table 10.1 Summary information on how the different faunal groups have been treated and studied in relation to wind farms, and the main effects and impacts that have been revealed.

Group (Chapter)	Treatment and level of study	Main effects	Impacts	Population-scale impacts
Seabed communities (Dannheim et al., Chapter 4)	Typically considered in EIA, and often monitored	Positive reef effect through provision of new hard 'bed to surface' substrate Benefit to natural soft-sediment communities with any restriction of commercial fishing activity Unsubstantiated risk of colonisation of alien species	Change in natural structure and function of communities, with possible trophic consequences	Increase in population size and range of colonising and associated species
Fish (Gill & Wilhelmsson, Chapter 5)	Typically considered in EIA, but functionally important groups are rarely monitored and poorly studied, with a very few exceptions	Noisy construction (especially pile driving) causing displacement and potential injury or mortality Positive reef effect for small 'hard-substrate' species and large predatory species Possible exclusion of pelagic species even into operation	Change in natural structure and function of communities, with possible trophic consequences for seabirds and marine mammals	Reduced population of a noise-sensitive species in one case Increase in population size and range of small hard-substrate species Potential for increases in large predatory species not yet demonstrated
Marine mammals (Nehls et al., Chapter 6)	Invariably considered in EIA and well studied in some cases	Noisy construction (especially pile driving) causing displacement and potential injury, particularly hearing loss Positive reef effect for seals	Increase in available habitat for some species, perhaps especially seals, and potential losses for others, including some cetaceans, although studies are limited	Possible negative (cetaceans) and positive (seals) impacts, especially in a cumulative context, but not yet demonstrated
Bats (Hüppop et al., Chapter 7)	Rarely, if ever, considered and poorly studied	Potential collision risk, especially of migratory species	Unknown	Unknown, but concern over cumulative effects in particular

Table 10.1 – *continued*

Group (Chapter)	Treatment and level of study	Main effects	Impacts	Population-scale impacts
Migratory non-seabirds (Hüppop *et al.*, Chapter 7)	Often considered in EIA but only monitored in a few cases	Potential collision risk, especially for passerines and raptors	Possible, especially at a local scale, but not demonstrated	Unknown, but concern over cumulative effects in particular
Seabirds: displacement (Vanermen & Stienen, Chapter 8)	Virtually always considered and often studied through comparative surveys	Displacement of some species leading to habitat loss, especially in passage and wintering areas Attraction of a few species Little knowledge of effects on breeding species	Possible and even likely in some cases, especially for species of conservation concern	Potential exhibited by modelling studies in some cases
Seabirds: collision (King, Chapter 9)	Virtually always considered in EIA but rarely monitored, with a few notable exceptions	Collision events only very rarely recorded	Possible, perhaps especially in the case of breeding species, but not demonstrated	Unknown, but concern over cumulative effects in particular

EIA: Environmental Impact Assessment.

Themes

Disturbance and displacement

The small footprint of turbines, coupled with their wide spacing (generally at least 600 m to 1 km or more), means that the amount of original habitat lost within a wind farm is generally very small, so that the potential for the original fauna present to be routinely displaced is low, even accepting changes in geomorphological conditions (Rees & Judd, Chapter 2) and ocean dynamics (Broström *et al.*, Chapter 3). For the more mobile and larger fauna in particular, low levels of disturbance may not lead to any detectable changes in their distribution or abundance. For birds, low levels of disturbance may be expressed as changes in behaviour such as increased vigilance, short redistributing flights or dives of birds from the sea surface, or subtle deviations of birds in flight. At higher levels of disturbance, a change in temporal or spatial use may be exhibited by some species, thereby leading to avoidance and ultimately displacement from otherwise suitable habitat within or close to a wind farm.

Throughout the life of a wind farm, vessels provide a source of disturbance for seabirds and marine mammals. The prospect of disturbance, through both visual and audible cues, is likely to be at its zenith during construction, when many vessels of different types and sizes may be permanently present and active, albeit often within clusters of activity (see Jameson *et al.*, Chapter 1). Even when the wind farm has been built, maintenance vessels generally make daily trips to the site.

However, no studies have managed to separate the effects of vessels from other activities such as noisy construction or the presence of the operational turbines, although vessels in isolation are known to be disruptive for some birds such as divers (*Gavia* spp.) and seaduck such as scoters (*Melanitta* spp.), which often flush several hundred metres away. Similarly, the general impression is that Harbour Porpoise *Phocoena phocoena*, unlike seals and bow-riding dolphins, tend to avoid vessels, and in the case of the Dutch wind farm Egmond aan Zee, the activity of Harbour Porpoise increased significantly inside the operational site relative to baseline conditions, which Scheidat *et al.* (2011) suggested was consistent with a refuge effect from nearby shipping lanes. Prey aggregation around turbine bases may also have been a factor, however (see *Colonisation and attraction*, below).

For marine mammals, construction noise, particularly in relation to pile driving, is the principal cause of disturbance and displacement (Nehls *et al.*, Chapter 6). Brandt *et al.* (2018) provide a neat summary of events for the first seven wind farms installed in the German Bight between 2010 and 2013. Here, declines in the number of detections of Harbour Porpoise were observed during piling exceeding 143 dB re 1 μPa^2 s (see Nehls *et al.*, Chapter 6 for an explanation of terminology) at up to 14 km distant, even using noise mitigation systems (17 km away without such systems). Without noise mitigation, declines of up to 50% were recorded at 10–15 km, with this reduced to an average of 17% where noise mitigation was applied. In all cases, however, porpoise detections declined at up to 2 km or so from the construction site for several hours before piling commenced, suggesting that the commotion of vessels and human activity was also a disturbing factor in the near field. At this distance, detections were reduced for 1–2 days after piling. Farther away, an effect was noted only during piling.

It is of note that while noise mitigation systems are routinely applied in Germany, they are not in the UK, where ramp-up procedures are used to encourage animals to disperse away from the construction area. Relatively few sites have also included specific monitoring of the response of marine mammals (MMO 2014). Results from Sheringham Shoal may

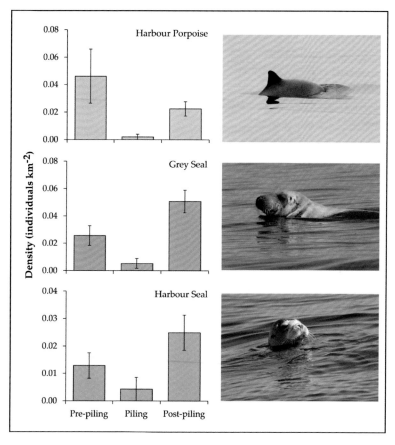

Figure 10.1 Trends in the relative abundance (mean±1 SE) of Harbour Porpoise *Phocoena phocoena*, Grey Seal *Halichoerus grypus* and Harbour Seal *Phoca vitulina* in boat-based surveys before (*n*=40), during (*n*=15) and after (*n*=66) piling activity at Sheringham Shoal, UK. Note the decline in abundance of all species during piling relative to values before piling, with increases following the completion of piling. (Martin Perrow)

be indicative of a general pattern, however. Here, boat-based surveys were conducted over 9 years, with 3 years of data before, during and after construction. This enabled a comparison of relative densities of animals incorporating distance-sampling corrections where possible, in the wind farm and 0–2 km buffer. Only data from the summer months, when most of the construction was undertaken and survey intensity was the highest, were included. Although the comparison was crude in that surveys were not tied to specific short-duration piling events, trends were still apparent as Harbour Porpoises declined during the general piling period relative to pre-piling (Figure 10.1). Partial recovery occurred thereafter, in the sense that numbers of sightings were still lower, albeit not significantly so, during the operational period compared to pre-construction. This is in line with Teilmann and Carstensen (2012), who noted that Harbour Porpoise populations had not fully recovered some 10 years after the construction of the Nysted wind farm in Denmark. Some change in the resource base in and around the wind farm for Harbour Porpoises is implied, perhaps linked to changes in the fish community according to the reef effect (see *Colonisation and attraction*, below) that do not favour the species preferred by porpoises.

The general response of seals to piling events, as exemplified by that observed at Sheringham Shoal (Figure 10.1), is characterised by short-term aversion, rapid recovery and the subsequent increase in the use of operational sites, which seem to provide enhanced prey resources (see *Colonisation and attraction*, below). However, the response of seals to piling activity is complicated by the fact that they have different sensitivities according to whether they are in and below water, or above water when hauled on sandbanks and rocks. In more detail, in their study of 24 tagged Harbour Seals *Phoca vitulina* in the Greater Wash, UK, Hastie *et al.* (2015) showed that the closest individual seals came to active pile driving of the Lincs site varied between 4.7 and 40.5 km. In a separate spatial analysis, Russell *et al.* (2016) also confirmed significant displacement of seals up to 25 km from the centre of the wind farm during piling. However, the recovery time of return to the impacted area was only 2 hours. Thus, the gaps in piling of a few hours or days at Lincs seemingly allowed unhindered travel and foraging from haul-outs at least 20 km away. Nevertheless, the auditory modelling of Hastie *et al.* (2015) predicted sound exposure levels resulting in a high chance of auditory damage, with all seals predicted to potentially suffer temporary threshold shifts and 50% to gain permanent threshold shifts on a number of occasions (see Nehls *et al.*, Chapter 6 for an explanation of terms), with potential to influence individual fitness and the ability to function normally. There is also evidence of species-specific differences between seal species from Scroby Sands in the UK. Here, Grey Seals *Halichoerus grypus* using a haul-out less than 2 km from the wind farm in the UK showed a year-on-year increase both during and after construction, in line with the increase in the wider area, in contrast to the only partial recovery of Harbour Seals after construction was completed (Skeate *et al.* 2012; Nehls *et al.*, Chapter 6).

In contrast to marine mammals, the response of fish to pile driving has been poorly studied and the impacts generally remain unquantified (Gill & Wilhelmsson, Chapter 5). In general, fish that possess organs to detect sound pressure are able to hear sounds in the range of 10–500 Hz (Slabbekoorn *et al.* 2010) and appear to be affected by construction noise. Some species, such as clupeids, with an anatomical link between their swim bladders and hearing apparatus, are particularly sensitive (Thomsen *et al.* 2006). Others, such as sandeels, without a swimbladder, appear to be less sensitive. The typical behavioural response may be simply to swim away from the noise source, but if this does not occur rapidly enough, then in extreme cases high exposure levels can lead to death as a result of the rupture of the swimbladder and other organs or internal bleeding. The author has personal experience of attending wind farms sites during or immediately after piling activity and recording large aggregations of birds, particularly gulls, Northern Gannet *Morus bassanus* and Sandwich Tern, seemingly attracted by injured or disorientated fish. The frequency of such events and the potential for impact upon populations remains unquantified, although it is clear from the events at Scroby Sands in relation to a local spawning population of Atlantic Herring *Clupea harengus* (see *Food-web and indirect effects*, below) that any effects should not always be assumed to be negligible. The results of current studies of Atlantic Cod *Gadus morhua* in Belgian waters according to construction activity (see van der Knaap & Reubens, Volume 4, Chapter 1) are eagerly awaited, although further work, especially on the small schooling species that comprise the prey base for many seabirds and marine mammals, is also badly needed.

In relation to seabirds, little consideration has generally been given to the effects of noise. However, a study of breeding African Penguins *Spheniscus demersus* showed that these birds avoided preferred foraging grounds during seismic activity (Pichegru *et al.* 2017). It therefore seems likely that noisy pile driving may have some adverse effect on the distribution and abundance of swimming and diving species that forage underwater, such as auks, divers and cormorants, where these are exposed.

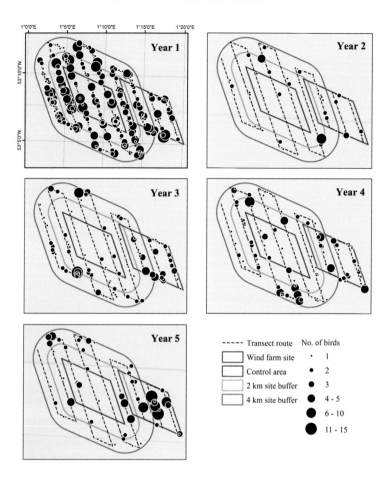

Figure 10.2 Records of Razorbill *Alca torda* within the wind-farm site, the control area and the 2 km and 4 km buffers during the five monitoring 'years' (February to January) from February 2009 to January 2014 inclusive. Construction was undertaken in 2010, 2011 and 2012 (years 2–4).

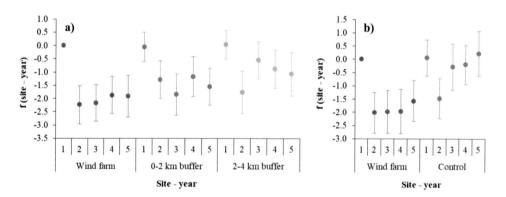

Figure 10.3 Generalised additive mixed model predictions with associated 95% confidence intervals of Razorbill *Alca torda* abundance in relation to (a) the site:buffers (before–after gradient) comparison and (b) the site:control (before–after control-impact) comparison. The monitoring years in both cases are: 1, 2009/10; 2, 2010/11; 3, 2011/12; 4, 2012/13; and 5, 2013/14. Construction was undertaken in years 2–4.

Figure 10.4 A number of studies have shown that Common Guillemots *Uria aalge* are displaced from wind farms at least in the non-breeding season, when they are dispersed throughout the North and Baltic Seas. (Johan Buckens)

Overall, there has been rather little research on the impacts of construction upon seabirds, partly as several key studies in European waters, such as those by Lindeboom *et al.* (2011), Leopold *et al.* (2013) and Vanermen *et al.* (2015a; 2015b), were unable to monitor during the construction phase. In contrast, there has been no restriction on monitoring during construction at some sites in the UK. At Sheringham Shoal, displacement of passing and wintering Razorbill *Alca torda* began during the initial construction phase (Figure 10.2 and Figure 10.3), affecting not only the site, but also the nearby control and buffer (to 4 km) areas (Figure 10.3). Thereafter, while there was no obvious recovery within the site, including into the operational period, numbers of birds recovered both in the control area and in the buffer areas, especially at 2–4 km from the site (Figure 10.3). The pattern at Sheringham Shoal mirrors the pre-construction to post-construction (operational) displacement of auks including Common Guillemot *Uria aalge* at several sites reviewed by Vanermen & Stienen in Chapter 8 (Figure 10.4). At Robin Rigg, however, numbers of Common Guillemot remained similar across all construction phases (Vallejo *et al.* 2017). The difference may be that here, birds were also present during the breeding season as a result of the proximity of breeding colonies. The fact that breeding birds are tied to central-place foraging restricts their options for foraging sites, from which they may not be readily displaced. Similarly, the breeding status of Sandwich Terns (Harwood *et al.* 2017) and Lesser Black-backed Gulls (Thaxter *et al.* 2018) may have contributed to their lack of strong or consistent avoidance of the wind farms routinely encountered on foraging trips from their colonies; although these aerial foraging species tend to show an indifferent or inconsistent response to wind farms.

The consent for wind farms within foraging range of a variety of breeding seabirds at large colonies such as Bempton Cliffs within the Flamborough Head and Filey Coast SPA, with connectivity to the Hornsea sites (Langston & Teuten 2018), and the four wind farms in the Firth of Forth, with connectivity to a number of colonies including those within the Forth Islands and Fowlsheugh SPAs (Cleasby *et al.* 2015), provides an invaluable opportunity to test the link between reproductive status and displacement for species such as Northern Gannet. Outside the breeding season, this species strongly avoids operational wind farms (see Vanermen & Stienen, Chapter 8), but if it is not displaced during the breeding season, it is likely to be vulnerable to collision, especially as an aerial foraging species that encroaches into the risk zone when foraging compared to commuting (Cleasby *et al*. 2015) (see *Collision with turbines*, below).

That at least some species are displaced from operational wind farms begs the question of what causes the negative response. Compared to construction, operational noise levels are much reduced, as is vessel traffic, thereby suggesting that the visual disturbance of the turbines is the dominant factor. However, it may not simply be the presence of spinning blades that is disruptive, as some studies suggest that the standing structures are as equally scary as the structures with moving parts, as was the case for Common Eider *Somateria mollissima* in the study by Larsen and Guillemette (2007). Similarly, Harwood *et al*. (2017) provide some evidence of the apparent avoidance of standing structures by Sandwich Tern, all the more surprising for an agile, fast-flying seabird.

In general, however, studies have not routinely distinguished between possible sources of disturbance from the turbines and thus the wind farm as a whole, including noise, shadow effect or movement of the blades. In other cases, the patterns observed may be caused indirectly through changes in the distribution and abundance of prey, rather than linked directly to the activity itself (see *Food-web and indirect effects*, below). Teasing apart causal and proximate factors remains a particular challenge bearing in mind that even with rigorous and intensive monitoring coupled with increasingly sophisticated analytical techniques, the nature of the zero-inflated and overdispersed data prevalent in seabird surveys makes it difficult to detect changes in abundance and distribution in the first place, particularly in relation to discrete short-term events. Good experimental design is crucial (Vanermen & Stienen, Chapter 8) and there is an increasing trend towards before–after gradient (BAG) designs rather than before–after control-impact (BACI) studies, largely

Figure 10.5 Appearances may be deceptive. Of these migrating flocks of Pink-footed Geese *Anser brachyrhynchus* at Lynn and Inner Dowsing in the Greater Wash, UK, radar studies by Plonczkier and Simms (2012) showed that few (5.54% of *n*=292 flocks predicted to enter the arrays) were actually at risk; that is, within the sites at the 'wrong' height. (Martin Perrow)

owing to the difficulties of establishing a true control. A combination of both approaches may be beneficial (Figure 10.2 and Figure 10.3).

Where the flightlines of birds undertaking regular commuting or foraging movements or more generally during dispersal or migration are disrupted, the resultant displacement is typically termed a barrier effect. There have been surprisingly few studies of barrier effects, perhaps simply as tracks of birds are required to demonstrate them, which, in turn, requires the use of telemetry (Vanermen & Stienen, Chapter 8, Box 8.3; Thaxter & Perrow, Volume 4, Chapter 4), visual tracking (Perrow *et al.* 2011b), video, radar or a combined sensor system (Skov *et al.* 2018; Molis *et al.*, Volume 4, Chapter 5). For breeding Sandwich Terns at Sheringham Shoal, the proportion of birds on a consistent flight path actually entering the site dropped from a baseline of 95% to 82.5% during initial construction and 65.1% during turbine installation (Harwood *et al.* 2017), highlighting both the scope for changes in the strength of any barrier effect according to the nature of works undertaken and potential individual variation (as a variable proportion of birds entered or deviated around the site) that may depend on a number of factors, including sex, reproductive status and season (Thaxter *et al.* 2015) (see *Collision with turbines*, below). Desholm & Kahlert (2005) documented the response of migrant waterfowl, mainly Common Eider with some geese, at Nysted wind farm. Here, the number of flocks entering the wind farm decreased by a factor of 4.5 from pre-construction to initial operation. The plot of tracks diverting around the wind farm, with only a low percentage flying down the rows of the wind farm, remains one of the most enduring images of wind-farm studies to date. Similarly, of the 292 flocks of migrating Pink-footed Geese *Anser brachyrhynchus* (Plonczkier & Simms 2012) predicted to enter the Lynn and Inner Dowsing wind farms in the Greater Wash, UK, 167 (57%) actually did so, with only 9 (9.68%) of the 93 flocks for which flight height information was recorded, flying at rotor-swept height while within the array (Figure 10.5). It is important to note that all of the studies described above benefited from studies of the movements of birds before the sites were built.

Avoidance distance is a well-developed concept and is routinely measured onshore (Hötker 2017). A key factor determining avoidance distance for birds temporarily resident in an area when breeding, on stopover or wintering is likely to be resource availability (Perrow 2017). Where resources are limited and there is no suitable habitat nearby, birds may be forced to tolerate various forms of human disturbance, including potentially any caused by wind farms regardless or whether of not this affects survival or reproductive success (Gill *et al.* 2001). In contrast, where resources are available elsewhere, birds may readily avoid disturbance and be displaced, perhaps without significant costs to fitness. A good example of different responses by the same population of birds according to the available resources is the strong avoidance of wind farms by wintering Barnacle Geese *Branta leucopsis* where resources are plentiful in Germany, compared to greater tolerance of wind farms on staging grounds in Sweden, where food resources are more limited (Percival 2005). There does not yet appear to be a good example of this differential response for offshore sites.

The fact that some birds show greater tolerance of wind farms in some locations is the basis of habituation, a phenomenon defined as a form of learning by an individual that ceases to respond to a stimulus after repeated exposure (Hötker 2017). Habituation can only be conclusively demonstrated where birds can be individually recognised, as in the current crop of Global Positioning System (GPS) telemetry studies (Thaxter *et al.* 2017; 2018), although it may be strongly implied for long-lived species, with what appear to be the same individuals using the same area. Increasing use of an area over time could also illustrate a cultural change in the population, driven, perhaps, by the progeny of birds that showed more tolerance. So far, the best example of possible habituation appears to be that of Petersen & Fox (2007), where up to several thousand Common Scoters began using the Horns Rev 1 wind farm 4 years after its construction. It was previously thought that Common Scoters were likely to be displaced from wind farms. It is possible that this was linked to an increasing food resource of bivalve molluscs (see *Colonisation and attraction*, below). This may also explain why other seaduck such as Common Eider, with a similar diet, show an inconsistent or indifferent response to operational wind farms (Vanermen & Stienen, Chapter 8).

Avoidance distance is an important concept, as where this is greater than half of the distance between turbines, 100% displacement would result and the abundance of birds within the wind farm should decline to zero. Equally, it may be possible to increase the porosity of a site to seabirds, especially those commuting to and from foraging grounds and colonies, by increasing turbine spacing to more than double avoidance distance. This could provide useful mitigation of any barrier effects (Harwood & Perrow, Volume 4, Chapter 8). Up until now, however, modelling has tended to conclude that barrier effects do not have energetic consequences and do not impact individual fitness. For example, Masden *et al.* (2009) calculated that small deviations around sites were inexpensive in energetic terms for a migratory species. In relation to breeding seabirds undertaking regular commuting flights, Masden *et al.* (2010) suggested that energetic costs were more important for those species with relatively high energetic costs of flight resulting from high wing loading, but that the costs of extra flight appeared much lower than those imposed by low food abundance or adverse weather conditions. Empirical studies of real situations, preferably with known individuals undertaking breeding attempts, are needed.

Colonisation and attraction

In principle, a proportion of the original habitat within a wind farm may be permanently lost as a result of the footprint of the turbines themselves, and that of any scour protection, which is typically much larger than the turbines. The area lost is generally thought to be a small proportion, at less than 1% (Petersen & Malm 2006).

Figure 10.6 Mobile fauna such as crabs benefit from the resources provided by colonising species, such as these anemones, favoured by the hard substrata of the wind-farm structures. (Roland Krone)

Figure 10.7 Long-spined Sea Scorpion *Tauralus bubalis*, an obligate hard-bottom dwelling small fish, which did not occur away from the turbine foundations at the *Alpha Ventus* wind-farm site in the North Sea.

Moreover, the area of new substrate gained in the form of the turbine tower and associated rocky scour protection may be 2.5 times greater than the area lost, and may provide 0.5–1.2 ha additional hard substrate per square kilometre of wind-farm area according to the type of foundations (monopile or gravity-base) installed (Dannheim *et al.*, Chapter 4).

The fact that the majority of wind farms in north-western Europe have been built using monopiles (80%, according to EWEA 2016) indicates that most wind farms have been installed over soft sediments and thus the hard surfaces introduced are new habitats available for colonisation by the veritable soup of larvae and propagules carried on the prevailing currents. Colonisation is rapid. For example, at the Alpha Ventus test site in German waters, after only a few months an average of five taxa were present on the turbines in each 0.12 m² area at depths of 1–10 m sampled by scientific diving. Within 2 years, this had trebled to around 15 taxa per sampled area (Gutow *et al.* 2014). However, Dannheim *et al.* (Chapter 4) make the point that the sessile macrofaunal community typically colonising the foundations seems impoverished relative to that expected on natural hard substrata in similar oceanographic conditions, with dominance of a few species such as the tube-building amphipod *Jassa herdmani*, anemones (Figure 10.6), hydroids *Tubularia* spp. and Blue Mussel *Mytilus edulis*, as well as a range of mobile species such as shrimps, hermit crabs and small fish (Figure 10.7).

Perhaps more notable than the list of species present, these 'biofouling', typically filter-feeding assemblages rapidly attain impressive biomass. At Alpha Ventus, the average biomass increased from around 1 kg/m² less than a year after foundation installation to around 25 kg/m² around 18 months later. This was, however, at 1 m below the surface, and at depths of 5–10 m the biomass appeared to quickly stabilise at 1 kg/m² or so.

Modelling by Adams *et al.* (2014) reinforced the notion that wind farms could operate as ecological stepping stones for hard-substratum species with pelagic larvae, thereby allowing range expansion including of 'climate migrants' responding to climate change. On a positive note, such species may include vulnerable native species that are currently range restricted, but with the downside that some invasive species may also be favoured, including those without pelagic larvae that may be deposited through the use of attendant vessels that may be inadvertently carrying such species, including in ballast water. Any concern about the spread of invasive species does not appear to have been realised as yet, although this is reliant on suitable, continued monitoring (Dannheim *et al.* Chapter 4; Dahlgren *et al.*, Volume 4, Chapter 1).

The natural soft-sediment communities, dominated by benthic deposit feeders such as polychaetes, bristleworms and sea urchins, that were originally present at the wind-farm site may be influenced on a local scale as a result of localised scour (Rees & Judd, Chapter 2), providing sinks of organic material with new microhabitats, and by the 'rain' of organic material from the dense epifauna on the turbine foundations (Dannheim *et al.*, Chapter 4). Otherwise, the number of taxa, total abundance and biomass may vary little between reference areas outside the wind farm and areas within the wind farm (Gutow *et al.* 2014). Dannheim *et al.* (Chapter 4) also make the important point that sediment communities within the wind farm may flourish and develop where commercial fishing, particularly the destructive attention of trawls, is restricted, typically as a result of health and safety concerns about towed fishing gear snagging cables and other structures. As a result, the regeneration of reef-building Ross Worms *Sabellaria spinulosa* and attendant species has been noted at a number of sites in the UK. Thus, wind farms may provide important refuges for otherwise vulnerable habitats and species.

The refuge principle extends to mobile demersal fauna such as large crabs and fish. At Alpha Ventus, commercially important Edible Crab *Cancer pagurus* reached densities several orders of magnitude higher on wind-farm foundations than on the surrounding sediments (Gutow *et al.* 2014). While this may provide a refuge from exploitation in most localities, at Sheringham Shoal, local pot fishers are allowed into the wind farm to take crabs and lobsters *Homarus gammarus*. Although the general restriction of commercial fishing may favour the aggregation of larger predatory fish, rod-and-line anglers may be attracted to wind farms in the same way as they are attracted to wrecks and rigs, as a

Figure 10.8 Great Cormorant *Phalacrocorax carbo* perched within the Thornton Bank wind farm in Belgian waters. Wind farms appear to have allowed Great Cormorants to expand their range farther offshore. (Hilbran Verstraete)

result of the prospect of large specimens of valuable sport species such as Atlantic Cod and European Bass *Dicentrarchus labrax*. Reubens *et al.* (Chapter 5, Box 5.1) showed the importance of wind farms as habitats for Atlantic Cod, with tagged individuals being highly sedentary and persisting throughout the summer and autumn foraging period, before moving away in winter, potentially to spawn. Unfortunately, at this time some individuals were 'recovered' by recreational fishermen and lost to the study, although others returned to the wind farm the following spring.

The presence of large fish and crustaceans within wind farms is attractive to apex predators such as marine mammals (Nehls *et al.*, Chapter 6). The example from Russell *et al.* (2014) of tracked Harbour Seals routinely attending Sheringham Shoal and systematically moving from foundation to foundation, consistent with foraging, is particularly compelling. In this part of the North Sea, Harbour Seals are known to take a range of prey, from small demersal fish to crabs (Hall *et al.* 1998). The latter are known to be abundant from the viable pot fishery that includes the wind farm (see previous paragraph), whereas the former are also likely to be concentrated in the wind farm, as described in other studies (Gill & Wilhelmsson, Chapter 5). Observations made during the boat-based monitoring programme that both Harbour and Grey Seals had increased within the wind-farm area after its construction (see Figure 10.1) reinforce the view from the telemetry study by Russell *et al.* (2014).

Box 10.2   The use of navigation buoys around the perimeter of an offshore wind farm by Sandwich Terns

Richard J. Berridge, Andrew J.P. Harwood and Martin R. Perrow

The visual tracking method established by Perrow *et al.* (2011b) was used to investigate the response of breeding Sandwich Tern *Thalasseus sandvicensis* to the construction and operation of Sheringham Shoal offshore wind farm (Greater Wash in Eastern England) from April to July from 2009 to 2015 inclusive. Sheringham Shoal is 19.5 km at its closest point from the Blakeney Point colony within the North Norfolk Coast Special Protection Area. Visual tracking involves following individual birds in a high-speed rigid-hulled inflatable boat, mirroring the birds' track and recording all behaviours.

Of the 2,705 birds tracked, a moderate proportion (15.7%) interacted with the eight navigation buoys distributed around the perimeter of the wind farm. Interactions included beginning the track from a buoy or the track was interrupted or ended as the bird landed on the buoy for a variable amount of time (Figure 10.9). Observations showed buoys were being used for resting and roosting (perhaps even overnight) and as a platform for social behaviours, including a full range of courtship displays such as posturing, food presentation and passing, as well as copulation. Preening and washing were frequently observed on and around buoys, and ten birds were found roosting on the water near the north-west buoy on one occasion. Food passes were also observed on the sea surface around marker buoys and display flights in the vicinity of buoys were commonplace in the pre-breeding season.

Combining visual tracking and boat-based surveys, Harwood *et al.* (2017) concluded that buoys were being used as staging posts by birds on extended foraging trips farther offshore and that buoys were responsible for the increased use of the

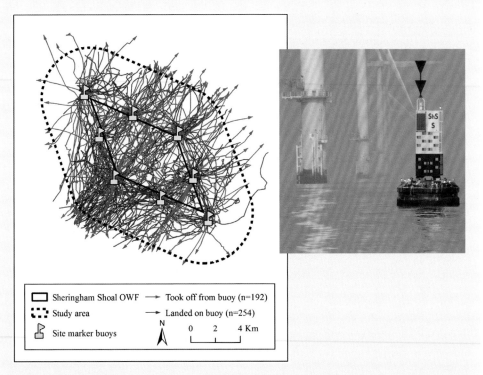

Figure 10.9 Flight lines of visually tracked (Perrow *et al.* 2011b) breeding Sandwich Tern *Thalasseus sandvicensis* interacting with the navigation buoys installed around Sheringham Shoal offshore wind farm (OWF) in eastern England. A total of 15.7% (*n*=424) of the 2,705 individuals tracked were recorded landing on or taking off from buoys, with *n*=22 of these interacting with a buoy more than once in the same track. Insert: Sandwich Terns *in situ* on a buoy. (Richard Berridge)

0–2 km buffer area in which the buoys were situated, with a resultant effect on the use of the wind farm.

In the 2014 breeding season, an effort was made to regularly record Sandwich Tern numbers and activity at each buoy during the tracking period (*n*=73 days). A total of *n*=330 counts were made over the breeding season, incorporating the pre-breeding, incubation, provisioning and post-breeding fledging periods. The south-west buoy was sampled most frequently (*n*=58) (Table 10.2), in accordance with this being the closest buoy to the inception of most tracks of birds from the Blakeney Point colony on a south-west to north-east flightline (Figure 10.9). The east buoy was the most frequently utilised overall, with the highest mean counts in three of the four periods and the highest mean count overall of 17.8 birds observed in the pre-breeding season. The highest single count of terns was 29, on the south-east buoy in mid-July, with ten Harbour Porpoises *Phocoena phocoena* feeding nearby, suggesting that birds were aggregating near a profitable foraging patch. The overall mean count (±1 SE) across the entire period and all buoys, was 5.73±0.37.

Peak mean counts were observed in the pre-breeding period on all buoys (Table 10.2), reflecting the propensity of local breeding birds to range farther at this time, while there is no pressure to incubate eggs or provision chicks. It is also possible that

Table 10.2 Number of Sandwich Terns *Thalasseus sandvicensis* recorded on navigation buoys in different locations around Sheringham Shoal wind farm during the different periods in the 2014 breeding season: pre-breeding in April, incubation in May, provisioning in June and post-breeding/fledging period in July.

Period	Location of buoy								
	North	North-east	East	South-east	South	South-west	West	North-west	Total
Pre-breeding	14.00±2.41 (8)	13.50±5.12 (4)	17.80±2.12 (10)	12.80±1.94 (10)	7.71±2.28 (7)	8.23±2.11 (13)	8.60±2.46 (5)	9.67±3.82 (6)	11.65±0.96 (63)
Incubation	5.33±1.42 (15)	9.80±2.16 (10)	10.26±1.49 (23)	5.40±3.89 (5)	7.29±1.52 (17)	3.00±0.81 (18)	4.29±2.52 (7)	6.00±1.28 (9)	6.76±0.63 (104)
Provisioning	0.91±0.39 (11)	1.92±0.90 (12)	3.00±1.47 (4)	2.75±2.75 (4)	0.69±0.29 (13)	1.13±0.54 (15)	1.14±0.65 (14)	2.60±1.29 (5)	1.42±0.28 (78)
Post-breeding	1.46±0.56 (13)	6.95±1.25 (19)	5.33±2.25 (9)	9.29±3.84 (7)	3.70±1.33 (10)	1.33±0.72 (12)	0.86±0.86 (7)	2.50±1.10 (8)	4.04±0.59 (85)

Data are shown as mean ± 1 SE, with the number of samples in parentheses. Highest to lowest numbers over the different periods for each buoy are shown by graded colours from red to orange, yellow and green.

birds on migration to more northerly colonies in the UK such as the Farne Islands or Coquet Island also use the buoys *en route*. Numbers of terns dropped during the incubation period to the lowest levels in the provisioning period, before rising again once chicks had fledged (Table 10.2). In the fledging and post-breeding period, the north-east and especially the south-east buoy held the greatest numbers of terns (Table 10.2), again possibly because of their proximity to foraging areas.

As well as Sandwich Terns, small numbers of Common Tern *Sterna hirundo* and Common Gull *Larus canus* were observed utilising buoys alongside Sandwich Terns on occasion. Large gulls, especially Great Black-backed Gull, were also regularly recorded, with these displacing terns from the buoys.

In 2015, the buoys were removed, presumably as the needs of navigational safety were met by what had become a well-lit operational site. While a considerable decline in the density of birds utilising the site (unpublished data) was observed, this could not be entirely attributed to the loss of the buoys, as a result of the coincidental failure of the Blakeney Point colony relative to the rise of the Scolt Head colony at much greater distance from Sheringham Shoal (closest point of 33.5 km).

Overall, if navigation buoys are to be used around wind farms, it is wise to consider their effect on the use of the area by local birds, especially if enhanced use of the wind-farm area seems likely to enhance the risk of collision.

For seabirds, the 'stand-out' species attracted to the resources within wind farms is Great Cormorant *Phalacrocorax carbo* (Lindeboom *et al.* 2011; Vanermen & Stienen, Chapter 8). As well as supplying suitable fish prey, wind-farm infrastructure provides suitable perches for cormorants to stand on and extend their wings to fully dry out their waterlogged feathers (Figure 10.8). The presence of wind farms appears to have allowed Great Cormorants to extend their relatively limited foraging range, from a mean maximum (± 1SD) of 25±10 km according to Thaxter *et al.* (2012), further into the offshore realm. It may be that some non-breeding individuals spend much of their time in wind farms, representing a form of colonisation as far as this is possible for a mobile species. Elsewhere, Sandwich Terns regularly associated with the eight navigation buoys installed around the perimeter of the Sheringham Shoal wind farm, which appeared to be responsible for the increased use of the wind farm and the 0–2 km buffer area in which the buoys were situated, compared to the wider area 2–4 km from the wind farm (Harwood *et al.* 2017; Box 10.2).

The only other seabirds attracted to wind farms are gulls, especially Great Black-backed Gull *Larus marinus* (Vanermen & Stienen, Chapter 8). This may be because the reasons of being disturbed and thus displaced (see *Disturbance and displacement*, above) outweigh the possible reasons for attraction, or simply that relatively few suitable prey resources are present. An obvious exception would be sessile fauna such as Blue Mussels and other bivalves that may be attractive to seaduck such as scoter (see *Disturbance and displacement*, above, in relation to habituation). Otherwise, many seabirds feed on small shoaling pelagic species such as sandeels (*Ammodytes* spp.) and clupeids including Atlantic Herring and European Sprat *Sprattus sprattus*. As yet, there have been relatively few studies on these functionally important species. An exception is the study of Harwood *et al.* (2017), who showed a significant reduction in foraging rate of Sandwich Terns within a wind farm compared to outside, implying that the availability of pelagic prey was lower inside. An alternative explanation is that as the birds crossing the site avoided the turbines they may have been focused on transit rather than foraging. Moreover, at Alpha Ventus,

Krägefsky (2014) did not detect significant differences in the abundance of pelagic species within compared to outside the wind farm both by day and at night, although the Atlantic Mackerel *Scomber scombrus* captured within the wind farm had significantly reduced stomach fullness compared to those outside, suggesting lower foraging activity. Sessile fauna attached to turbine foundations were of negligible interest to the pelagic fish studied, which at least provides a reason to remain outside the wind farm even if this was not directly demonstrated in the study. The presence of other larger predatory fish and marine mammals such as seals providing a risk of predation is another reason for pelagic species to possibly avoid wind farms (see *Food-web and indirect effects*, below).

Among migratory non-seabirds, work by Skov *et al.* (2016) showed that migrant raptors crossing the Baltic Sea were significantly attracted to a wind farm during adverse wind conditions, akin to the 'island effect' behind the attraction of landbirds to small islands (or boats on a very small scale), whereby birds seek rest and shelter. Hüppop *et al.* (Chapter 7) also suggest that nocturnally migrating passerines in particular may be attracted to lit wind farms, in the same manner as documented for a variety of illuminated structures around the coast and onshore. But whether migrant bats are as attracted to wind farms as they are on land, for a variety of possible reasons (Barclay *et al.* 2017), remains unclear as yet, largely as a result of a lack of studies. The attraction of any volant species to wind farms carries an increase in collision risk (see *Collision with turbines*, below).

Collision with turbines

Collision with turbines is seen as the most serious effect of wind farms as it almost invariably involves the death of the individual concerned. Birds generally collide with moving blades, although collision with the tower has been recorded (see Box 11.1 in Perrow 2017). Bats suffer both collisions with blades and barotrauma, whereby internal organs and blood vessels are ruptured as a result of the pressure differences associated with rapidly moving blades (Barclay *et al.* 2017). Onshore, collision studies are advanced and aided by the potential for the recovery of collision victims and an understanding of the factors influencing that recovery, which allows estimates of collision rates of both birds and bats to be made with reasonable confidence (Huso *et al.* 2017). This has revealed large variation in collision rates of zero to 125 birds per turbine per year, with this distilled down to an average of around four birds per turbine per year from 24 intensive studies across the world (de Lucas & Perrow 2017). The rate for bats is similar, at two bats per turbine per year in Europe (sample of 40 sites) and 3.7 bats per turbine per year in North America (Barclay *et al.* 2017). The numbers of bats estimated to be killed by turbines globally per year runs into the millions and is concentrated on relatively few species, suggesting that at least some species of bats are particularly vulnerable to population impacts. A similar case has been made for some raptors among birds (Perrow 2017). Part of the vulnerability of bats lies in that they are often attracted to turbines within wind farms for a range of possible reasons (Barclay *et al.* 2017). It is not yet clear whether migratory bats especially are attracted to offshore turbines in a similar way (Hüppop *et al.*, Chapter 7).

Offshore, there is, as yet, no prospect of collecting victims at sea, and empirical evidence for collision has relied on direct observation (for example, Petersson 2005) and an increasingly sophisticated range of remote thermal devices, cameras and radars (Desholm *et al.* 2006; Skov *et al.* 2018; Molis *et al.*, Volume 4, Chapter 5). Such studies are invariably expensive in terms of equipment, the demanding logistics of working offshore and the intensive resources required, particularly in relation to analysis. A lot of effort is required to detect what we already know from onshore to be rare events. For example, at Nysted OWF in Denmark, Petersen *et al.* (2006) calculated that 0.018–0.02% of waterfowl passing by would collide with turbines. With such a low probability of collision it was predicted that the thermal animal detection system employed at one turbine would fail to detect a waterfowl collision, and this was indeed the case even with 2,445 hours of effort. At Alpha Ventus, using a video camera on a single turbine at night aided by strong infrared illuminators and two thermal imaging cameras, Hill *et al.* (2014) recorded hundreds of birds flying through or near the blades over a 3 year period, but no direct collisions.

Thus, in essence, detecting collisions becomes a 'numbers game', although even a 'typical' wind farm may see hundreds of thousands, if not millions of movements of seabirds and other species through the wind farm annually. At Nysted, for example, 235,000 waterfowl passed by the 72-turbine site annually. At the similarly sized 88-turbine Sheringham Shoal in the UK, density data gathered during boat-based surveys before the site was built suggested that 348,716 individuals of five species seen to be at risk of collision would pass through the site area on an annual basis (SCIRA Offshore Energy 2006). The number of passages may be much higher for larger sites at other locations such as migration bottlenecks or near breeding colonies, especially large ones. Of course, birds have to fly through a rotor-swept area to be at risk from collision and the chances of this are low based on the proportion of the airspace this occupies, even without high avoidance rates.

It is, therefore, not surprising that even where intensive studies have been undertaken, there is a dearth of records of actual collisions (King, Chapter 9). At Yttre Stengrund, Pettersson (2005) observed one collision event whereby the tail of a flock of 310 Common Eider snaked into a rotor. Four birds were knocked to the water, but three flew away and only one was (probably) killed. At Nysted, Petersen *et al.* (2006) recorded one collision of a small bird (presumably a passerine). The recent Offshore Renewables Joint Industry Programme (ORJIP) study at Thanet off the Kent coast seemingly holds the record for the highest number of collisions in a single study, with six collision victims from November 2014 to June 2016 (Skov *et al.* 2018). These were all gulls, comprised of one adult Black-legged Kittiwake *Rissa tridactyla*, one Lesser or Great Black-backed Gull, two unidentified large gulls and two unidentified gulls. A further key value of the ORJIP study, particularly for the offshore industry and the estimation of collision risk using collision risk modelling within an EIA, has been the calculation of empirical avoidance rates (with SD) for key species (Northern Gannet, Black-legged Kittiwake, Herring Gull, Great Black-backed Gull and Lesser Black-backed Gull), with values ranging from 0.996±0.011 to 0.999±0.003. These are an improvement on previous estimates of likely avoidance rates for the same species derived mainly from the mortality of birds at onshore wind farms in the case of gulls, including of ecologically dissimilar small gulls as proxy for Black-legged Kittiwake (Cook *et al.* 2018).

However, without the benefit of knowing the movements of birds *before* the site was built, it is difficult to judge whether the presence of birds outside, and thus apparently avoiding the wind farm, is linked directly to the presence of the wind farm itself and not to other factors such as fishing boats, which tend to attract scavenging gulls and may not be allowed within the wind farm (see *Disturbance and displacement*, above, for further discussion of the benefits of *before* and *after*). Moreover, in the ORJIP study (Skov *et al.* 2018), the birds present appeared to be mostly non-breeding individuals that may be more readily displaced depending on resource availability (see *Disturbance and displacement*, above) compared to breeding birds. As the authors recognise, there is also always the limitation of this being a specific site with specific (mainly benign) weather conditions.

Furthermore, even with high avoidance rates, provided there are enough passages, collisions may build up over time. For example, after 7 years, a total of 49 White-tailed

Eagles *Haliaeetus albicilla* had been killed at the 68-turbine Smøla wind farm (see Box 8.4 in de Lucas & Perrow 2017). It is not inconceivable that some similarly *k*-selected seabirds with low reproductive output over long lifespans may be at risk, and this may only be determined by detailed monitoring of key species at specific sites. The choice of which species to concentrate on is often framed within first principles of the type of bird involved and its likely behaviour, especially flight behaviour, vulnerability to additional mortality and level of conservation concern. These considerations spawned the sensitivity index (King, Chapter 9), first with Garthe & Hüppop (2004) and subsequently with Furness *et al.* (2013). The latter, in particular, produced ranked lists of birds susceptible to both collision and displacement separately, producing different arrangements of species. This links with the fact that collision and displacement are more or less mutually exclusive, in that displaced birds cannot collide.

Thus, the evidence to date suggests that, as onshore (Hötker 2017), naturally wary species showing a high degree of aversion, such as waterfowl, are at little risk (see *Disturbance and displacement*, above). In the study by Desholm and Kahlert (2005), less than 1% of the flocks of migrating Common Eider and geese were close enough to the turbines to be at any risk of collision. Raptors represent the other extreme to waterfowl in that they are fearless and often indifferent to the presence of turbines (de Lucas & Perrow 2017), although there is also evidence of displacement in some circumstances, perhaps especially on migration (Johnston *et al.* 2014). At sea, however, migrating raptors may be attracted to wind farms, perhaps as they may appear to offer refuge during the challenges of crossing the open sea (Skov *et al.* 2016). The potential for collision of raptors is further exacerbated by how their eyes work (reviewed in Perrow 2017). For example, scavenging *Gyps* vultures have a small binocular region and large blind areas above, below and behind their heads (Martin *et al.* 2012). A typical head position when foraging would allow scanning of the ground as they search for food and extensive lateral coverage to check out conspecifics in flight, and would shield the eyes from the sun, but with the key disadvantage that the birds would be blind in the direction of travel (Martin *et al.* 2012).

Several piscivorous or rather generalist predatory (and/or kleptoparasitic) seabirds, including Northern Gannet, terns, gulls, and skuas, are the equivalent of raptors at sea, with all of these species spending much time visually scanning the sea surface for foraging opportunities. Martin (2011) noted the potential similarities between terns (specifically Gull-billed Tern *Gelochelidon nilotica*), and eagles which both turn their heads systematically to look both laterally and downwards (forward pitch of the head by 60º) when searching for prey on the water or ground below, which may make them vulnerable to collision with artificial objects such as wind turbines. In general, however, it appears that we know relatively little about the visual systems of aerial foraging birds likely to collide with turbines in comparison to groups such as auks (Martin & Wanless 2015) or procelliiformes (Martin & Prince 2001). Nevertheless, at least Northern Gannet appears to be able to see and perceive the clutter of rotating turbines as a threat and avoids wind farms (Vanermen & Stienen, Chapter 8). Sandwich Terns also show clear awareness of the location of the turbines in order to maximise the distance from them by generally passing down the centre of the rows (Harwood *et al.* 2017). Similarly, although the GPS-tagged Lesser Black-backed Gulls in the study by Thaxter *et al.* (2018) frequently entered wind farms at risk height, the overlap with the three-dimensional rotor-swept volume was significantly lower than expected from a random distribution, thereby indicating meso-scale avoidance.

However, birds will make mistakes. These may be exacerbated by inclement weather conditions, although understanding of the conditions during which collision may occur remains limited (King, Chapter 9). After modelling flight distribution in accordance with weather conditions, Skov *et al.* (2012) suggested that the risk of collision for Northern Gannet and large gulls increased with tail or side winds, with increased risk in winds of intermediate strength for Northern Gannet and low wind speed for gulls, perhaps partly as gulls increased flight altitude with decreasing wind speed. Fog or mist is often seen to be problematic, especially for nocturnal migrating passerines, which may be brought down to turbine height by such conditions (Hill *et al.* 2014), with these becoming disorientated and 'trapped' among lighted turbines (Hüppop *et al.*, Chapter 7). There appears to be less of an issue for waterfowl, as Pettersson (2005) commented that few waterfowl migrated under fog and mist (5–6%), broadly corresponding with the periods during which these conditions occurred (5–9%), and the flocks migrated in the same way as they did under good conditions. Plonczkier & Simms (2012) ran radar for 24 hours a day over a range of weather conditions for 38–46 days in each of 4 years during the autumn migration period. They noted that 85% of the 979 flocks of geese (571 visually confirmed as Pink-footed Geese comprising 39,957 individuals) migrated by day. Of these, only 14 diurnal flights were recorded when visibility was perceived to be poor. Although these were at lower height (100–150 m) than normal (250–300 m) under such conditions and thus at more risk of turbine strike, Plonczkier & Simms (2012) concluded that this represented a negligible increase in the risk to the population as a whole. For nocturnally migrating waterfowl, Desholm & Kahlert (2005) had previously demonstrated that although flocks of migrating Common Eider and geese were more likely to enter the Nysted wind farm at night, they counteracted any risk by increasing the distance to individual turbines, demonstrating that they could see and/or sense the presence of the turbines and react accordingly.

The behaviour of individuals is thought to be a key factor contributing to collision, and Smallwood (2017), mainly in relation to raptors, concluded that collision is often linked to territorial clashes and general distraction by conspecifics and any mobbing individuals. The sex-biased mortality of Common Tern *Sterna hirundo* during the egg-laying and incubation periods at Zeebrugge was thought to be the result of the higher foraging activity of males in order to provision their mates during these periods, thereby exposing them to wind turbines (Stienen *et al.* 2008; Everaert & Stienen, Chapter 9, Box 9.1). Ultimately, Smallwood (2017) suggested that surveys are best focused on behaviour rather than conventional utilisation rate. The latter distilled to the number of passages (flux) through the area of risk is generally the key descriptor of collision risk modelling (Smales 2017). This may partly explain why EIA based on predictions of collision risk modelling may be only weakly correlated with actual collision rate once the site is built (Ferrer *et al.* 2012), although there are other cases where the two are comparable (Smales 2017).

Finally, individual habitat use coupled with individual behaviour may increase the risk of collision (Cleasby *et al.* 2015; Thaxter *et al.* 2015; 2018). In the study by Thaxter *et al.* (2015), the use of proposed, consented and operational wind farms by Lesser Black-backed Gulls varied between years and reproductive state, with little use during incubation and greater use in the early chick provisioning period, and males making significantly more use of wind-farm areas compared to females on top of individual differences. However, judging from the spatial use maps, little use was actually made of operational sites, and it will be interesting to see how this changes as more sites become operational. Moreover, the propensity for risky behaviour within wind farms may increase as the breeding season unfolds and the pressure on adult seabirds to provision chicks increases. Much clearly remains to be learned about the nature of bird collision with turbines. For bats, even more needs to be understood.

Food-web and indirect effects

In the UK at least, indirect or food-web effects are generally not included in EIA, although with some exceptions, notably in the Greater Wash (for example SCIRA Offshore Energy 2006). For birds in particular, this is in keeping with the notion that the principal effects are limited to (1) collision, (2) displacement, and (3) barrier effects (MMO 2014), even though (3) is really a subset of (2). In contrast, the review of wind-farm effects on birds by Gove *et al.* (2013) included 'indirect impacts' as well as 'habitat loss or damage' as effects. Unfortunately, the section on OWFs in this review missed what appears to be the only published study of an indirect effect of a wind farm on a seabird through its fish prey (Perrow *et al.* 2011a). This study was the subject of the apposite comment from Furness *et al.* (2013, p64): "complex and indirect effects such as alteration of fish and benthic invertebrate prey abundance by wind farms is something that is extremely difficult to predict, so caution is needed in interpretations and collecting post-construction data will be important".

However, there is little prospect of detecting effects where monitoring is not conducted in an integrated manner across trophic levels. The review of monitoring in the UK by the Marine Management Organisation (MMO 2014) revealed that up to that point, monitoring had tended to be undertaken on a case-by-case basis and more or less specifically on birds, benthos and fish. For birds, this is in keeping with this group often being the focus of EIA and/or HRA. The additional focus on the benthos is less clear considering that negative effects are generally not predicted in EIA, although a few benthic studies did attempt to link with monitoring of benthic-feeding birds, specifically Common Scoters. In contrast, while fish were often monitored, this was almost invariably linked to benthic or benthopelagic species with the use of small beam trawls. Although some sites were concerned with possible effects on spawning clupeids, sampling of functionally important pelagic forage fish Alpha Ventus was not undertaken and linked to the abundance of seabirds or marine mammals in the manner of Certain *et al.* (2011). At Alpha Ventus, Krägesky (2014)

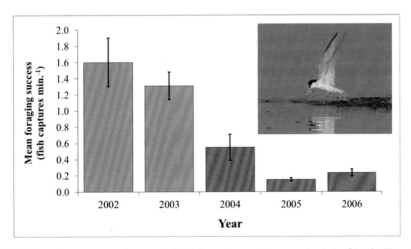

Figure 10.10 Foraging success (mean±1 SE fish capture rate per minute) of Little Tern *Sternula albifrons* observed offshore at North Denes Great Yarmouth Special Protection Area within the breeding season from 2002 to 2006. Construction of the Scroby Sands wind farm was undertaken from November 2003 to February 2004, and thus 2002 and 2003 represent 'before' (green) and 2004–2006 represent 'after' (red) construction. Significant differences are shown by the different colours corresponding exactly to 'before' and 'after'. (Redrawn from Perrow *et al.* 2011a.) Insert: Little Tern capturing a small fish. (Martin Perrow)

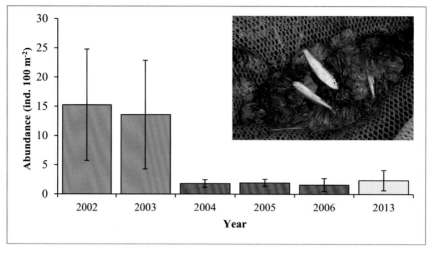

Figure 10.11 Inter-annual variation in the abundance (mean±1 SE individuals per 100 m²)of young clupeids, mainly locally spawned young-of-the-year (YOY) Atlantic Herring *Clupea harengus*, captured by a surface tow net within foraging range of Little Tern *Sternula albifrons* breeding at the North Denes Great Yarmouth Special Protection Area from 2002–2006 and 2013. The net (0.92 × 0.3 m opening) was towed over a distance of 1 km at each of four inshore sites on six to eight occasions in each breeding season from May to August inclusive. Data from 2002 contain a back-calculation for missing early-season samples derived from the seasonal trend before and after 12 June 2003. Construction of the Scroby Sands wind farm was undertaken from November 2003 to February 2004, and thus 2002 and 2003 represent 'before' (green) and 2004–2006 represent 'after' (red) construction. Data from 2013 (yellow) are from follow-up work for Natural England on the impact of the wind farm on Little Terns (Perrow *et al.* unpublished report, 2013). Insert: Catch of young clupeids including translucent YOY among Sea Gooseberries *Pleurobrachia pileus* in the surface trawl. (Martin Perrow)

demonstrated that sampling pelagic fish may be achieved using hydroacoustic methods supported by pelagic trawls. Admittedly, the integrated sampling of seabirds and their prey is not easy, and it may be more efficacious to sample oceanographic variables that account for the abundance of fish and link these to the distribution of seabirds and marine mammals (Scott *et al.* 2010), rather than sample fish themselves. Overall, by the time of the MMO (2014) overview report, few studies had been completed with fully analysed results, with others suffering from issues of survey design and analyses. Thus, few conclusions on the success of monitoring and lessons learned could be reached, thereby providing no impetus to further understand indirect and food-web effects.

Elsewhere in European waters, and specifically at demonstration sites such as Alpha Ventus (BSH & BMU 2014) in Germany and Egmond aan Zee (Lindeboom *et al.* 2011) in the Netherlands, where a range of trophic levels were monitored, it is clear that indirect and food-web effects are highly likely to occur as a result of wind-farm installation. For example, Lindeboom *et al.* (2011) reported that within 2 years, the site was operating as a new type of habitat with a higher biodiversity of benthic organisms, with a possible increased use of the area by benthic invertebrates, fish, marine mammals and some bird species, and a decrease in use by several other bird species.

The case of Little Tern breeding at the Great Yarmouth North Denes SPA (Perrow *et al.* 2011a), then the largest breeding colony in the UK, containing more than 11% of the population, shows that indirect effects may be initiated during construction. Here, significant declines in the prey capture rate of breeding birds (Figure 10.10) coincided perfectly with declines in the abundance of young clupeids, mainly young-of-the-year

Box 10.3 Beneficial effects of the re-emergence of the Scroby Sands sandbank on breeding terns

The Scroby Sands sandbank, around 3 km offshore of Great Yarmouth, Norfolk, in eastern England, is known to have supported important local colonies of breeding Common Tern *Sterna hirundo*, Sandwich Tern *Thalasseus sandvicensis* and Little Tern *Sternula albifrons* at least from the late 1940s to the mid-1960s (Table 10.3). In this period, the bank was mostly emergent at all states of tide, but with some risk of overtopping by high tides, especially in conjunction with strong onshore winds, with disastrous consequences for the breeding birds. In the mid-1960s, it seems that the bank started to shrink and was more frequently overtopped, and apparently remained submerged at anything other than low water after 1976.

Table 10.3 Estimated numbers of pairs and estimated productivity (numbers of fledglings in parentheses) of different tern species nesting on Scroby Sands from 1948 to 2018 inclusive. Brief notes on any important factors determining the success of the colony are provided.

Year	Species			Notes
	Sandwich Tern	Little Tern	Common Tern	
1948	170 (0)	27 max. (?)	368 (0)	
1949	185 (0)		50–360 (?)	Submerged under high tides
1950	450 (?)			Submerged several times during season
1951				
1952	100 (0)			Recorded in Taylor *et al.* (1999)
1953	87 (0)			High tides
1954	16 (0)			High tides and strong north-westerly winds
1955	400 (51)	15 max. (?)		
1956	350 (124)			Predation by Great Black-backed Gulls
1957	327 (?)			
1958	~45 (20)			Strong south-westerly wind and high tide caused losses of chicks
1959	320 (0)			Taken by Great Black-backed Gulls
1960	95 (0)			Abnormally high tides: 149 chicks ringed, but none survived tides and north-westerly gale on 1 July
1961	53 (?)			North-westerly gale caused high tide
1962	110 (119)			All nests washed away, although some young survived
1963	160 (73)			
1964	275 (62)			Human interference: 50 eggs removed
1965	2 (0)			Adverse weather, high tides, shrinking size of Scroby
1971			50–360 (?)	
1972	60 (0)			Submerged under high tides
1975	200 (?)			Submerged several times during season
1976	2 (?)	15 (?)		

Table 10.3 – *continued*

Year	Species			Notes
	Sandwich Tern	Little Tern	Common Tern	
2010		200+ (?)	40+ (0)	Washed out by high tide 18/19 June. 80 pairs Little Tern re-nested by 5 July but outcome unknown
2011		180–200 (80+)	80–100 (?)	
2012		35 (7)	160 (50)	Peter Allard count on 23 July
2013	2 (?)	~70 (~9)	174–200 (100+)	ECON study for Natural England and Peter Allard count on 15 July separately
2014	250 (?)	50+ (?)	100 (?)	Exceptionally strong winds on 8/9 July
2015	40 (?)	~35 (0)	~40 (~50)	High tides and storm surge mid-July
2016	~10 (?)	30 (?)	100–150 (?)	Benign tides and Little Terns at least may have done well
2017	?	100 (0)	200–250 (0)	High tides probably washed out all nests. RSPB data
2018	>3 (?)	>40 (?)	~200 (?)	RSPB data

Data from annual Norfolk Bird and Mammal Reports published by Norfolk and Norwich Naturalists' Society (http://www.nnns.org.uk), Taylor *et al.* (1999), the Royal Society for the Protection of Birds (RSPB) and ECON Ltd. Accurate counts are generally prevented by the inability to land and infrequency of visits.

In the summer of 2004, the main bank was seen to be emergent at all states of tide, coincident with the presence of a subsidiary sandbar within the Scroby Sands wind farm observed on aerial images used to determine seal haul-out patterns (Skeate *et al.* 2012) and by workers undertaking bird surveys, especially tracking of Little Terns (Perrow *et al.* 2006; 2015) (Figure 10.12). The wind farm had been constructed a few months earlier from November 2003 to February 2004.

The construction of the wind farm in a highly dynamic environment produced some of the largest scour pits around turbines yet recorded, at up to 5.8 m (1.38 times monopile diameter) deep and up to 100 m in diameter (Whitehouse *et al.* 2011; Rees & Judd, Chapter 2). Attempts were made to counteract scour with *post hoc* dumping of 900 m³ rock armour per foundation in 2004 and subsequently with the use of tyre-filled nets around five foundations in 2015 (Offshore Wind 2016). While detailed geophysical investigation was undertaken within the wind farm (Rees & Judd, Chapter 2), the far-field effects of the mobilisation of fine materials and patterns of erosion and deposition were not studied. However, the presence of the sandbar in the wind farm (Figure 10.12) is consistent with the mobilisation of material derived from scour and what appears to be the main vector of sediment transport in a southerly direction towards the main Scroby Sands sandbank 2–3 km away.

In the absence of definitive monitoring, the underlying cause of the re-emergence of the main bank for the first time in around 30 years cannot be attributed to the wind farm, particularly as the effect of storms redistributing material is not understood

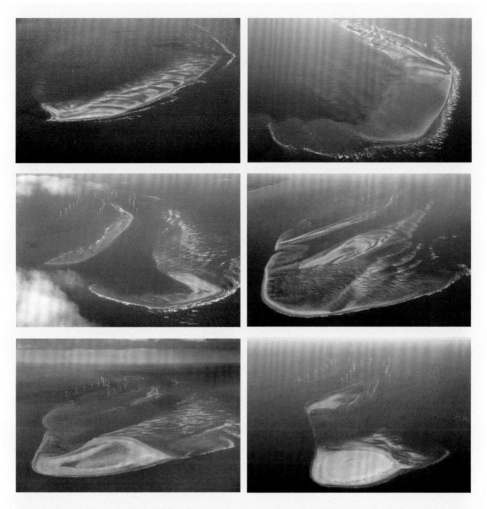

Figure 10.12 Development of Scroby Sands as a time series before wind-farm construction in 1999 (top left) and 2003 (top right), in 2004 immediately following wind-farm construction (middle left) and thereafter in 2007 (middle right), 2012 (bottom left) and 2018 (bottom right). Note that the bank was submerged at high water before 2003, but after 2004 was emergent at all states of the tide, excepting the potential for overtopping on the highest tides and in storm events. A subsidiary sandbar emerged in 2004, which coalesced with the main bank over time. The main bank also appears to have also migrated north, closer to the wind farm. The large increase in relative height of the bank is shown by the dryness of the sand by 2018, which persists even on the highest tides. (Mike Page aerial photography)

and cannot be separated, in what is a highly dynamic environment. Whatever the cause, a time series of aerial photographs shows that the main (and subsidiary) bank has expanded, dried with increasing height and perhaps migrated northward, to the major benefit of breeding terns, which recolonised their former breeding grounds, beginning with Little and Common Terns in 2010 and Sandwich Terns in 2013 (Table 10.3, Figure 10.13). What may have been a pair of Roseate Terns *Sterna dougallii*, the UK's rarest breeding seabird, was observed in 2013 (Figure 10.13), with another seen in 2018 (RSPB, personal communication). Data from the Royal Society for the Protection

Figure 10.13 Part of the Common Tern *Sterna hirundo* colony at Scroby Sands in 2013 that included a possible pair of Roseate Terns *Sterna dougallii* (middle and right in flight). Up to 200 pairs of Little Terns *Sternula albifrons* have bred since 2010, and Sandwich Terns *Thalasseus sandvicensis* recolonised in 2013 after an absence of 37 years. (Martin Perrow)

of Birds (RSPB) now suggest that the Common Terns breeding as a qualifying species at the nearby Breydon Water Special Protection Area are now entirely located on Scroby Sands.

In addition to breeding terns, the main bank supports impressive numbers of terns in the autumn migration period, with 3,694 counted from photographs taken on 31 July 2013 (1,936 Common Terns, 1,616 Sandwich Terns, 141 Little Terns and one Roseate Tern) (Perrow & Eglington 2014). The bank is also a stopover for several species of migrant wading birds and forms part of the habitat for a range of locally breeding and non-breeding gulls and Great Cormorant *Phalacrocorax carbo*; as well as being a haul-out for Grey Seal *Halichoerus grypus* and Harbour Seal *Phoca vitulina* (Skeate *et al*. 2012).

Atlantic Herring (Figure 10.11), linked to pile driving during the construction of the Scroby Sands wind farm, encompassing the entire spawning period of the local herring population. Furthermore, the decline in foraging success was linked to the unprecedented abandonment of eggs in both 2004 and 2005, such that not a single chick hatched in 2004, which could not be linked to other factors such as predators, human disturbance or high tides. Sampling 9 years after the event indicated only slight recovery of the herring stock, and it is of note that the numbers of Little Terns breeding in the immediate area are much reduced, although birds are now present on Scroby Sands themselves (Box 10.3). As Little Tern has the smallest foraging range of any seabird in the UK (Thaxter *et al*. 2012), meaning that it is vulnerable to local effects, it remains possible that the situation at North Denes was exceptional. Nevertheless, given that construction of large wind farms may take several years to complete and some key species of fish are highly sensitive to noise (see *Disturbance and displacement*, above), it should not be assumed that construction impacts on fish would invariably be of short duration and have no effects on seabirds.

Once sites are built, there are a number of potential ecological pathways that may ultimately affect seabirds as well as other apex predators, which stem from changes in physical processes and the physical environment. From the available evidence, Rees & Judd (Chapter 2) concluded that any physical effects may occur predominantly in the near field for tides and sediment transport, although modelling studies suggest that impacts can extend beyond this for waves. The effects of changes in physical processes are likely to be site specific, and the situation observed at Scroby Sands in a highly dynamic environment provides the unconfirmed prospect of possible sediment transport effects over a greater distance, ultimately with benefits for wildlife (Box 10.3). Otherwise, scour around wind-turbine structures, particularly monopiles in sandy sediments, is typically controlled by the use of rock foundations, which contribute to the hard substrate provided by the turbines themselves, influencing habitat diversity and changing the nature of faunal assemblages (Dannheim *et al.*, Chapter 4). Gill & Wilhelmsson (Chapter 5, Figure 5.5,) speculate that a shift in the nature of ecological interactions will occur, with increased benthic relative to pelagic productivity and perhaps benthopelagic coupling expressed as a relative increase in the biomass of demersal relative to pelagic fish. The extent to which any effect extends from the wind farm is unknown, although the migration of fish to other habitats at some times of year (Reubens *et al.*, Chapter 5, Box 5.1) could be argued to represent a considerable extension of the wind-farm effect. In turn, an increased presence of piscivorous fish (Gill & Wilhelmsson, Chapter 5) may directly compete with seabirds for shared prey (Jakobsen 2014) as well as changing the dynamics of that shared prey, including its redistribution as a result of predation risk (Heithaus *et al.* 2008).

At the same time, large wind farms in particular may conceivably cause an increase in planktonic production through the wind-wake effect, whereby changes in Ekman transport and convergence/divergence of water masses cause downwelling and upwelling effects, which theoretically increase nutrient transport and availability to increase primary production (Broström *et al.*, Chapter 3). The wind-wake effect may affect large areas of surrounding sea, perhaps ten times the area of the wind farm. Broström *et al.* (Chapter 3) outline the possible scale of increased production of small 'forage fish' such as clupeids or sandeels within the planktonic pelagic food chain that supports a wide range of species and culminates in many seabirds and marine mammals (Cury *et al.* 2012). But an increase in primary production may conceivably be tempered by the colonisation of wind structures by filtering ecosystem engineers such as Blue Mussels. Modelling work by Slavik *et al.* (2018), using knowledge of the distribution of operational and proposed sites coupled with the likely colonisation patterns of Blue Mussels in the southern North Sea, suggests non-negligible effects on regional annual primary production of up to a few per cent, and larger changes (±10%) in the phytoplankton stock and thus in water clarity during blooms. The importance of these potential pathways cannot be underestimated, but, with no current research on the topic, the scope and scale of the resultant ecological and even ecosystem effects remain unknown.

As well as interactions between predators and prey at a number of trophic levels, competitive interactions between species are expected to change as a result of the new habitats available, the changing resource base and even the different sensitivities to any effects. For example, Skeate *et al.* (2012) suggested that the greater tolerance of Grey Seal to construction disturbance could have led to invasion and rapid occupation of the available haul-out space and local resources by this colonising species, thereby depressing the recovery of the long-standing Harbour Seal population. It is not difficult to imagine or even predict a wealth of shifts in the fortunes of one species relative to another (or others) among invertebrates, fish or birds, with some of these being of functional significance at the ecosystem level. In some cases, the increases in exploitable (edible) species, such as large fish or crabs or lobsters, due to wind farms may also be of socio-economic importance for human populations (see *Colonisation and attraction*, above).

In-combination and cumulative effects

Conceptually, in-combination and cumulative effects across wind farms may magnify the effect of a factor or factors at a local scale. The definition of in-combination effects used here is both the combined effect of different factors within a wind farm, such as the effect of noise added to the effect of the provision of new habitat, and the combined effect of wind farms with other factors operating in the marine environment, such as pollution. Thus, in any EIA or HRA process, as well as the in-combination synergistic and antagonistic effects of all aspects of the same development, the in-combination effects with other marine activities need to be considered.

For sediment communities, Dannheim *et al.* (Chapter 4) record sand and gravel extraction, trawling, oil and gas activities as the focus of in-combination effects among the many human activities and pressures, whereas for marine mammals, Nehls *et al.* (Chapter 6) note incidental by-catch, prey depletion, further sources of anthropogenic noise such as sonar, boat traffic and pollution, as well as other renewable energies such as tidal turbines as being of particular importance. At the same time, Nehls *et al.* (Chapter 6) identify a clear need to better understand the in-combination interaction of wind farms with other different pressures and stressors. For fish, Gill & Wilhelmsson (Chapter 5) further state that the knowledge base is currently inadequate to assess in-combination effects, or indeed the cumulative effects of the installation of multiple wind farms (see later in this section).

The assessment of in-combination effects for birds reveals some of the inherent issues encountered when undertaking assessment. Here, the two principal effects, collision and displacement (but see *Food-web and indirect effects*, above), are often thought to be more or less mutually exclusive (see *Disturbance and displacement*, above). This is unlikely to be completely true, however, as a result of individual variation in response according to a number of factors. Thus, part of the population may be displaced while the remainder is potentially at risk of collision. Any failure to consider in-combination effects can then be linked to the difficulty of combining effects as a result of the different metrics produced. For example, collision results in a loss of individuals from the population, whereas displacement is judged as an effect on a proportion of the population. To allow integration, the effect of displacement needs to be translated into relevant population parameters such as survival or reproductive output. By making some assumptions, particularly in relation to prey distribution and energy intake and provisioning rate, Searle *et al.* (2014) were able to develop models to determine the population consequences of displacement from the effects upon individuals, which were later used in Appropriate Assessment (one stage of the HRA process) of the Forth and Tay wind farms (Marine Scotland 2014). Humphreys *et al.* (2015) further outline the options in integrating the impacts of collision and displacement in an ecologically and statistically appropriate manner for seabirds.

Cumulative effects, herewith described as the combined effect of one wind farm with others, were raised as being of importance in all the taxa-based chapters (see Table 10.1). The more obvious cumulative impacts are associated with the readily definable reef effect of turbines. For example, Slavik *et al.* (2018) were able to quantify the biomass and thus filtering capacity of Blue Mussels likely to be produced by current and proposed wind farms in the North Sea. A similar exercise could conceivably be undertaken for other colonising fauna and even for some fish species such as Atlantic Cod that show distinct patterns of distribution, if not abundance, as a result of association with wind farms (see Reubens *et al.*, Chapter 5, Box 5.1). Moreover, although the cumulative impact of numerous large wind farms as a result of upwelling from the wind-wake effect is not known, in theory it is likely to be considerable in terms of ocean productivity (Broström *et al.*, Chapter 3), with positive impacts of increased production of forage fishes that feed seabirds and marine mammals. Both the wind-wake effect and particularly the reef effect may even be of socio-economic importance to humans.

In contrast, there is growing concern over cumulative negative impacts of displacement for a number of bird species, especially divers, which have a clear negative response to wind farms (Vanermen & Stienen, Chapter 8). Studies cited by Vanermen & Stienen predicted cumulative effects of up to a 5% loss in habitat, up to a 5% decline in breeding success and up to a 2% decline in adult survival. Although quantifying the impact of direct losses of seabirds through collision may require fewer modelling assumptions, the considerable uncertainty of the predictions of collision risk modelling, which in turn results from the lack of empirical evidence of collisions to allow avoidance rates to be clearly quantified (see King, Chapter 9), means that cumulative effects remain particularly difficult to assess. Advances in the use of modelling to allow decision making with confidence are outlined by Cook & Masden (Volume 4, Chapter 6).

Broadly similar modelling approaches have been developed in relation to the impacts of pile-driving noise on birth and survival rates of individual marine mammals (see Nehls *et al.*, Chapter 6), namely the Interim Population Consequences of Disturbance (iPCOD) stage-structured stochastic population models (Harwood *et al.* 2016) and the Disturbance Effects on the Harbour Porpoise Population in the North Sea (DEPONS) (van Beest *et al.* 2015). Both models can be used to determine both cumulative effects of multiple developments and in-combination effects including mortality associated with incidental by-catch. To date, no significant cumulative impacts appear to have been recorded, although the extent to which both models have been used remains unclear.

According to Vanermen & Stienen (Chapter 8), an ideal cumulative impact assessment would include all existing and new wind-farm developments within the year-round distribution range of the species' population under study. This may be very large for mobile species such as seabirds or many marine mammals, as well as migratory birds and bats. The range of the population under investigation may also be extremely difficult to define in terms of a biogeographic unit. Thus, in practice, cumulative impact assessments are generally performed on smaller geographical scales such as legislative units associated with national boundaries, or according to particular protected sites or a suite of sites, such as SPAs for birds or Special Areas of Conservation for marine mammals. Overall, much work remains to be done to truly understand the full extent of cumulative impacts.

Population-scale impacts

It is first important to note that any discussion of 'population' impacts is invariably hampered by the difficulty of defining the 'population' adequately. Onshore, the National Wind Coordinating Committee (2003) suggested that it may be more appropriate to determine the biological significance of impacts, bearing in mind that any cumulative increase in anthropogenic 'take' could be argued to be of biological significance. In general, a focus on specific populations of defined protected areas or on national or international populations, as tends to occur in EIA and especially Appropriate Assessment in Europe, simplifies the issues somewhat.

Population-level impacts stem from effects on components of individual fitness such as reproductive output and the prospects of survival. Even onshore, where more studies have been conducted, and for birds, probably the taxon with the most available information, very few studies have demonstrated that wind farms affect individual fitness through displacement, and therefore there is little prospect of population-scale impacts (Hötker 2017). There are, however, a few examples of population-scale impacts as a result of collision mortality of individuals in species such as White-tailed Eagle at Smøla in Norway (see Dahl, Box 8.4 in de Lucas & Perrow 2017) and Golden Eagle *Aquila chrysaetos* at Altamont in California (Katzner *et al.* 2016). The emerging prospect of population-scale impacts onshore is perhaps because a number of rigorous long-term studies have now been conducted. This contrasts with the situation offshore, where there are currently no examples of population-scale impacts of collision mortality, perhaps stemming from the difficulty of measuring and quantifying collision (see *Collision with turbines*, above, and King, Chapter 9), coupled with the relative immaturity of the industry.

However, the coastal wind farm at Zeebrugge has had detectable effects upon several species of terns at the scale of the regional population (Everaert & Stienen, Chapter 9, Box 9.1). This reinforces that even a single wind farm or development area could have an impact on a particular population, as has been seen onshore. Nevertheless, the prospect of population-scale impacts intuitively increases as in-combination and cumulative effects grow (see *In-combination and cumulative effects*, above), as was clear from the few modelling studies cited by Vanermen & Stienen (Chapter 8) demonstrating the potential for population-scale impacts of displacement upon sensitive birds such as divers.

For migratory non-seabirds of a wide range of groups as well as bats making crossings over the open sea, the overall risk to individual populations may ultimately be low (Hüppop *et al.*, Chapter 7), although the evidence base on which to make this judgement is currently small to non-existent. For the better studied migratory waterfowl such as geese and ducks, which tend to display strong aversion to turbines, there seems little prospect of population impacts, even on small populations of conservation concern. But for raptors, the attraction to a wind farm observed by Skov *et al.* (2016) provides the basis for concern. Passerines may suffer high collision rates in some cases, especially as a result of attraction to illuminated structures under adverse weather conditions, but even relatively high losses locally seem unlikely to be significant for these typically *r*-selected species with high reproductive turnover and relatively low survival rates. For bats, the real risk to populations is unknown, especially considering the considerable mortality of bats seen onshore (Barclay *et al.* 2017). Important factors could be the proportion of the population crossing the sea in areas with a high density of turbines, and whether offshore turbines are as attractive to bats as they are onshore.

In relation to marine mammals, Nehls *et al.* (Chapter 6) stress that species with small populations compared to widely dispersed species with large populations, are more likely to suffer effects from wind farms, although no study has yet demonstrated a population impact. In a similar vein to seabirds, a number of modelling approaches are now available to determine potential population impacts from effects on individual vital rates, typically as a result of noise from pile driving (see *In-combination and cumulative effects*, above). The iPCoD model (Harwood *et al.* 2016) includes two species of seals and three species of cetacean, which is particularly useful considering that the empirical evidence base of effects on cetaceans is mostly limited to Harbour Porpoise, which typically suffers short-term displacement during noisy construction, but with a more variable and ambiguous response to constructed sites. The potential for opposing effects is also neatly illustrated by seals. In one population of Harbour Seals, Hastie *et al.* (2015) estimated that 50% of individuals would experience permanent threshold shifts in hearing with consequences for foraging ability and social interactions, whereas individuals in the same population were also shown to intensively exploit constructed wind farms in the area as foraging grounds (Russell *et al.* 2014) (Figure 10.1), and presumably benefit as a result.

Finally, it is clear from the pages of this volume that there are many examples of positive increases in populations of many species as a result of wind farms, largely tied to the reef effect. The scale of the prospective population increase in Blue Mussels in the southern North Sea is predicted to be so large as to be of ecosystem-scale functional significance (Slavik *et al.* 2018). Similarly, many species from anemones to crabs to small benthic fishes (Figure 10.6 and Figure 10.7) will have undergone or will undergo population increases. Although wind farms may be important habitats for fish such as Atlantic Cod and Pouting *Trisopterus luscus* (Gill & Wilhelmsson, Chapter 5 in this volume), it remains unclear whether wind farms have the potential to significantly increase the populations of such species, rather than simply redistribute and aggregate the larger adults in particular. Population-scale effects may depend on the enhanced survival of significant numbers of adults to influence population fecundity.

Concluding remarks

Even in isolation, an OWF may have a range of what may be seen as negative, neutral and positive effects on wildlife. As development continues, any effects will magnify in a cumulative context. In terms of generally positive effects, wind farms may exert a strong artificial reef effect by creating new habitat for colonising hard-substratum species and, in turn, providing prey and habitat for mobile invertebrates and fish. As such, wind farms may become 'biodiversity hotspots' (Dannheim *et al.*, Chapter 4). This provides a case to retain the bases even following decommissioning of a site in a 'renewables to reefs' initiative, although retention is also favoured by the technical difficulty of removing bases in the first place. Conceptually, the value of wind farms as reefs may be further enhanced where they extend pre-existing hard-bottom habitats with diverse fauna vertically upwards to the water surface. However, placing wind farms in hard substrate depends on the use of currently expensive gravity bases as alternatives to driven monopiles (see Jameson *et al.*, Chapter 1). Perhaps especially where fishing activity is restricted, there is even the prospect of wind farms becoming Marine Protected Areas, either in their own right or as part of a wider area.

The installation of wind farms provides considerable scope for cascading effects through the food web as a result of the new trophic system set up around bases, with potential to create resources for apex predators, as well as affecting primary production of surrounding seas through the filtering capacity of a large biomass of Blue Mussels. In addition, the wind-wake effect at large wind farms leading to upwelling and increasing nutrient transport may ultimately manifest as an increase in the biomass of forage fishes such as clupeids or sandeels, perhaps to the benefit of seabirds and marine mammals.

As it stands, while there is some indication that seals benefit from developing wind farms, only a few seabird species, notably Great Cormorant, appear to have proved able to exploit the resources contained within. Far more seabird species appear to be disturbed and thus displaced, although there is potential for habituation as time passes. The fundamental reasons for displacement remain unclear, although fear of novel structures, awareness of collision risk, disturbance by vessels, the lack of suitable pelagic forage fish, and competition with predatory fish and/or seals for forage fish all remain possibilities. There is some suggestion that displacement begins during construction for some species, which could link with the displacement of fish during noisy construction in a similar manner as occurs for Harbour Porpoise. Unfortunately, construction effects on fish have not yet been adequately studied.

Such is the strength of displacement by some seabirds such as divers that there is clear potential for population-scale impacts through effects on individual fitness (survival and reproductive output), especially in a cumulative context as wind-farm expansion continues and large areas of what may be preferred habitat are lost. Further empirical data gathering is urgently required to support theoretical modelling studies.

The collision of seabirds, migratory birds and bats remains of concern, but with little empirical basis. However, the technical issues involved mean that relatively few studies have been undertaken and only a handful of collision events have ever been observed. The question of collision is rarely taken on at individual sites and is generally seen to be an industry-scale issue with demonstration projects such as ORJIP (Skov et al. 2018). The more vulnerable groups appear to include less risk-averse species such as many gulls, especially large Larus spp., and possibly migratory raptors and passerines in adverse weather conditions. For passerines, it remains questionable whether losses, even in a cumulative context, could be significant at a population level in the presence of other limiting factors, although in real terms not enough is yet known. The same could be said for bats, with the proviso that these may be more vulnerable as a result of their population dynamics, and as yet, we have barely begun to understand what is out there. For seabirds, the key issues may be where wind farms are developed in foraging grounds associated with breeding colonies, where birds may take more 'risks' in order to provision chicks. In general, without better focus on key sites or suites of sites and key species, we may continue to chase the elusive 'phantom of collision risk', perhaps diverting attention away from more tangible impacts, including displacement and indirect effects.

Indeed, the review of information in this volume points to the high likelihood of indirect effects cascading throughout the food web. But, rather than being given the attention required, indirect effects seem to have generally been placed in the 'too difficult box' and have largely been ignored. As such, the true scope and scale of ecological effects remain virtually completely unknown and integrated ecosystem-scale research, particularly as the expansion of wind farms continues, is urgently required. A model for rigorous data collection for different faunal groups as well as physical effects and ocean dynamics already exists in the form of the StUKplus programme by BSH and BMU (2014), which could be adapted to better understand biophysical and trophic linkages.

New wind farms near seabird breeding colonies (Cleasby et al. 2015; Marine Scotland 2014; Langston & Teuten 2018) are seen to be a key area of research on the indirect effects of wind farms upon seabirds through their fish prey, which will almost invariably change over time through short-term displacement of fish during noisy construction, into an ecologically different trophic system with new fish 'actors' as the operational reef effect develops. The vulnerability of seabirds to anthropogenic change affecting their prey stocks is already well known (Frederiksen et al. 2004; Daunt et al. 2008; Cury et al. 2012).

Given the difficulties of quantifying wind-farm losses, it would seem prudent to also closely monitor those seabird colonies with the potential to be affected as an early warning of any impacts, bearing in mind that the loss of relatively few adults to the breeding population, either through collision or through failure to breed as a result of displacement from preferred areas, may be important for these k-selected species. This is notwithstanding that it will always be difficult to disentangle any impact from wind farms from the plethora of other factors limiting seabird populations (e.g. Frederiksen *et al.* 2007; Cury *et al.* 2012), perhaps especially on what are generally long-distance migratory species that inhabit different geographical areas subject to their own pressures at different times of year (e.g. Ramos *et al.* 2012). The same case for monitoring could be made for seals and cetaceans, although seals may be at less risk from wind farms in general, whereas monitoring of cetaceans is especially challenging as their entire lives are spent at sea.

Significant plans are now also in place in numerous other countries in the world for rapid expansion of wind farms, in their surrounding seas with different trophic conditions and a different set of dependent wildlife (Jameson *et al.*, Chapter 1). The lessons learned from north-western Europe must be incorporated into planning, using the monitoring and modelling tools and potential mitigation strategies outlined in Volume 4 of this series. However, until further research fills the knowledge gaps, the uncertainty of impacts should be embraced by avoiding conflict through careful selection of sites and even turbine locations bearing in mind that some turbines typically have greater impact than others (Perrow 2017). This will be vital to achieve Kiesecker *et al.*s' (2011) vision of a 'win–win' for wind energy and wildlife.

Acknowledgements

I am indebted to all the other 27 chapter authors and 3 additional box authors for their knowledge and hard work in preparing their chapters. The considered comments of Nicolas Vanermen of INBO on two earlier drafts of this manuscript are also greatly appreciated. I am deeply grateful to SCIRA Offshore Energy for funding the (as yet) unpublished work in relation to Sheringham Shoal, excerpts of which are presented here, and to E.ON for funding the previously published work in relation to Scroby Sands. Andrew Harwood kindly redrew Figures 10.1, 10.2, 10.3, 10.10 and 10.11, as well as co-authoring Box 10.2 with Richard Berridge, both of ECON Ecological Consultancy Ltd. Philip Pearson and Daniel Hercock of the RSPB kindly provided data gathered by RSPB staff and volunteers on recent counts of terns in Box 10.3.

References

4C Offshore (2018) Events of Cirrus Shell Flat Array. Retrieved 14 May 2018 from https://www.4coffshore.com/windfarms/project-dates-for-cirrus-shell-flat-array-uk24.html

Adams, T.P., Miller, R.G., Aleynik, D. & Burrows, M.T. (2014) Offshore marine renewable energy devices as stepping stones across biogeographical boundaries. *Journal of Applied Ecology* 51: 330–338.

Anonymous (2009) Move to protect sea bird colonies. BBC News. Retrieved 14 May 2018 from http://news.bbc.co.uk/1/hi/wales/8382216.stm

Barclay, R.M.R., Baerwald, E.F. & Rydell, J. (2017) Bats. In Perrow, M.R. (ed.) *Wildlife and Wind Farms, Conflicts and Solutions. Volume 1. Onshore: Potential effects.* Exeter: Pelagic Publishing. pp. 191–221.

Brandt, M.J., Dragon, A.-C., Diederichs, A., Bellmann, M.A., Wahl, V., Piper, W., Nabe-Nielsen, J. & Nehls, G. (2018) Disturbance of harbour porpoises during construction of the first seven offshore wind farms in Germany. *Marine Ecology Progress Series* 596: 213–232.

BSH & BMU (2014) *Ecological Research at the Offshore Windfarm alpha ventus: Challenges, results and perspectives.* Federal Maritime and Hydrographic

Society of Germany (BSH) and Federal Ministry for the Environment, Nature Conservation and Nuclear Safety (BMU). Weisbaden: Springer Spektrum.

Buckland, S.T., Burt, M.L., Rexstad, E.A., Mellor, M., Williams, A.E. & Woodward, R. (2012) Aerial surveys of seabirds: the advent of digital methods. *Journal of Applied Ecology* 49: 960–967.

Certain, G., Masse, J., van Canneyt, O., Petitgas, P., Doremus, G., Santos, M.B. & Ridoux, V. (2011) Investigating the coupling between small pelagic fish and marine top predators using data collected from ecosystem-based surveys. *Marine Ecology Progress Series* 422: 23–39.

Cleasby, I.R., Wakefield, E.D., Bearhop, S., Bodey, T.W., Votier, S.C. & Hamer, K.C. (2015) Three-dimensional tracking of a wide-ranging marine predator: flight heights and vulnerability to offshore wind farms. *Journal of Applied Ecology* 52: 1474–1482.

Cook, A.S.C.P., Humphreys, E.M., Bennet, F., Masden, E.A. & Burton, N.H.K. (2018) Quantifying avian avoidance of offshore wind turbines: current evidence and key knowledge gaps. *Marine Environmental Research*. https://doi.org/10.1016/j.marenvres.2018.06.017.

Cury, P.M., Boyd, I.L., Bonhommeau, S., Anker-Nilssen, T., Crawford, R.J.M., Furness, R.W., Mills, J.A., Murphy, E.J., Österblom, H., Paleczny, M., Piatt, J.F., Roux, J.-P., Shannon, L. & Sydeman, W.J. (2012) Global seabird response to forage fish depletion – one-third for the birds. *Science* 334: 1703–1706.

Daunt, F., Wanless, S., Greenstreet, S.P.R., Jensen, H., Hamer, K.C. & Harris, M.P. (2008) The impact of the sandeel fishery closure in the north-western North Sea on seabird food consumption, distribution and productivity. *Canadian Journal of Fisheries and Aquatic Sciences* 65: 362–381.

de Lucas, M. & Perrow, M.R. (2017) Birds: collision. In Perrow, M.R. (ed.) *Wildlife and Wind Farms, Conflicts and Solutions. Volume 1. Onshore: Potential effects.* Exeter: Pelagic Publishing. pp. 155–190.

Department of Energy & Climate Change (DECC) (2012) Record of the Appropriate Assessment undertaken for applications under Section 36 of the Electricity Act 1989. Projects: Docking Shoal Offshore Wind Farm (as amended), Race Bank Offshore Wind Farm (as amended), Dudgeon Offshore Wind Farm. London: DECC. Retrieved 20 August 2018 from https://www.og.decc.gov.uk/EIP/pages/projects/AAGreaterWash.pdf

Desholm, M. & Kahlert, J. (2005) Avian collision risk at an offshore wind farm. *Biology Letters* 1: 296–298.

Desholm, M., Fox, A.D., Beasley, P. & Kahlert, J. (2006) Remote techniques for counting and estimating the number of bird–wind turbine collisions at sea: a review. *Ibis* 148: 76–89.

European Wind Energy Association (EWEA) (2016) The European offshore wind industry – key trends and statistics 2015. Retrieved 16 May 2018 from https://www.ewea.org/fileadmin/files/library/publications/statistics/EWEA-European-Offshore-Statistics-2015.pdf

Ferrer, M., de Lucas, M., Janss, G.F.E., Casado, E., Muñoz, A.R., Bechard, M.J. & Calabuig, C.P. (2012) Weak relationship between risk assessment studies and recorded mortality in windfarms. *Journal of Applied Ecology* 49: 38–46.

Frederiksen, M., Wanless, S., Harris, M.P., Rothery, P. & Wilson, L.J. (2004) The role of industrial fishery and oceanographic change in the decline of North Sea black-legged kittiwakes. *Journal of Applied Ecology* 41: 1129–1139.

Frederiksen, M., Edwards, M., Mavor, R.A. & Wanless, S. (2007) Regional and annual variation in black-legged kittiwake breeding productivity is related to sea surface temperature. *Marine Ecology Progress Series* 350: 137–143.

Furness, R.W., Wade, H.M. & Masden, E.A. (2013) Assessing vulnerability of marine bird populations to offshore wind farms. *Journal of Environmental Management* 119: 56–66.

Garthe, S. & Hüppop, O. (2004) Scaling possible adverse effects of marine wind farms on seabirds: developing and applying a vulnerability index. *Journal of Applied Ecology* 41: 724–734.

Gill, J.A., Norris, K. & Sutherland, W.J. (2001) Why behavioural responses may not reflect the population consequences of human disturbance. *Biological Conservation* 97: 265–268.

Goodship, N., Caldow, R., Clough, S., Korda, R., McGovern, S., Rowlands, N. & Rehfisch, M. (2015) Surveys of red-throated divers in the Thames Estuary SPA. *British Birds* 108: 506–513.

Gove, B., Langston, R.H.W., McCluskie, A.J., Pullan, D. & Scrase, I. (2013) Wind farms and birds: an updated analysis of the effects of wind farms on birds, and best practice guidance on integrated planning and impact assessment. In *Convention on the Conservation of European Wildlife and Natural Habitats*. Report prepared by BirdLife International on behalf of the Bern Convention. Sandy: RSPB/BirdLife International. Retrieved

3 June 2014 from http://www.birdlife.org/sites/default/files/attachments/201312_BernWindfarmsreport.pdf

Green, R.E., Langston, R.H.W., McCluskie, A., Sutherland, R. & Wilson, J.D. (2016) Lack of sound science in assessing wind farm impacts upon seabirds. *Journal of Applied Ecology* 53: 1635–1641.

Gutow, L., Teschke, K., Schmidt, A., Dannheim, J., Krone, R. & Gusky, M. (2014) Rapid increase of benthic structural and functional diversity at the alpha ventus test site. In BSH & BMU (eds) *Ecological Research at the Offshore Windfarm alpha ventus: Challenges, results and perspectives*. Federal Maritime and Hydrographic Society of Germany (BSH) and Federal Ministry for the Environment, Nature Conservation and Nuclear Safety (BMU). Weisbaden: Springer Spektrum. pp. 67–81.

Hall, A.J., Watkins, J. & Hammond, P. (1998) Seasonal variation in the diet of harbour seals in the south-western North Sea. *Marine Ecology Progress Series* 170: 269–281.

Harvey, F. (2014) Migrating birds halt expansion of London Array. Retrieved 14 May 2018 from https://www.theguardian.com/environment/2014/feb/19/migrating-birds-london-array-offshore-windfarm

Harwood, J., King, S., Booth, C., Donovan, C., Schick, R.S., Thomas, L. & New, L. (2016) Chapter 49. Understanding the population consequences of acoustic disturbance for marine mammals. In Popper, A. & Hawkins, A. (eds) *The Effects of Noise on Aquatic Life II*. Advances in Experimental Medicine and Biology, Vol. 875. New York: Springer. pp. 417–423.

Harwood, A.J.P., Perrow, M.R., Berridge, R., Tomlinson, M.L. & Skeate, E.R. (2017) Unforeseen responses of a breeding seabird to the construction of an offshore wind farm. In Köppel, J. (ed.) *Conference on Wind Energy and Wildlife Interactions. Presentations from the CWW2015 conference*. Cham: Springer International Publishing. pp. 19–41.

Hastie, G.D., Russell, D.J.F., McConnell, B., Moss, S., Thompson, D. & Janik, V.M. (2015) Sound exposure in harbour seals during the installation of an offshore wind farm: prediction of auditory damage. *Journal of Applied Ecology* 52: 631–640.

Heithaus, M.R., Frid, A., Wirsing, A.J. & Worm, B. (2008) Predicting ecological consequences of marine top predator declines. *Trends in Ecology & Evolution* 23: 202–210.

Hill, R., Hill, K., Aumüller, R., Schulz, A., Dittmann, T., Kulemeyer, C. & Coppack, T. (2014) Of birds, blades and barriers: detecting and analysing mass migration events at *alpha ventus*. In BSH & BMU (eds) *Ecological Research at the Offshore Windfarm alpha ventus: Challenges, results and perspectives*. Federal Maritime and Hydrographic Society of Germany (BSH) and Federal Ministry for the Environment, Nature Conservation and Nuclear Safety (BMU). Weisbaden: Springer Spektrum. pp. 111–131.

Hötker, H. (2017) Birds: displacement. In Perrow, M.R. (ed.) *Wildlife and Wind Farms, Conflicts and Solutions. Volume 1. Onshore: Potential effects*. Exeter: Pelagic Publishing. pp. 119–154.

Humphreys, E.M., Cook, A.C.P. & Burton, N.H.K. (2015) Collision, displacement and barrier effect concept note. BTO Research Report No. 669. Thetford: British Trust for Ornithology. Retrieved 11 July 2018 from https://www.bto.org/file/336011/download?token=k5RcgmA8

Huso, M., Dalthorp, D. & Korner-Nievergelt, F. (2017) Statistical principles of post-construction fatality monitoring. In Perrow, M.R. (ed.) *Wildlife and Wind Farms, Conflicts and Solutions. Volume 2. Onshore: Monitoring and mitigation*. Exeter: Pelagic Publishing. pp. 84–102.

Intergovernmental Panel on Climate Change (IPCC) (2014) *Climate Change 2014: Impacts, adaptation, and vulnerability. Part A: Global and sectoral aspects. Contribution of Working Group II to the Fifth Assessment Report of the Intergovernmental Panel on Climate Change* [Field, C.B., Barros, V.R., Dokken, D.J., Mach, K.J., Mastrandrea, M.D., Bilir, T.E., Chatterjee, M., Ebi, K.L., Estrada, Y.O., Genova, R.C., Girma, B., Kissel, E.S., Levy, A.N., MacCracken, S., Mastrandrea, P.R. & White, L.L. (eds)]. Cambridge: Cambridge University Press.

Jakobsen, S.E. (2014) Puffin chicks die of hunger. ScienceNordic. Retrieved 10 July 2018 from http://sciencenordic.com/puffin-chicks-die-hunger

Johnston, N.N., Bradley, J.E. & Otter, K.A. (2014) Increased flight altitudes among migrating Golden Eagles suggest turbine avoidance at a Rocky Mountain wind installation. *PLoS ONE* 9 (3): e93030. doi: 10.1371/journal.pone.093030.

Joint Nature Conservation Committee (JNCC) (2017) Liverpool Bay/Bae Lerpwl SPA. Peterborough: JNCC. Retrieved 27 July 2018 from http://jncc.defra.gov.uk/page-7507

Kaiser, M.J., Galanidi, M., Showler, D.A., Elliot, A.J., Caldow, R.W.G., Rees, E.I.S., Stillman, R.A. & Sutherland, W.J. (2006) Distribution and behaviour of common scoter *Melanitta nigra* relative to prey resources and environmental parameters. *Ibis* 148: 110–128.

Katzner, T.E., Nelson, D.M., Braham, M.A., Doyle, J.M., Fernandez, N.B., Duerr, A.E., Bloom, P.H., Fitzpatrick, M.C., Miller, T.A., Culver, R.C.E., Braswell, L. & DeWoody, J.A. (2016) Golden eagle fatalities and the continental-scale consequences of local wind-energy generation. *Conservation Biology* 31: 406–415.

Kiesecker, J.M., Evans, J.S., Fargione, J., Doherty, K., Foresman, K.R., Kunz, T.H., Naugle, D., Nibbelink, N.P. & Niemuth, N.D. (2011) Win–win for wind and wildlife: a vision to facilitate sustainable development. *PLoS ONE* 6 (4): e17566. doi: 10.1371/journal.pone.0017566

Krägefsky, S. (2014) Effects of the *alpha ventus* offshore test site on pelagic fish. In BSH & BMU (eds) *Ecological Research at the Offshore Windfarm alpha ventus: Challenges, results and perspectives*. Federal Maritime and Hydrographic Society of Germany (BSH) and Federal Ministry for the Environment, Nature Conservation and Nuclear Safety (BMU). Weisbaden: Springer Spektrum. pp. 83–94.

Langston, R.W. & Tueten, E. (2018) Ranging behaviour of northern gannets. *British Birds* 111: 131–143.

Larsen, J.K. & Guillemette, M. (2007) Effects of wind turbines on flights behaviour of wintering common eiders: implications for habitat use and collision risk. *Journal of Applied Ecology* 44: 516–522.

Leopold, M.F., van Bemmelen, R.S.A. & Zuur, A. (2013) Responses of local birds to the offshore wind farms PAWP and OWEZ off the Dutch mainland coast. Report C151/12. Report commissioned by Prinses Amaliawindpark. Texel: Imares. Retrieved 1 September 2017 from http://library.wur.nl/WebQuery/wurpubs/443769

Lindeboom, H.J., Kouwenhoven, H.J., Bergman, M.J.N., Bouma, S., Brasseur, S., Daan, R., Fijn, R.C., de Haan, D., Dirksen, S., van Hal, R., Hille Ris Lambers, R., ter Hofstede, R., Krijgsveld, K.L., Leopold, M. & Scheidat, M. (2011) Short-term ecological effects of an offshore wind farm in the Dutch coastal zone; a compilation. *Environmental Research Letters* 6: 035101.

Lovich, J.E. & Ennen, J.R. (2017) Reptiles and amphibians. In Perrow, M.R. (ed.) *Wildlife and Wind Farms, Conflicts and Solutions. Volume 1. Onshore: Potential effects*. Exeter: Pelagic Publishing. pp. 97–118.

Marine Management Organisation (MMO) (2014) Review of post-consent offshore wind farm monitoring data associated with licence conditions. A report produced for the Marine Management Organisation. MMO Project No. 1031. Retrieved 11 July 2018 from https://assets.publishing.service.gov.uk/government/uploads/system/uploads/attachment_data/file/317787/1031.pdf

Marine Scotland (2014) Appropriate Assessment: Marine Scotland's consideration of a proposal affecting designated Special Areas of Conservation ('SACs') and Special Areas of Protection ('SPAs'). Glasgow: Marine Scotland. Retrieved 18 November 2014 from http://www.gov.scot/Resource/0046/00460542.pdf

Martin, G.R. (2011) Understanding bird collisions with man-made objects: a sensory ecology approach. *Ibis* 153: 239–254.

Martin, G.R. & Prince, P.A. (2001) Visual fields and foraging in Procellariiform seabirds: sensory aspects of dietary segregation. *Brain, Behavior and Evolution* 57: 33–38.

Martin, G.R. & Wanless, S. (2015) The visual fields of common guillemots *Uria aalge* and Atlantic puffins *Fratercula arctica*: foraging, vigilance and collision. *Ibis* 157: 798–807.

Martin, G.R., Portugal, S.J. & Murn, C.P. (2012) Visual fields, foraging and collision vulnerability in *Gyps* vultures. *Ibis* 154: 626–631.

Masden, E.A., Haydon, D.T., Fox, A.D., Furness, R.W., Bullman, R. & Desholm, M. (2009) Barriers to movement: impacts of wind farms on migrating birds. *ICES Journal of Marine Science* 66: 746–753.

Masden, E.A., Haydon, D.T., Fox, A.D. & Furness, R.W. (2010) Barriers to movement: modelling energetic costs of avoiding marine wind farms amongst breeding seabirds. *Marine Pollution Bulletin* 60: 1085–1091.

National Wind Coordinating Committee (2003) How is biological significance determined when assessing possible impacts of onshore wind power facilities? Meeting of the Wildlife Workgroup of the National Wind Coordinating Committee, 17 November 2003, Washington, DC. Retrieved 3 November 2016 from https://www.nationalwind.org/wp-content/uploads/assets/wildlife_workgroup_ meetings/Biological_Significance_Meeting_Summary.pdf

Offshore Wind (2016, 4 February) Norfolk Marine's TFN scour remediation system installed at Scroby Sands. Retrieved 21 June 2018 from https://www.offshorewind.biz/2016/02/04/norfolk-marines-tfn-scour-remediation-system-installed-at-scroby-sands/

Offshore Wind (2017, 15 March) World's first offshore wind farm passes into history. Retrieved 27 July 2018 from https://www.offshorewind.biz/2017/03/15/worlds-first-offshore-wind-farm-passes-into-history/

Orloff, S. & Flannery, A. (1992) Wind turbine effects on avian activity, habitat use, and mortality in Altamont Pass and Solano County wind resource areas 1989–1991. Sacramento, CA: Planning Departments of Alameda, Contra Costa and Solano Counties and the California Energy Commission. Retrieved 27 July 2018 from https://tethys.pnnl.gov/sites/default/files/publications/Orloff-1992.pdf

Percival, S.M. (2005) Birds and wind farms: what are the real issues? *British Birds* 98: 194–204.

Perrow, M.R. (2017) A synthesis of effects and impacts. In Perrow, M.R. (ed.) *Wildlife and Wind Farms, Conflicts and Solutions. Volume 1. Onshore: Potential effects.* Exeter: Pelagic Publishing. pp. 241–276.

Perrow, M.R. & Eglington, S.M. (2014) Securing a future for East Norfolk's little terns. *British Wildlife* 25: 237–245.

Perrow, M.R., Skeate, E.R., Lines, P., Brown, D. & Tomlinson, M.L. (2006) Radio telemetry as a tool for impact assessment of wind farms: the case of little terns *Sterna albifrons* at Scroby Sands, Norfolk, UK. *Ibis* 148 (Suppl. 1): 57–75.

Perrow, M.R., Gilroy, J.J., Skeate, E.R. & Tomlinson, M.L. (2011a) Effects of the construction of Scroby Sands offshore wind farm on the prey base of little tern *Sternula albifrons* at its most important UK colony. *Marine Pollution Bulletin* 62: 1661–1670.

Perrow, M.R., Skeate, E.R. & Gilroy, J.J. (2011b) Novel use of visual tracking from a rigid-hulled inflatable boat (RIB) to determine foraging movements of breeding terns. *Journal of Field Ornithology* 82: 68–79.

Perrow, M.R., Harwood, A.J.P., Skeate, E.R., Praca, E. & Eglington, S.M. (2015) Use of multiple data sources and analytical approaches to derive a marine protected area for a breeding seabird. *Biological Conservation* 191: 729–738.

Petersen, I.K. & Fox, A.D. (2007) Changes in bird habitat utilisation around the Horns Rev 1 offshore wind farm, with particular emphasis on common scoter. NERI Report commissioned by Vattenfall A/S. National Environmental Research Institute, Ministry of the Environment (Denmark). Retrieved 26 June 2018 from https://corporate.vattenfall.dk/globalassets/danmark/om_os/horns_rev/changes-in-bird-habitat.pdf

Petersen, J. & Malm, T. (2006) Offshore windmill farms: threats to or possibilities for the marine environment. *Ambio* 35: 75–80.

Petersen, I.K., Christensen, T.K., Kahlert, J., Desholm, M. & Fox, A.D. (2006) Final results of bird studies at the offshore wind farms at Nysted and Horns Rev, Denmark. NERI Report commissioned by DONG Energy and Vattenfall A/S. National Environmental Research Institute, Ministry of the Environment (Denmark). Retrieved 19 June 2018 from https://tethys.pnnl.gov/sites/default/files/publications/NERI_Bird_Studies.pdf

Pettersson, J. (2005) The impact of offshore wind farms on bird life in southern Kalmar Sound, Sweden. A final report based on studies 1999–2003. Report to the Swedish Energy Agency. Retrieved 19 June 2018 from https://tethys.pnnl.gov/sites/default/files/publications/The_Impact_of_Offshore_Wind_Farms_on_Bird_Life.pdf

Pichegru, L., Nyengera, R., McInnes, A.M. & Pistorius, P. (2017) Avoidance of seismic survey activities by penguins. *Scientific Reports* 7: 16305.

Plonczkier, P. & Simms, I.C. (2012) Radar monitoring of migrating pink-footed geese: behavioural responses to offshore wind farm development. *Journal of Applied Ecology* 49: 1187–1194.

Ramos, R., Granadeiro, J.P., Nevoux, M., Mougin, J.-L., Peixe Dias, M. & Catry, P. (2012) Combined spatio-temporal impacts of climate and longline fisheries on the survival of a trans-equatorial marine migrant. *PLoS ONE* 7 (7): e40822. doi: 10/1371/journal.pone.0040822.

Russell, D.F., Brasseur, S.M.J.M., Thompson, D., Hastie, G.D., Janik, V.M., Aarts, G., McClintock, B.T., Matthiopoulos, J., Moss, S.E.W. & McConnell, B. (2014) Marine mammals trace anthropogenic structures at sea. *Current Biology* 24: R638–R639.

Russell, D.F., Hastie, G.D., Thompson, D., Janik, V.M., Hammond, P.S., Scott-Hayward, L.A.S., Matthiopoulos, J., Jones, E.L. & McConnell, B.J. (2016) Avoidance of wind farms by harbour seals is limited to pile driving activities. *Journal of Applied Ecology* 53: 1642–1652.

Scheidat, M., Tougaard, J., Brasseur, S., Carstensen, J., van Polanen Petel, T., Teilmann, J. & Reijnders, P. (2011) Harbour porpoises (*Phocoena phocoena*) and wind farms: a case study in the Dutch North Sea. *Environmental Research Letters* 6: 025102.

SCIRA Offshore Energy (2006) Sheringham Shoal offshore wind farm environmental statement. London: SCIRA Offshore Energy. Retrieved 30

July 2018 from http://sheringhamshoal.co.uk/downloads/Offshore%20environmental%20statement.pdf

Scott. B.E. (2013) Seabirds and marine renewables: are we asking the right questions about indirect effects? *BOU Proceedings – Marine Renewables and Birds*. Retrieved 17 August 2018 from https://www.bou.org.uk/bouproc-net/marine-renewables/scott.pdf

Scott, B.E., Sharples, J., Ross, O.N., Wang, J., Pierce, G.J. & Camphuysen, C.J. (2010) Sub-surface hotspots in shallow seas: fine scale limited locations of marine top predator foraging habitat indicated by tidal mixing and sub-surface chlorophyll. *Marine Ecology Progress Series* 408: 207–226.

Scott, B.E., Langton, R., Philpott, E. & Waggitt, J.J. (2014) 7. Seabirds and marine renewables: are we asking the right questions? In Shields, M.A. & Payne, A.I.L. (eds) *Marine renewable energy technology and environmental interactions*. Humanity and the Sea DOI10.1007/978-94-017-8002-5_7. Dordrecht, Springer Science+Business Media, pp.81-91.

Searle, K., Mobbs, D., Butler, A., Bogdanova, M., Freeman, S., Wanless, S. & Daunt, F. (2014) Population consequences of displacement from proposed offshore wind energy developments for seabirds breeding at Scottish SPAs (CR/2012/03). Report to Marine Scotland Science, Scottish Government. Retrieved 1 September 2017 from http://www.gov.scot/Topics/marine/marineenergy/Research/SB7

Skeate, E.R., Perrow, M.R. & Gilroy, J.J. (2012) Likely effects of construction of Scroby Sands offshore wind farm on a mixed population of harbour *Phoca vitulina* and grey *Halichoerus grypus* seals. *Marine Pollution Bulletin* 64: 872–881.

Skov, H., Leonhard, S.B., Heinänen, S., Zydelis, R., Jensen, N.E., Durinck, J., Johansen, T.W., Jensen, B.P., Hansen, B.L., Piper, W. & Grøn, P.N. (2012) Horns Rev 2 monitoring 2010–2012. Migrating birds. Orbicon, DHI, Marine Observers and Biola. Report commissioned by DONG Energy. Retrieved 19 June 2018 from https://tethys.pnnl.gov/sites/default/files/publications/Horns_Rev_2_Migrating_Birds_Monitoring_2012.pdf

Skov, H., Desholm, M., Heinänen, S., Kahlert, J.A., Laubek, B., Jensen, N.E., Żydelis, R. & Jensen, B.P. (2016) Patterns of migrating soaring migrants indicate attraction to marine wind farms. *Biology Letters* 12: 20160804.

Skov, H., Heinänen, S., Norman, T., Ward, R., Méndez-Roldán, S. & Ellis, I. (2018) ORJIP bird collision and avoidance study. Final report – April 2018. London: Carbon Trust. Retrieved 20 June 2018 from https://www.carbontrust.com/media/675793/orjip-bird-collision-avoidance-study_april-2018.pdf

Slabbekoorn, H., Bouton, N., van Opzeeland, I., Coers, A., ten Cate, C. & Popper, A.N. (2010) A noisy spring: the impact of globally rising underwater sound levels on fish. *Trends in Ecology & Evolution* 25: 419–427.

Slavik, K., Lemmen, C., Zhang, W., Kerimoglu, O., Klingbeil, K. & Wirtz, K.W. (2018) The large scale impact of offshore wind farm structures on pelagic primary productivity in the southern North Sea. *Hydrobiologia*. https://doi.org/10.1007/s10750-018-3653-5

Smales, I. (2017) Modelling collision risk and populations. In Perrow, M.R. (ed.) *Wildlife and Wind Farms, Conflicts and Solutions. Volume 2. Onshore: Monitoring and mitigation*. Exeter: Pelagic Publishing. pp. 58–83.

Smallwood (2017) Monitoring birds. In Perrow, M.R. (ed.) *Wildlife and Wind Farms, Conflicts and Solutions. Volume 2. Onshore: Monitoring and mitigation*. Exeter: Pelagic Publishing. pp. 1–30.

Stienen, E.W.M., Courtens, W., Everaert, J. & Van de Valle, M. (2008) Sex-biased mortality of common terns in wind farm collisions. *The Condor* 110: 154–157.

Taylor, M., Seago, M., Allard, P. & Dorling, D. (eds) (1999) *The Birds of Norfolk*. Robertsbridge: Pica Press.

Teilmann, J. & Carstensen, J. (2012) Negative long term effects on harbour porpoises from a large scale offshore wind farm in the Baltic – evidence of slow recovery. *Environmental Research Letters* 7: 045101.

Teilmann, J., Tougaard, J., Carstensen, J., Dietz, R. & Tougaard, S. (2006) Summary on seal monitoring 1999–2005 around Nysted and Horns Rev Offshore Wind Farms. Report to Energi E2 A/S and Vattenfall A/S. Report No. 2389313244. National Environmental Research Institute. Retrieved 17 August 2018 from https://www.researchgate.net/profile/Jonas_Teilmann/publication/267985927_Summary_on_seal_monitoring_1999-2005_around_Nysted_and_Horns_Rev_Offshore_Wind_Farms_-_Technical_Report_to_Energi_E2_AS_and_Vattenfall_AS/links/54b7969e0cf24eb34f6ebbb3/Summary-on-seal-monitoring-1999-2005-around-Nysted-and-Horns-Rev-Offshore-Wind-Farms-Technical-Report-to-Energi-E2-A-S-and-Vattenfall-A-S.pdf

Thaxter, C.B., Lascelles, B., Sugar, K., Cook, A.S.C.P., Roos, S., Bolton, M., Langston, R.H.W. & Burton,

N.H.K. (2012) Seabird foraging ranges as a tool for identifying candidate marine protected areas. *Biological Conservation* 156: 53–61.

Thaxter, C.B., Ross-Smith, V.H., Bouten, W., Rehfisch, M.M., Clark, N.A., Conway, G.J. & Burton, N.H.K. (2015) Seabird–wind farm interactions during the breeding season vary within and between years: a case study of lesser black-backed gull *Larus fuscus* in the UK. *Biological Conservation* 186: 347–358.

Thaxter, C.B., Ross-Smith, V.H., Bouten, W., Masden, E.A., Clark, N.A., Conway, G.J., Barber, L., Clewley, G.D. & Burton, N.H.K. (2018) Dodging the blades: new insights into three-dimensional space use of offshore wind farms by lesser black-backed gulls *Larus fuscus*. *Marine Ecology Progress Series* 587: 247–253.

Thompson, P.M., Lusseau, D., Barton, T., Simmons, D., Rusin, J. & Bailey, H. (2010) Assessing the responses of coastal cetaceans to the construction of offshore wind turbines. *Marine Pollution Bulletin* 60: 1200–1208.

Thompson, P.M., Hastie, G.D., Nedwell, J., Barham, R., Brookes, K.L., Cordes, L.S., Bailey, H. & McLean, N. (2013) Framework for assessing impacts of pile-driving noise from offshore wind farm construction on a harbour seal population. *Environmental Impact Assessment Review* 43: 73–85.

Thomsen, F., Lüdemann, K., Kafemann, R. & Piper, W. (2006) Effects of offshore wind farm noise on marine mammals and fish. Hamburg: Biola, on behalf of COWRIE Ltd. Retrieved 20 August 2018 from https://tethys.pnnl.gov/sites/default/files/publications/Effects_of_offshore_wind_farm_noise_on_marine-mammals_and_fish-1-.pdf

Tougaard, J., Carstensen, J., Teilmann, J., Skov, H. & Rasmussen, P. (2009) Pile driving zone of responsiveness extends beyond 20 km for harbor porpoises (*Phocoena phocoena* (L.)). *Journal of the Acoustical Society of America* 126: 11–14.

Vallejo, G.C., Grellier, K., Nelson, E.J., McGregor, R.M., Canning, S.J., Caryl, F.M. & McLean, N. (2017) Responses of two marine top predators to an offshore wind farm. *Ecology and Evolution* 2017: 8698–8708.

van Beest, F.M., Nabe-Nielsen, J., Carstensen, J., Teilmann, J. & Tougaard, J. (2015) Disturbance Effects on the Harbour Porpoise Population in the North Sea (DEPONS): status report on model development. Scientific Report from DCE – Danish Centre for Environment and Energy No. 140. Retrieved 25 July 2018 from http://dce2.au.dk/pub/SR140.pdf

Vanermen, N., Onkelinx, T., Courtens, W., Van de walle, M., Verstraete, H. & Stienen, E.W.M. (2015a) Seabird avoidance and attraction at an offshore wind farm in the Belgian part of the North Sea. *Hydrobiologia* 756: 51–61.

Vanermen, N., Onkelinx, T., Verschelde, P., Courtens, W., Van de walle, M., Verstraete, H. & Stienen, E.W.M. (2015b) Assessing seabird displacement at offshore wind farms: power ranges of a monitoring and data handling protocol. *Hydrobiologia* 756: 155–167.

Whitehouse, R., Harris, J., Sutherland, J. & Rees, J. (2011) The nature of scour development and scour protection at offshore windfarm foundations. *Marine Pollution Bulletin* 62: 73–88.

Wind Europe (2018) Offshore wind in Europe: key trends and statistics 2017. February 2018. Retrieved 27 July 2018 from https://windeurope.org/wp-content/uploads/files/about-wind/statistics/WindEurope-Annual-Offshore-Statistics-2017.pdf

Index

Page numbers in *italics* denote figures and in **bold** denote tables.